Assessing Revolutionary and Insurgent Strategies

HUMAN FACTORS CONSIDERATIONS
OF UNDERGROUNDS
IN INSURGENCIES

I0095707

SECOND EDITION

Paul J. Tompkins Jr., USASOC Project Lead

Nathan Bos, Editor

United States Army Special Operations Command

and

The Johns Hopkins University Applied Physics Laboratory

National Security Analysis Department

CONFLICT RESEARCH GROUP

About Conflict Research Group

Conflict Research Group (CRG) is a non-profit think tank based in the United Kingdom, dedicated to advancing understanding of the art and science of Unconventional Warfare. With a focus on the academic study of guerrilla warfare, revolutionary warfare, asymmetric warfare, Fourth Generation Warfare, Fifth Generation Warfare, and political unrest, CRG's work sheds light on the complexities and nuances of modern conflicts. By bringing critical and key works into print, the organization serves as a vital resource for academics, policymakers, and military professionals seeking in-depth knowledge in these specialized fields.

At the heart of CRG's mission is the belief that a comprehensive understanding of Unconventional Warfare is essential for addressing contemporary security challenges. The group's research and publications delve into historical and contemporary case studies, exploring the strategies, tactics, and implications of irregular warfare. Through this rigorous analysis, CRG contributes to the development of more effective and adaptable strategies for dealing with non-traditional threats.

One of the key aspects of CRG's work is its publishing arm, which is dedicated to bringing into print seminal works on Unconventional Warfare. The group's publications cover a wide range of topics, from historical accounts of guerrilla movements to theoretical analyses of contemporary conflict dynamics and reprints of government publications. By making these works accessible to a broader audience, CRG aims to enrich the discourse on Unconventional Warfare and contribute to the development of more nuanced and effective approaches to resolving conflicts and disrupting, degrading and defeating unconventional threats.

CRG's research is characterized by its interdisciplinary approach, drawing on insights from military history, political science, sociology, and international relations. This holistic perspective allows the organization to address the multifaceted nature of unconventional warfare, considering not only military tactics but also the granularity of the political, social, and economic dimensions of conflicts. Through this comprehensive approach, CRG provides a deeper understanding of the root causes and long-term implications of irregular warfare.

Published by Conflict Research Group.

First published by USASOC in 2013

ISBN: 978-1-925907-19-3

ASSESSING REVOLUTIONARY AND INSURGENT STRATEGIES

The Assessing Revolutionary and Insurgent Strategies (ARIS) series consists of a set of case studies and research conducted for the US Army Special Operations Command by the National Security Analysis Department of The Johns Hopkins University Applied Physics Laboratory.

The purpose of the ARIS series is to produce a collection of academically rigorous yet operationally relevant research materials to develop and illustrate a common understanding of insurgency and revolution. This research, intended to form a bedrock body of knowledge for members of the Special Forces, will allow users to distill vast amounts of material from a wide array of campaigns and extract relevant lessons, thereby enabling the development of future doctrine, professional education, and training.

From its inception, ARIS has been focused on exploring historical and current revolutions and insurgencies for the purpose of identifying emerging trends in operational designs and patterns. ARIS encompasses research and studies on the general characteristics of revolutionary movements and insurgencies and examines unique adaptations by specific organizations or groups to overcome various environmental and contextual challenges.

The ARIS series follows in the tradition of research conducted by the Special Operations Research Office (SORO) of American University in the 1950s and 1960s, by adding new research to that body of work and in several instances releasing updated editions of original SORO studies.

VOLUMES IN THE ARIS SERIES

Casebook on Insurgency and Revolutionary Warfare, Volume I: 1933–1962 (Rev. Ed.)
Casebook on Insurgency and Revolutionary Warfare, Volume II: 1962–2009
Undergrounds in Insurgent, Revolutionary, and Resistance Warfare (2nd Ed.)
Human Factors Considerations of Undergrounds in Insurgencies (2nd Ed.)
Irregular Warfare Annotated Bibliography

FUTURE STUDIES

The Legal Status of Participants in Irregular Warfare
Case Studies in Insurgency and Revolutionary Warfare—Colombia (1964–2009)
Case Studies in Insurgency and Revolutionary Warfare—Sri Lanka (1976–2009)

SORO STUDIES

Case Study in Guerrilla War: Greece During World War II (pub. 1961)
Case Studies in Insurgency and Revolutionary Warfare: Algeria 1954–1962 (pub. 1963)
Case Studies in Insurgency and Revolutionary Warfare: Cuba 1953–1959 (pub. 1963)
Case Study in Insurgency and Revolutionary Warfare: Guatemala 1944–1954 (pub. 1964)
Case Studies in Insurgency and Revolutionary Warfare: Vietnam 1941–1954 (pub. 1964)

Published by:

The United States Army Special Operations Command

Fort Bragg, North Carolina

25 January 2013

Second Edition

First Edition published by Special Operations Research Office, American University, December 1965

Comments correcting errors of fact and opinion, filling or indicating gaps of information, and suggesting other changes that may be appropriate should be addressed to:

United States Army Special Operations Command

G-3X, Special Programs Division

2929 Desert Storm Drive

Fort Bragg, NC 28310

All ARIS products are available from USASOC at www.soc.mil under the ARIS link.

LETTERS OF INTRODUCTION

The foreword to Special Warfare's 1966 *Human Factors Considerations of Undergrounds in Insurgencies* notes, "in the desire to understand the broad characteristics and societal impact of revolutionary movements we often neglect the study of the human element involved." "To understand the individual, his reasons, his behavior, and the pressures that society places upon him is at the heart of the problem of social change." The earlier study and this updated edition represent part of our intellectual investment in understanding the human domain. Understanding the human domain remains critical for future Special Warfare operations.

Since the inception of the United States Army Special Forces, understanding indigenous individuals and the human domain in which they exist has been a persistent Army Special Operations Forces cornerstone. Relationships with indigenous individuals enable Special Warfare. Understanding why individuals choose to join an underground movement, why law-abiding citizens are tempted to lead a dangerous underground life, why individuals stay in underground organizations, and what behaviors individuals use to survive are key questions that will reveal insights into the individuals that may be our partners. Special Warfare's leverage of and reliance on indigenous forces offers a unique capability. This Special Warfare capability offers our nation's leaders necessary and different strategic options. Our Special Warfare mission necessitates our continued educational and intellectual commitment to studying human factors. Our endeavor must include institutional and individual commitments. This updated volume offers a beginning, and the text will be integrated into our schoolhouse curriculums. The schoolhouse introduction represents only the starting point for each Army Special Operations Forces member's continued learning. Our nation requires a Special Warfare capability. The Special Warfare capability requires intellectual investment and continuous evolution to understand the people that the human domain represents. I encourage each member to read, analyze, debate, and challenge this work as we endeavor to remain the premier Special Warfare capability in the world.

LTG Charles T. Cleveland
Commanding General, U.S. Army Special Operations Command

The first time I saw the original version of this book was when it was presented to me in 1979 at the Special Forces Qualification Course, and I have reread it multiple times in the years since. The original edition of this book, and the remainder of the series produced by Special Operations Research Office (SORO),[a] constitute some of the foundational references necessary to a full understanding of Special Forces operations and doctrine. Thirty-three years later, I still use it routinely to teach and mentor new and experienced Special Forces personnel.

This book is intended as an update to the original text. Basic human nature does not change, and much has remained the same throughout the history of human political conflict. The new edition therefore includes the enduring principles and methods from the original. The sections of the book discussing techniques, causes, and methods, however, required updating to reflect the many changes in the world and experiences of Special Forces in the last fifty years.

While the need to resist perceived oppression has not changed, the ways in which oppressed societies express this need through culture have changed significantly. Whereas insurgencies prior to the publication of the first edition of this book in 1963 were predominantly Communist-inspired, modern rebellions have been inspired by a greater number of factors. This increase in the causes of insurgency began with the fall of the Soviet Union, and accelerated with the attacks of September 11, 2001, on the United States.

In addition, technology has changed the world significantly over the last fifty years. Modern communications, the Internet, global positioning and navigation systems, and transportation have all introduced different dynamic pressures on how insurgencies develop and operate. As a result of the cultural and technological shifts of the past fifty years, we decided that the original series by SORO needed revision and updating. Like the original series, this update is written by sociologists, this time from The Johns Hopkins University Applied Physics Laboratory.

Plenty of histories and military analyses have been written on the different revolutions that have taken place over the past fifty years, but this series—including the new, second volume of the *Casebook*—hopefully provides a useful perspective. Since rebellion is a sociopolitical issue that takes place in the human domain, a view through the nonmilitary lens broadens the aperture of learning for Special Forces soldiers.

---

[a] They include: *Undergrounds in Insurgent, Revolutionary, and Resistance Warfare* and the *Casebook on Insurgency and Revolutionary Warfare*, which includes case studies on Cuba, Algeria, Guatemala, and Vietnam, and a study of guerrilla warfare in Greece during World War II.

This project would not have been possible without the support of COL Dave Maxwell, who was the USASOC G3 (my boss) and fought to get the project approved. Also I must thank Michael McCran, a long-time friend who encouraged me to continue to seek the approval of this project.

Paul J. Tompkins Jr.
USASOC Project Lead

PREFACE TO THE FIRST EDITION, 1966

Human Factors Considerations of Undergrounds in Insurgencies is the second product of Special Operations Research Office (SORO) research on undergrounds. The first, *Undergrounds in Insurgent, Revolutionary, and Resistance Warfare*, was a generalized description of the organization and operations of underground movements, with seven illustrative cases. The present study provides more detailed information, with special attention to human motivation and behavior, the relation between the organizational structure of the underground and the total insurgent movement, and Communist-dominated insurgencies.

Because an understanding of the general nature of undergrounds is necessary to more detailed considerations, some of the information from the earlier study of undergrounds has been included in this report. Wherever possible, material from insurgency situations since World War II has been used. Occasionally, however, it was necessary to use information from studies of World War II underground movements in order to fill gaps about certain operations.

In the methodological approach it was assumed that confidence could be placed in the conclusions if data on underground operations and missions and similar data could be found in other insurgencies. An attempt was made to base conclusions on empirical information and actual accounts rather than theoretical discussions, and upon data from two or more insurgencies. An effort was made to find internal consistencies within the information sources. For example, if units were organized and trained to use coercive techniques for recruiting, and defectors described having been recruited in this manner, the conclusion that people were coerced into the movement can be made. Because of this approach there is a good deal of redundancy within and among the various chapters.

While the main emphasis in this report has been on underground organization, many characteristics can be understood only in relation to overt portions of the subversive organization. Therefore, discussions of guerrilla forces, the visible outgrowth of undergrounds, and of Communist structures, which often inspire, instigate, and support subversive undergrounds, have been included. The report is designed to provide the military user with a text to complement existing training materials and manuals in counterinsurgency and unconventional warfare, and to provide helpful background information for the formulation of counterinsurgency policy and doctrine. As such, it should be particularly useful for training courses related to the counterinsurgency mission.

The authors wish to express thanks to a number of persons whose expertise and advice assisted substantially in the preparation of this report. Mr. Slavko N. Bjelajac, Director of Special Operations for the Office of the Deputy Chief of Staff for Operations, Department of the Army, on the basis of his personal experiences and special interest in the study of underground movements, contributed guidelines and concepts to the study.

Four men reviewed the entire report: Dr. George K. Tanham, Special Assistant to the President of the Rand Corporation of Santa Monica, California, made many helpful suggestions based upon his firsthand experiences and study of Communist insurgency; Dr. Jan Karski, Professor of Government at Georgetown University, Washington, D.C., whose personal experience as a former underground worker is combined with a talent for thorough, constructive criticism, also helped the final manuscript; Dr. Ralph Sanders of the staff of the Industrial College of the Armed Forces, Washington, D.C., offered a careful and useful critique of the manuscript and helpful suggestions; Lt. Col. Arthur J. Halligan of the U.S. Army Intelligence School, Fort Holabird, Maryland, provided valuable suggestions based upon his experience in Vietnam.

Within SORO, Dr. Alexander Askenasy, Brig. Gen. Frederick Munson (Ret.), Mr. Phillip Thienel, Mr. Adrian Jones, Dr. Michael Conley, Mrs. Virginia Hunter, and Mrs. Edith Spain contributed to the end product.

Andrew Molnar

PREFACE TO THE SECOND EDITION

This book, *Human Factors Considerations of Undergrounds in Insurgencies*, is the second edition to the 1966 book of the same name. The first edition of this book was produced by the Special Operations Research Office (SORO) at American University in Washington, DC. SORO was established by the U.S. Army in 1956. During the 1950s through the mid-1960s, SORO social scientists and military personnel researched relevant political, cultural, social, and behavioral issues occurring within the emerging nations within Asia, Africa, and Latin America.[1] The researchers conducted analyses, sometimes for the first time, on the effects of propaganda and psychological operations and the roles of the military in developing countries, and provided large bibliographies of unclassified materials related to counterinsurgency and unconventional warfare. The Army had a particular interest in understanding the processes of violent social change in order to be able to cope directly or indirectly through assistance and advice with revolutionary actions. In 1962, SORO published the *Casebook on Insurgency and Revolutionary Warfare*; in 1963, it published *Undergrounds in Insurgent, Revolutionary, and Resistance Warfare*; and in 1966, it published *Human Factors Considerations of Undergrounds in Insurgencies*—each of these publications remained in the Special Operations training curricula for subsequent generations. The work of SORO, whose scholarship grew more ambitious and controversial, and later the Center for Research in Social Systems (CRESS), has endured. Some of the reports under Project PROSYMS (Propaganda Symbols) have served as examples of incorporating rigorous social science research methods into psychological operations, and the CRESS report on the subversive manipulation of crowds is still widely used as a training resource. Relevant components of many of SORO and CRESS papers are referenced in this edition.

The American experience in Afghanistan, Iraq, the Philippines, the Horn of Africa, and other locales in the twenty-first century reaffirmed the need for cooperation between the academic and operational communities. In 2009, the U.S. Army Special Operations Command (USASOC) G-3X sought to recreate some of the capability SORO had provided. They turned to, among other institutions, The Johns Hopkins University Applied Physics Laboratory (JHU/APL). Under a project entitled Assessing Revolutionary and Insurgent Strategies (ARIS), researchers at JHU/APL have been engaged in understanding how social movements such as insurgencies and revolutionary groups are created; how they grow, spread, and sustain themselves; and how they succeed or fail.[2] The first product of that effort is the forthcoming *Casebook on Insurgency and Revolutionary Warfare, Volume II: 1962–2009*. Similar to the first volume, the second volume is intended to provide a foundational understanding of insurgent warfare by presenting cases using a common analytical framework and format. The second and third major products under ARIS are the second editions of *Undergrounds in Insurgent, Revolutionary, and Resistance Warfare* and *Human Factors Considerations of Undergrounds in Insurgencies.*

Most of the text in this edition is new. Some large sections of the first edition are retained verbatim, mostly in Chapter 3's study of Communist organizations and sections of Chapter 5 on recruitment and retention, but also in smaller sections of the other chapters as well. We preserved much of the overall structure, although not the specific chapters, and strove to answer many of the same underlying questions. Material from the first edition is used without citation. Material from other SORO studies is referenced like any other source.

Intended as a complement to the second edition of *Undergrounds in Insurgent, Revolutionary, and Resistance Warfare*, this book delves deeper into theory and further into background materials and focuses less on operational details. We provide numerous chapter cross-references to the second edition of *Undergrounds*. We also drew heavily on the new, second volume of the *Casebook*; these cases are cited in the normal way. We also provide a table of contents at the beginning of every chapter to make the book more useful as a reference.

The first edition of *Human Factors* was an important synthesis of a poorly understood topic and has proved to have some remarkable staying power, with much still relevant even in the edition's fifth decade. An update to the first edition is needed, however, simply because the world has changed since the 1960s.

CHANGES SINCE THE ORIGINAL EDITION

The 1966 *Human Factors* edition focused on the contemporary threat of Maoist insurgencies, particularly in Southeast Asia, and also drew extensively on World War II resistance movements in Europe. Much of this information is still relevant and has been retained and integrated. In the post-Cold War world, the most important insurgencies tend to be ethnic and religious. Long-simmering conflicts, sometimes with roots in colonial policies, have become prominent; examples include the Liberation Tigers of Tamil Eelam (LTTE) in Sri Lanka, Euskadi Ta Askatasuna (Basque Homeland and Freedom or ETA) in Spain, the Hutu-Tutsi genocides, the Ushtia Çlirimtare e Kosovës (Kosovo Liberation Army, or KLA), and the Provisional Irish Republican Army (PIRA). Battle lines in these conflicts are often drawn along ethnic lines, even when land or politics are the immediate issues in contention. The other important new category is extremist religious movements, most prominently

Islamic groups, including regional insurgent movements like Hizbollah and Ḥarakat al-Muqāwamah al-'Islāmiyyah (Islamic Resistance Movement, or HAMAS) and global movements like Al Qaeda. These present a different profile of ideology, organizational forms, and psychology than either Cold War Maoists or post-colonial ethnic insurgencies (although the Palestinian cause could be considered a post-colonial issue).

Globalization has also changed underground operations in numerous ways. Insurgencies, enabled by low-cost transportation, Internet-based communications, and other information technologies, can more easily recruit, communicate, and operate across borders. It is correspondingly much more difficult to contain an insurgency in a region. Global media has led to development of new tactics, in particular new types of terrorism, designed to capture worldwide attention.

Compared with what was available in the 1960s, there are orders of magnitude more academic research available relevant to this study's topics. We were able to draw on more recent work in psychology, political science, economics, sociology, organizational studies, and communications studies. Readers of this edition will, over the course of eleven chapters, get a wide exposure to basic concepts from a number of disciplines. This breadth also presented a challenge in trying to evaluate and synthesize many competing explanatory frameworks while keeping focus on what is relevant (sometimes directly, sometimes as background) to the study. The authors strove to accurately synthesize relevant background material but avoided wading into inconsequential details or academic debates; readers can make their own judgment as to how effectively the latter part was accomplished.

ACKNOWLEDGMENTS OF ORIGINAL AUTHORS AND ARIS CONTRIBUTORS

Each chapter in this edition does have author credits at the beginning; chapters that borrow significant sections of the original edition list "SORO authors" as authors. (At the beginning of this effort, the co-authors for this edition, along with project lead Chuck Crossett, had a brief social meeting with Andrew Molnar, first author of the original study, who is now retired and living a short drive from our campus in Maryland. Dr. Molnar, for his part, was quite amenable to his work being used in an updated edition.)

The ARIS project is the result of the joint vision and initiative of Paul Tompkins Jr., chief USASOC, G-3X Special Programs Division, and our colleague Chuck Crossett of JHU/APL.

John Shissler, Senior Researcher, JHU/APL, served as quality reviewer for this edition and made many substantive contributions. This effort also benefitted from the knowledge and efforts of our JHU/APL colleagues Jerome Conley, Angela Hughes, Robert Leonhard, Kelly Livieratos, and Summer Newton.

Nathan Bos, Ph.D., and Jason Spitaletta
Laurel, Maryland, March 2012

ENDNOTES

[1]   Joy E. Rohde, "The Social Scientists War: Expertise in a Cold War Nation" (Ph.D. diss., University of Pennsylvania, 2007).

[2]   Chuck Crossett and Ronald J. Buikema, "Analysis of Social Movements in Warfare," *Johns Hopkins APL Technical Digest* 30, no. 1 (2011): 5–12.

TABLE OF CONTENTS

LIST OF ILLUSTRATIONS

LIST OF TABLES

CHAPTER 1.

INTRODUCTION

CHAPTER CONTENTS

Nathan Bos and Jason Spitaletta

Warfare is an inherently human activity, fraught with the idiosyncrasies of human behavior experiencing existential stress. The complexity of human factors in warfare is perhaps best articulated by Clausewitz's wondrous trinity, which describes war as "more than a true chameleon that slightly adapts its characteristics to the given case. As a total phenomenon its dominant tendencies always make war a paradoxical trinity—composed of primordial violence, hatred, and enmity, which are to be regarded as a blind natural force; of the play of chance and probability within which the creative spirit is free to roam; and of its element of subordination, as an instrument of policy, which makes it subject to pure reason."[1]

An understanding of human factors is arguably more critical in irregular warfare than in conventional warfare and especially critical in complex civil conflicts that extend over time into a protracted war for "hearts and minds." The Director of Central Intelligence Directive (DCID) 7/3 defines human factors as, "The psychological, cultural, behavioral, and other human attributes that influence decision-making, the flow of information, and the interpretation of information by individuals and groups at any level in any state or organization." This definition is appropriate but somewhat cold. It fails to communicate the simultaneously fascinating and frustrating part of human factors in insurgencies, which is that such conflicts push common behavioral and social dynamics to their extremes. Organizational design, leadership, social influence, mental health, and the other topics discussed in this book are standard topics in the behavioral sciences. But these choices take on existential significance when they involve an insurgent organization weighing trade-offs between reliability and flexibility, a strategic choice whether to embrace violence or nonviolence, or an individual's judgments about whether to trust an ideology, a social contact, or a charismatic leader. Understanding individual choices regarding underground movements requires knowledge of mental health, the dynamics of hate, and the power of social influence in extremis. Understanding a population's support or rejection of such movements requires understanding of a broad set of political, economic, and social factors, and often requires an understanding of how individuals respond to oppression, violence, or terrorism.

This book is divided into two background chapters and three main sections. This structure follows the basic organization and flow of the first edition. Previews of the sections and chapters are provided below.

UNDERLYING CAUSES

Why do violent insurgencies appear in some countries and persist for decades, while countries that appear culturally, politically, or economically similar experience no such events? Understanding where and why civil violence is likely to occur is of great interest to policy makers, of course, and is also relevant to soldiers on the ground inasmuch as it helps explain the causes and internal dynamics of a conflict. Chapter 2 delves into the political and social factors that seem to predict civil violence.

Poverty contributes to instability, but not in as direct a way as might be supposed. People's perception of *relative deprivation*, which is failure to meet economic expectations or keep up with visible peer groups, tends to be more important than the absolute level of deprivation. Poor governance combined with poverty is an even more dangerous combination. The legitimacy of the government in the eyes of the populace can insulate governments in poor countries or endanger them even in wealthier ones; legitimate governments provide security, justice (especially related to corruption), and economic development and have ideological legitimacy conveyed to them by other powerful institutions.

Marginalization or persecution of ethnic or religious groups within the country is one of the strongest factors predicting violent resistance. The importance of social identity-based conflicts in understanding insurgencies is a common theme in this book. Countries in conflict-filled "neighborhoods," or with a history of domestic conflict, are much more likely to see future conflict because of the ready availability of weapons and trained personnel, and because the population may be desensitized to violence. Countries experiencing a demographic "youth bulge" seem to be particularly vulnerable to instability; such population bulges often lead to feelings of relative deprivation if young people are unemployed at high levels.

Insurgencies need funding, and those that do not have external sponsors must find funding domestically; the presence of an exploitable commodity resource that can be smuggled, stolen, or extorted, such as diamonds, drugs, or oil, seems to increase the likelihood of an ongoing insurgency. The terrain itself may be a contributor, as dense forests or mountainous terrain provide cover for insurgents. Chapter 2 of this work reviews the evidence surrounding these factors and details the mechanisms by which these factors may lead to violent resistance.

PART I: UNDERGROUNDS AS ORGANIZATIONS

Chapter 3: Organizational Structure and Function and *Chapter 4: Leadership* complement the similar chapters in the companion to this work, the second edition of *Undergrounds in Insurgent, Revolutionary, and Resistance Warfare.* In Chapter 3, we examine six organizational challenges faced by all insurgent groups: command and control, aligning of structure with strategy, secrecy and compartmentalization, evolution and growth of organizations, underground and aboveground connections, and criminal connections. To understand how these factors play out in a contrasting set of cases, we examine these aspects of the Provisional Irish Republican Army (PIRA) in Ireland, of prototypical Communist organizations, and of Al Qaeda. In terms of command structure, the PIRA's military-type of command structure is an interesting contrast to Al Qaeda's "franchise" model, which is different again from the Communist practice in which nonmilitary decision-making bodies, at least in theory, maintain control over armed components. Navigating the aboveground and underground connections presents challenges for every insurgent group, and the PIRA's rejection, then gradual embrace of, political engagement (via Sinn Féin) highlights these issues. The international Communist movement, although well past its peak of influence, is still a fascinating organizational study in its complex, ingenious (in concept) interlocking coordination of infiltrated civil organizations, Communist front organizations, national and international party leadership, and military components.

Public component activities
Armed component activities
Underground activities

Negotiated settlement

International strategic communications

Large-scale military and
paramilitary actions

Minor military and paramilitary actions

Shadow governance activities

Overt

Shadow governance activities

Clandestine

Increased political violence, terror, and sabotage

Negotiations with government representatives

Intense sapping of morale of government,
administration, police, and military

Increased underground activities to demonstrate
strength of revolutionary organization

Sabotage and terror to demonstrate weakness of government

Overt and covert pressures against government; strikes,
riots, and disorders

Intensification of propaganda, increase in disaffection,
psychological preparation for revolt

Expansion of and coordination among resistance networks

Establishment of formalized resistance elements; appeal to
extraterritorial support infrastructure

Spreading subversive organizations into all sectors of life in
a country/region

Penetration into professional, social, and political organizations and into
all parts of society

Recruitment of like-minded individuals and others; indoctrination and use
of these for organizational purposes

Infiltration of foreign agents and agitators, and foreign propaganda material,
money, weapons, and equipment

Increased agitation, unrest, and disaffection, infiltration of administration, police,
and military and national organizations, and slowdowns and strikes

Assassination, forming favorable public opinion (advocating national cause), creation
of distrust of established institutions

Creation of atmosphere of wider discontent through propaganda, lies, and political and
psychological effort: discrediting government, police, and military authorities

Dissatisfaction with political, economic, social administrative, and/or other conditions; national
aspiration (independence) or desire for ideological and other changes

Preparation of revolutionary cadres and of masses for revolution

Preparation of revolutionary cadres and of masses for revolution

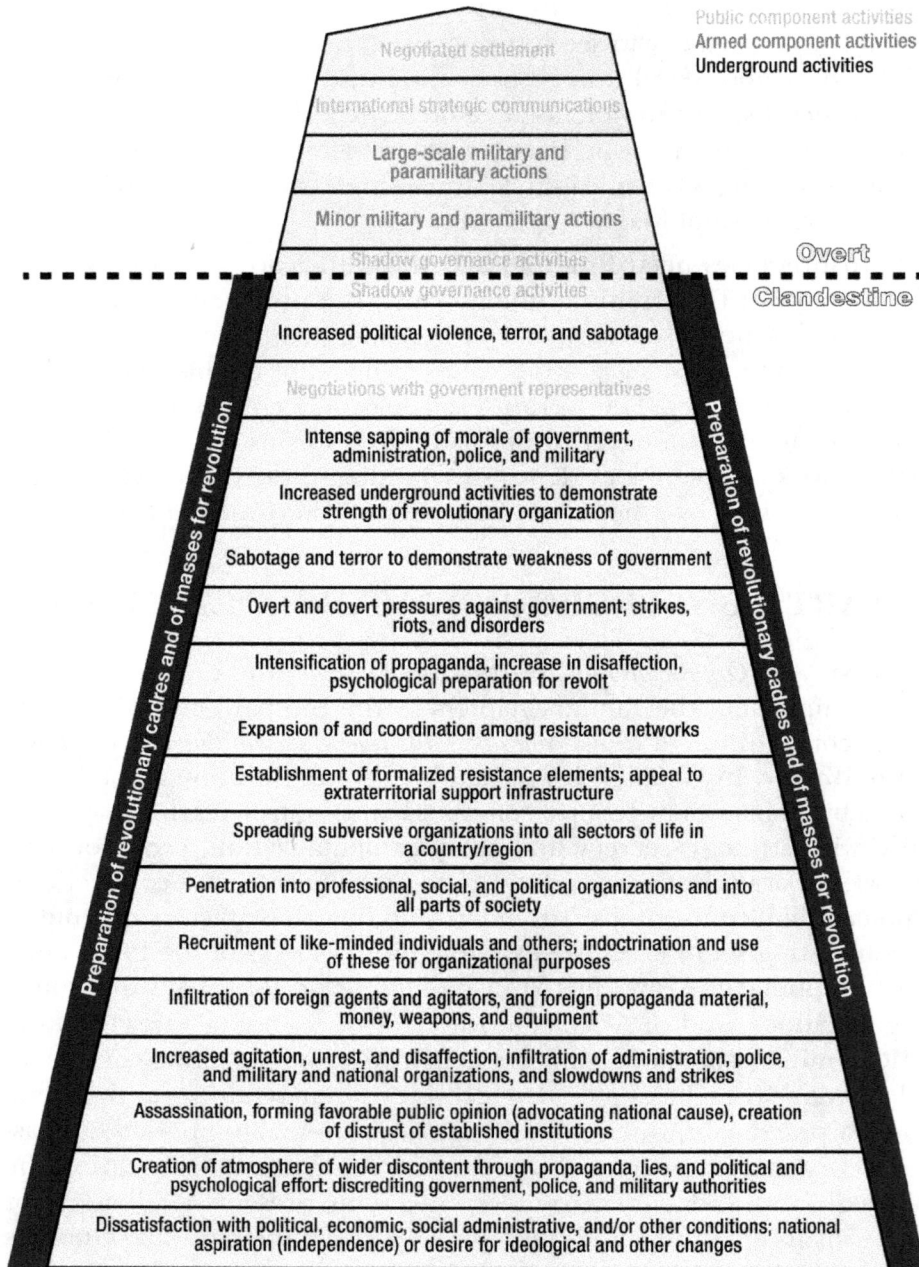

Figure 1-1. Covert and overt functions of an underground.

Figure 1-1, also shown in Chapter 3, shows the range of organizational functions performed by an underground. A figure very similar to this was included in the first edition and has been reproduced somewhat widely since; this update to the figure includes more aboveground functions of insurgent public components.

In addition to organizational structure, the personalities of key leaders can also have a strong influence on the operations of insurgent organizations, particularly during their early stages. What kind of personality is required to recruit and lead friends, relatives, and strangers in such a dangerous undertaking? Upstart insurgencies are often headed by leaders who are charismatic, a style with critical strengths and marked limitations, particularly when groups become larger. Several other strong personality traits, some bordering on psychopathologies, may be associated with insurgent leadership and may affect, hinder, or sometimes help a would-be underground leader. Chapter 4's study of leadership includes an example profile of former Egyptian Islamic Jihad (EIJ) and current Al Qaeda leader Dr. Ayman Al-Zawahiri; the profile was drawn from an unclassified integrated personality profile[2] using the Integrated Political Personality Profile method that was created, published, and taught by Dr. Jerrold M. Post.

PART II: MOTIVATION

The middle section of this book addresses the psychological and social variables important to understanding insurgent membership. Chapter 5 reviews what is known about joining, staying in, and leaving underground organizations, particularly foot soldiers in armed struggles. Social connections are always a key factor in recruitment. Ideology seems to play a more important role in early recruitment of true believers and a lesser role in mass recruitments during later-phase armed struggles. In these later phases, some groups turn to coercion, although this is usually paired with some subtlety with other incentives or manipulations. In Chapter 5, the recruitment strategies of rural insurgents (mostly Viet Minh and Phillipino Hukbalahap) are compared with the very different recruitment strategies used by the PIRA during "The Troubles" in Ireland and the recruitment of lonely and vulnerable college students by Egypt's EIJ/Egyptian Islamic Group (EIG).

Ideology is a necessary although not sufficient element that affects both recruitment and retention. Ideology provides intellectual common ground for a movement, reduces uncertainty, justifies violent actions, and motivates members to persist through difficulty. One modern example of a well-developed ideology that has grounded and inspired radical and insurgent groups is Qutbism. Based on the writings of Sayyid Qutb, this ideology condemned current Arab leadership and called Muslims to a different kind of engagement with the outside world. Its influential role in shaping first modern Egyptian and ultimately global Salafist jihad movements is discussed in some detail.

Research on retention in underground movements is sparse, but there seem to be good parallels to soldiers in conventional military units in losing efforts, who often remain loyal to military units because of their social commitments and quality of leadership. There is also emerging research on why people leave insurgent groups, including a number of interesting (but largely unproven) tactics for "deradicalization" that are reviewed in Chapter 5.

Individual behavior can rarely be understood without understanding the social influences that surround that person, and the focus of Chapter 6 is on group dynamics. Academic research on in-group/out-group conflicts started with bewildered researchers trying to explain the horrors of the Holocaust and other atrocities of World War II. These classic works remain relevant to today's ethnic and religious conflicts, such as the escalation of tensions between Sinhalese and Tamils in Sri Lanka that fed the Tamil Tiger insurgency.

A recent paper on mechanisms of radicalization[3] provides a framework for discussing how group dynamics and outside pressures can lead groups to "radicalize," or cross the line from nonviolent to violent actions. Some of these mechanisms are fairly easy to understand, such as radicalization under threat, but others are less intuitive, such as radicalization of groups in competition for the same base of support. The steps by which hate groups form and desensitize members to increasingly violent actions are also described.

Chapter 7 delves deeper into the individual psychological factors that may affect underground involvement. The mental stability of insurgents, terrorists, and suicide attackers is examined, and although the profile of the "psychopathic terrorist" is mostly a myth, there are some disorders that are helpful in understanding violence, suicidality, and related phenomena. A number of other more circumstantial influences on a person's psychological state can affect the decision to become involved in a violent movement, including emotional vulnerability, experience of grievances such as abuse by government or military forces, vicarious experiences of grievances such as through Internet-based recruiting materials, and the perception that one's religious or ethnic group has experienced humiliation at the hands of an opponent.

PART III: UNDERGROUND PSYCHOLOGICAL OPERATIONS

Insurgent groups never succeed as purely military operations; they must also effectively engage in battles of persuasion and influence. While the term "psychological operations" is no longer used by the

U.S. military (in favor of Military Information Support Operations or MISO), the phrase is an appropriate label for the functions that underground groups can and do perform. The information domain's impact on the radicalization process in modern insurgency cannot be overstated. The seemingly ubiquitous availability of information, including ideological narratives and success stories and even the presence of tactics, techniques, and procedures, can have a profound influence on cognitive processes.

Chapter 8 describes the use of media to reach a variety of audiences, to seek support from internal and external sympathizers, and to intimidate or demoralize internal opposition and the enemy. The basic functions of communication remain the same whether the medium is handbills or websites. But two recent historical technological developments did seem to be game changers for insurgent groups and those that oppose them. The first was the rise of global broadcast media, particularly satellite-linked television, which encouraged spectacular terrorist attacks such as the Munich Olympic kidnappings. The second technological breakthrough is still developing—use of the Internet. The Internet has lowered the cost and extended the reach for every important insurgent communications activity, including publicity, recruitment, training, fundraising, and command and control.

Insurgent groups, and those who oppose them, must also understand the basic principles of influence and its use in recruitment, persuasion, negotiation, and coercion. Chapter 9 is a more academic review of research in the area of influence, with relevant examples in the military domain. When influence is exercised one-on-one or one-to-many, charisma and nonverbal communication are critical factors. As influence extends throughout a population, understanding social networks and the roles of key individuals in a network becomes paramount; Everett Rogers's model of how innovations diffuse in a population via "change agents," "innovators," and "opinion leaders" is relevant for political movements as well. The content of messages themselves also matters; careful use of evidence, "inoculation" messages, personal relevance, and other factors are what make certain messages resonate while others are ignored. It has also long been understood that the combination of words and actions can be more powerful than either by itself. The psychological mechanism of "cognitive dissonance" underlies some models for how this can be accomplished. Chapter 9 also includes a study of narratives as influence mechanisms, matching types of narrative arguments with different types of resistance movements.

Mastery of nonviolent tactics, as detailed in Chapter 10, is also critical for undergrounds facing stronger opponents. Some movements rely on nonviolent tactics exclusively; some integrate them with military actions (although this does tend to detract from the legitimacy of the nonviolent aspects). Many have developed ways to subversively manipulate nonviolent actions such as street protests to provoke violent confrontations, justify retaliatory attacks, or divert attention from other actions. Nonviolent actions can be classified into three broad categories: attention-getting devices such as street protests and street performance art; noncooperation techniques such as boycotts and work slowdowns; and civil disobedience campaigns such as civil rights sit-ins and Gandhi's salt protests. Cyber activism and so-called Hacktivism are two newer developments in this area and are described in Chapter 10.

Terrorism, described in Chapter 11, has proven to be one of the more effective forms of psychological warfare. Chapter 11 discusses not only the individual and social psychological effects of terrorism but also the planning and justification processes behind the decision to use this tactic. The resultant anxiety and dysphoria associated with acts of terror create not only an increased fear but also awareness of death. This leads individuals to affiliate with those of similar worldviews and to be more willing to sacrifice their civil liberties to charismatic (and authoritarian) leaders. The chapter goes on to discuss some of the considerations and risks of terrorism as an insurgent tactic as well as some generalized patterns of behavior under the threat of terrorism.

ENDNOTES

1. Andreas Herberg-Rothe, "Clausewitz's 'Wondrous Trinity' as a Coordinate System of War and Violent Conflict," *International Journal of Conflict and Violence* 3, no. 2 (2009): 204–219.

2. Jason Spitaletta, "Integrated Personality Profile: Dr. Ayman Mohammed Rabie Al-Zawahiri" (unpublished paper submitted to The George Washington University, 2009).

3. Clark McCauley and Sophia Moskalenko, "Mechanisms of Political Radicalization: Pathways Toward Terrorism," *Terrorism and Political Violence* 20, no. 3 (2008): 415–433.

CHAPTER 2.

UNDERLYING CAUSES OF VIOLENCE

CHAPTER CONTENTS

Nathan Bos

Why do violent insurgencies appear in some countries and persist for decades, while countries that appear culturally, politically, or economically similar experience no such events? Many people would, of course, like to be able to predict where violence is likely to occur before it happens, as well as understand what regimes are vulnerable and when. Political scientists have for decades been trying to understand what factors increase the probability of terrorism, insurgency, or other forms of political violence appearing in a country or region. This chapter will review the set of economic, political, geographic, and social variables that have the best evidence of being linked to the likelihood of violence.

In recent years, this question has been addressed with statistical analysis of large historical datasets. Without delving too far into the mathematics of these techniques, it is worth describing qualitatively how these techniques use past data to try to identify the most important factors influencing violence.

A good example of such a study was recently sponsored by the Central Intelligence Agency (CIA) and conducted by a very accomplished group of political scientists.[1] This study sought to build a statistical model that could forecast instability events with as small a number of variables as possible. They produced a model that could "predict" 80 percent of instability events based on just four variables, which will be described later in this chapter.

Models to predict the future are necessarily based on past data, so this group, drawing on several well-known datasets, assembled data to construct a cross-national time-series dataset covering the period from 1955 through 2003 for all countries with a population of more than 500,000. Each record (row) in this dataset represented one country year (e.g., United States, 1965). The dataset included 7,500 total country years.

Their definition of an instability event included three types of events: civil wars, adverse regime changes, and genocides/politicides. (Precise definitions of each of these are included in the paper.) They identified 141 instability events and then began analysis of what factors might have predicted these.

Modern studies almost always use some form of multivariate regression, which is a statistical technique that allows one to propose a set of predictor variables (e.g., infant mortality, unemployment, civil wars in neighboring countries, climate) and an outcome variable (instability event) and determine which of the variables are most strongly associated with violent events while controlling for the other variables. The

13

ability to control for the influence of some variables and isolate the influence of others is the most important power of regression compared with other techniques.

Different combinations of variables are tried to see which ones produce the strongest model. Multivariate regression forces predictor variables to "compete" with each other as explanatory variables within a given model. So one could enter both "infant mortality" and "unemployment" into the same model and might find that when infant mortality was available as an explanatory variable, unemployment became unimportant. The model could also show the reverse or show that they were both important independent of each other. These models can also reveal more complex relationships between three or more predictor variables.

Researchers will often try out dozens of combinations of variables over the course of a study, adding and removing variables to understand how they interact. (Because of the mathematical assumptions of noncollinearity in regression, it is impermissible to include predictor variables that are strongly correlated with each other, so researchers cannot simply include every variable in every model.) In published papers they may present just one best model or may present a series of perhaps four or five best alternative models, allowing the reader to also use their own judgment.

One of the most difficult aspects of interpreting statistical studies of past events is separating correlations from causality. There may be a strong statistical relationship between infant mortality and civil war, but does this mean that people are more likely to revolt because of high infant mortality (and associated poverty and poor health care), does it simply mean that health care services tend to be disrupted during wartime, or is there another explanation? An important additional tool is the use of time-lagged analysis. If infant mortality is also lower in the two years before the onset of a civil war, this shows that disruption in health care due to war is not the only reason for the relationship. This does not completely answer the question of how infant mortality is involved in causing instability, of course, but it does eliminate some possible explanations.

In the CIA study mentioned earlier, the authors presented a four-factor model which "predicted" 80 percent of instability events. The four factors were (1) infant mortality (perhaps acting as a proxy for development), (2) presence of armed conflict in four or more bordering states, (3) regime type, and (4) presence of state-led discrimination. These will each be discussed below as risk factors for civil violence. The risk factors identified in this chapter are supported by similar

14

regression-based studies performed over several decades of research with a variety of datasets and theoretical assumptions.

This chapter will discuss eight risk factors for political violence. Six of these are human factors: economic deprivation, poor governance, lack of legitimacy, marginalization and persecution of identity groups, history of conflict in the country or conflict in nearby countries, and unfavorable demographics such as a "youth bulge." We will also discuss two risk factors that are not human factors: presence of a primary commodity resource and terrain type.

Each of these factors has been identified in at least one large-scale study relating it to political instability, political violence, or civil war. It is important to note that there is not universal agreement on any of these factors. For each of these factors, one can find at least one study or model that does *not* find a statistical relationship between that factor and political violence. However, absence of evidence is not necessarily evidence of absence, and we have chosen to err on the side of inclusiveness to present a broader set of possible factors.

ECONOMIC DEPRIVATION

Political violence is more likely to occur in countries with lower levels of economic development and less likely to occur in prosperous countries, making economic deprivation a risk factor. Overall level of economic development has been shown to be a significant predictor of civil war. Different authors have used different variables as indicators or proxies for development. Each of these has been shown to have a relationship with political violence: energy usage per capita,[2] per-capita income,[3] infant mortality,[4] and level of male secondary schooling.[5]

The simplest explanation for these relationships is probably wrong: insurgencies do not spring up solely because of a population's anger about poverty or associated deprivations such as lack of education, health care, or employment. Deprivation may lead individual poor people to participate in "bread riots" or to commit property crimes, but it does not directly lead to organized, sustained insurgencies.

It is also true that political opposition occurs in all settings, including more developed countries; and revolutions are usually led by wealthier and more educated members of society. Deprivation alone is rarely used by non-Communist groups as a narrative to explain their actions, although it is may be given as a justification after the fact of a successful revolution.[6]

Poverty is generally considered to be an indirect contributing factor and not a primary cause of political violence. Poverty may lead young military-age men to feel that they have fewer options in life and less to lose, making joining an insurgency a more attractive option. Poverty does tend to increase property crimes, which may desensitize populations to lawlessness and violence, create a criminal class and illegal markets to support insurgencies, and undermine the government's legitimacy.

Successful Communist revolutionaries understood these dynamics very well. Their goal was to convince peasants over time that their poverty and deprivation were the fault of the elite and the government, but they saw little value in entering a village and immediately exhorting the poorest villagers to rise up against their oppressors. Before seeking recruits for a class struggle, Maoists did a great deal of groundwork learning about very specific local grievances and teaching villagers to blame these things on wealthy landowners and the government. Equally important, they worked to build the complex social, organizational, and ideological infrastructure needed for an effective insurgency, as will be described in the following chapters of this book.

Theda Skocpol's well-regarded comparative study of three revolutions illustrates these points.[7] The three revolutions were France 1789 (the French Revolution), Russia 1917 (resulting in the takeover of Leninists), and China 1911–1949 (resulting in the takeover of Maoists). Each of these featured peasant revolts was motivated in the long term by economic exploitation and in the short term by food shortages or other economic failures. But it is the author's view that

> In all agrarian bureaucracies at all times, peasants have had grievance enough to warrant, and periodically spur, rebellions. . . . Economic crises (which are endemic in semi-commercial agrarian economies anyway) . . . might substantially enhance the likelihood of rebellions at particular times. But such events ought to be treated as short-term precipitants of peasant unrest, not fundamental underlying causes.[8]

In order to foment revolution, each of the three cases required other elements to be in place, specifically (1) outside military and economic pressure from other countries that were modernizing faster and undermined the ruling bureaucracies and (2) parallel movements led by "marginal elites" for their own reasons. These political elites, such as the Maoist Communists in China, might assist the peasantry in forming a cohesive identity, organizing across distance, and mobilizing in a coordinated fashion. In the French and Russian cases, there was less direct coordination between elites and peasants, but there was coincident timing and mutual inspiration between peasant anti-landlord movements and urban anti-government movements. Other details relating to the structure of the central bureaucracy, loyalties of the army, and external pressures also play important roles in determining the course of events. Poverty and deprivation laid the groundwork for revolution, but other instigating factors were needed.

Ted Gurr's influential 1970 book *Why Men Rebel*[9] proposed relative deprivation as another related factor. Relative deprivation is a mismatch between peoples' level of expectation and their economic reality. Impoverished peasants with no expectations for improvement have no reason to risk what little they have on an insurgency. But groups that have experienced sudden changes of fortune, are jealous of peer groups who are prospering more, or have had expectations raised for other reasons may experience more discontent. This discontent, in the right circumstances, can lead to organized rebellion. The revolutions of France, Russia, and China studied by Skocpal[10] fit this pattern, because in each case the nation was failing to modernize as fast as their neighbors, leading to widespread dissatisfaction among elites who were in a position to observe their comparative failures. Also, fitting with the theory of relative deprivation, the thwarted ambitions of educated and underemployed youths in Middle Eastern countries such as Tunisia and Egypt have been blamed for high levels of political discontent (see the *Demographic Youth Bulge* section).

Subsequent research questioned whether relative deprivation is a reliable predictor of revolution,[11,12] although it may be an important component of the political grievances of marginalized subgroups. Like economic deprivation, relative deprivation is considered to be a general risk factor but only an indirect cause of political violence.

POOR GOVERNANCE

A country's system of governance can, not surprisingly, be an important risk factor. But what kinds of governments are the most at risk for violent insurgencies? One might expect the most violent and repressive governments to be the most at risk, but the answer is more complex than this.

There is general agreement that the most democratic governments are the least vulnerable to violent insurgency. This does not mean they are free from opposition or dissent—far from it. But open democracies offer nonviolent channels for opposition, including elections, public protests, and many other forms of free speech. Most opposition groups realize that they have better options than trying to take on the central government with military force, and radicalized groups have difficulty rallying citizens to their cause when other, safer options are available.

Highly repressive regimes offer no such avenues. The most repressive regimes prevent opposition groups from forming, even informally, and prevent alternative messages from being heard. Established regimes have effective secret police embedded throughout the population who are not held in check by any legal restrictions on intelligence gathering, interrogation, or arbitrary arrest. Most dissent is quashed before it can gain any momentum. Modern North Korea is an example of a highly repressive but stable regime.

Most modern states fall somewhere in between full democracy and the most repressive dictatorships. These blended regimes are sometimes called hybrid regimes or anocracies. These regimes may allow opposition political parties but rig elections so that the ruling party is never seriously challenged; they may restrict political freedom but allow a great deal of economic freedom; or, as in recent times, they may restrict the media but allow relatively free Internet access.

Researchers in this area often make use of the Polity project dataset,[13] which assigns democracy ratings to every country with more than 500,000 people for every year since 1800. Every country-year has been assigned a "Polity" score between −10 and 10, with −10 being the most repressive regime and 10 being the most democratic. Polity scores take into account these subcomponents: competitiveness of political participation (are elections really contested?), regulation of political participation (who can vote?), competitiveness of executive recruitment (are high-level positions really contested?), openness of executive recruitment (who can be appointed?), and constraints on chief executive (checks and balances).

One interesting but controversial finding is an inverted U-shaped relationship between a nation's Polity score and the likelihood of political violence in that country.[14, 15,a] The theory is that countries at the far ends of the continuum—the most autocratic and the most democratic—are the most stable, because democracies allow nonviolent opposition and complete autocracies effectively quash dissent. Anocracies, which have a mix of democratic and autocratic policies, may be the most unstable. The reason for this would be that the hybrid governments allow enough freedom of speech and assembly to allow opposition groups to form but are autocratic enough that challengers are often put down forcefully, causing opposition groups to believe they must resort to violence to achieve their aims. These hybrid governments "possess

---

[a] It should be noted that this inverted "U" finding has been challenged recently. These sources argue that the presence of violence influenced the assignment of Polity scores, particularly the influential "regulation of political participation" subscore, and therefore Polity should not be used in models as a predictor of violence. Similar criticism has been made of the separate dataset used by Muller, who made a similar finding.

inherent contradictions . . . (they) are partly open yet somewhat repressive, a combination that invites protest, rebellion, and other forms of civil violence. Repression leads to grievances that induce groups to take action, and openness allows for them to organize and engage in activities against the regime."[16]

LACK OF GOVERNMENT LEGITIMACY

Beyond style of governance, the perceived legitimacy of the government is an important predictor of its ability to govern especially in the face of challenges; legitimacy has become a core concept in counter-insurgency theory. A government has legitimacy when it is perceived as having both the right to rule and the competency to fill expected functions of government. These are the most important factors affecting legitimacy:

- **Security.** People who experience threats to their physical safety tend to lose faith in their government. This is particularly true when threats are internal, from crime, insurgency, or terrorism, rather than external threats, which tend to evoke a unifying reaction. (Not surprisingly, governments tend to blame internal security problems on "outside agitators" or external manipulation whenever they can.) Terrorism is often an attempt to undermine a government's legitimacy by undermining people's sense of security.

- **Justice.** Governments are expected to settle disputes fairly and quickly. Widespread corruption in the judicial system undermines legitimacy. Many countries with widespread corruption rely on alternate judicial systems, such as the Shura system in Afghanistan and adoption of Shari'a law in a number of states in northern Nigeria; these work-arounds undermine the legitimacy of federal governments.[b]

- **Economic needs.** Governments are expected to make sure people of the nation are fed and to meet their other basic needs, which could include fuel, roads and utilities, health care, education, and employment. Expectations for what services a government should provide vary widely between cultures and nations and are closely tied to both prior conditions and conditions of immediate neighboring countries. Widespread corruption, by which employment, health care, and other services can only

---

[b] Establishing "shadow government" services can be an effective delegitimizing tactic for an insurgent group. See the *Shadow Governments* chapter in the companion to this work, the second edition of *Undergrounds in Insurgent, Revolutionary, and Resistance Warfare.*

be obtained through bribes or connections, can undermine legitimacy (although judicious use of patronage can in some circumstances increase it).

- **Ideological legitimacy.** Cultures also have idiosyncratic expectations for what constitutes a legitimate government. Religious leaders may undermine a government by withholding sanction or declaring the government illegitimate. The Catholic Church in the past held such power over many European states (Henry VIII founded the Church of England because he could not obtain legitimization by the Roman Catholic Church). Modern-day Islamists often direct their most vehement criticism at secular leaders of Muslim nations who do not meet their standards for Islamic rulers. Nonreligious ideologies also matter; governments may forfeit legitimacy for violating strongly held ideals of freedom and democracy or other values that a population feels to be ideologically nonnegotiable.

Legitimacy is ultimately a subjective judgment in the eyes of the governed. Regimes that provide poorly for their people by the standards of developed democracies may nevertheless enjoy a high standing with their own people. However, exposure to information from outside, particularly related to peer nations, may be enough to engender dissatisfaction. Information and communication with the outside world are therefore important determiners of legitimacy and something that underperforming governments will have good reason to try to control. Attacking the government's legitimacy is almost always a central theme of the war of words between insurgents and the government.

MARGINALIZATION OR PERSECUTION OF IDENTITY GROUPS

Perhaps the strongest and most immediate risk factor for radicalization is the systematic marginalization or persecution of identity groups in a country.[17] Politically, the most important identity groups tend to be based on ethnicity but can also be tribal, religious, political, ideological, regional, or economic.

A number of researchers have asked whether ethnically diverse countries are, in general, more prone to civil violence. Earlier models included measures such as Ethno-Linguistic Fractionalization, which is a simple measure of how diverse a country is.[18] But ethnic diversity taken by itself does not appear to be a risk factor;[19, 20] many countries have peacefully coexisting religions and ethnic groups.

Risk of political violence is great, however, when diversity is combined with some form of economic or political exclusion. Violence is particularly likely when a minority group controls the central state government and excludes other groups,[21] even if those groups have some level of local autonomy. State-led discrimination of other forms is also a strong predictor of political violence.[22]

Systematic discrimination against minority groups can endanger otherwise democratic societies. Groups usually turn to organized, violent resistance only when they believe they have no nonviolent alternatives. When a minority group believes it is systematically excluded from the democratic process, violent resistance becomes a more likely option. The Irish Republican Army (IRA) persisted within the democratic nation of Great Britain because of their Irish Republican feelings of exclusion. The civil rights movement in the United States may be seen as another example. The United States has always taken great pride in its democratic institutions, but the systematic exclusion of African Americans from these institutions led to violence during the civil rights movement and to the brink of what might have become organized insurgent-type activity.

To understand the typical types and ranges of discrimination, it is interesting to look at the criteria for discrimination used by the ongoing Minorities at Risk (MAR) project.[23] Data from this project are often used in high-level studies of political violence and discrimination. This effort monitors the treatment of minority groups identified as being at risk for discrimination. As of 2011, the project is monitoring 283 at-risk ethnic groups across the world; archived data go back to 1945. MAR examines a wide range of types of discrimination and rates each type on scales that increase from 0 to 4 or 5. Paid raters working for this project scour news archives and other sources to find evidence of different kinds of discrimination to inform these ratings. Criteria for ratings of government repression, political discrimination, and economic discrimination categories is listed in Table 2-1.

Table 2-1. Code guidance for three discrimination categories from the MAR dataset.

Rating	Government repression	Political discrimination	Economic discrimination
1	Surveillance, e.g., domestic spying, wiretapping, etc.	**Neglect/remedial polices** Substantial underrepresentation in political office and/or participation due to historical neglect or restrictions. Explicit public policies are designed to protect or improve the group's political status.	**Neglect/remedial polices** Significant poverty and underrepresentation in desirable occupations due to historical marginality, neglect, or restrictions. Public policies are designed to improve the group's material well-being.
2	Harassment/containment, e.g., saturation of police/military presence, militarized checkpoints targeting members of group, curfews, states of emergency	**Neglect/no remedial policies** Substantial underrepresentation due to historical neglect or restrictions. No social practice of deliberate exclusion. No formal exclusion. No evidence of protective or remedial public policies.	**Neglect/no remedial policies** Significant poverty and underrepresentation due to historical marginality, neglect, or restrictions. No social practice of deliberate exclusion. Few or no public policies aim at improving the group's material well-being.
3	Nonviolent coercion, e.g., arrests, show-trials, property confiscation, exile/deportation	**Social exclusion/neutral policy** Substantial underrepresentation due to prevailing social practice by dominant groups. Formal public policies toward the group are neutral or, if positive, inadequate to offset discriminatory social practices.	**Social exclusion/neutral policy** Significant poverty and underrepresentation due to prevailing social practice by dominant groups. Formal public policies toward the group are neutral or, if positive, inadequate to offset active and widespread discrimination.

Rating	Government repression	Political discrimination	Economic discrimination
4	**Violent coercion**, short of killing, e.g., forced resettlement, torture	**Exclusion/repressive policy** Public policies (formal exclusion and/or recurring repression) substantially restrict the group's political participation by comparison with other groups. (Note: This does not include repression during group rebellions. It does include patterned repression when the group is not openly resisting state authority.)	**Exclusion/repressive policy** Public policies (formal exclusion and/or recurring repression) substantially restrict the group's economic opportunities in contrast with other groups.
5	**Violent coercion**, killing, e.g., systematic killings, ethnic cleansing, reprisal killings		

Besides the three variables listed in columns here, MAR also records the history of violent conflicts between groups (not necessarily involving the government), grievances expressed by group leaders, and protests and rebellion by group members.

The social psychology of relationships between social identity groups ("in-groups" versus "out-groups") will be discussed extensively in *Chapter 6: Group Dynamics and Radicalization*, and elsewhere in this work.

HISTORY OF CONFLICT IN THE COUNTRY OR CONFLICT IN NEARBY COUNTRIES

Countries with a history of violence are more likely to experience violence in the future.[24] The same is true for countries whose geographic neighbors have experienced violence. There are both psychological and non-psychological reasons for this. The simplest cause may be the available supply of weapons and people trained to use them, either in-country or nearby. When one conflict ends or dies down, both weapons suppliers and soldiers may be unemployed and have few other skills; they may return to their home countries or cross borders as mercenaries. A second reason for the bleed-over of violence across borders may be large numbers of refugees or other displaced persons. These refugees may strain the resources of new areas, leading to violence. Or the refugees may hold claims on their prior land (such as displaced Palestinians) or have other grievances (e.g., lost relatives and friends) to be redressed with violence in a new location.

There are also psychological processes at work that create second-order effects of regional violence. Populations can become desensitized to violence, which leads to more violence.

Wartime crime rates give some evidence for this. American sociologists Dane Archer and Rosemary Gartner[25] were among the first to show that homicide rates increased during wars and continued to be higher after the wars ended. They found that homicide rates were higher in countries that were involved in World War II than in nations that were not; in Italy, the homicide rate more than doubled. Similarly, the homicide rate increased in all six nations involved in the Vietnam War. In the United States, despite the fact that none of the fighting took place on U.S. soil and that large numbers of the demographic group responsible for homicides (young men) were fighting overseas, the homicide rate increased every year of the war, peaking in 1973 at a rate 107 percent higher (more than double) than the prewar rate. Of course, other dramatic changes were taking place in the United States at that time, including anti-war protests and the civil rights movement, but the effect holds statistically across many other nations in different situations. The increase in homicide rates holds across both victorious and defeated nations, and is, in fact, higher in victorious nations. Countries with more battle deaths experience subsequently higher homicide rates. These increases in violence cannot be attributed solely to violent or battle-scarred veterans; more violent crimes are also committed by women and older age groups during and after wars.

Most humans have natural inhibitions toward violence, and all cultures have norms and practices designed to prevent and manage violence. Wars and political conflicts can desensitize people to violence on a large scale, making many kinds of violence more likely.

Desensitization has been demonstrated in many different settings. Most of the relevant research has been done with children. Children who witness adults behaving aggressively, for example, by pummeling a stuffed animal, tend to imitate that aggression. Children who watch violent television or play violent video games have more aggressive thoughts and behaviors, less empathy toward victims, and lower physiological reactions when witnessing violence.[26] And children exposed to actual violence show a range of negative stress reactions that persist long after the events.[27] Similar results have been found with adults.

Being witness to violence does not universally cause violence. No amount of watching violent television or playing violent video games will make a child violent if they are not predisposed. Exposure to violence will lead some to commit more acts of violence, through desensitization or simple imitation, and desensitization on a large scale can affect how quickly people intervene or punish incidents, and generally weaken cultural mores that prevent violence.

There is also speculation that some national and regional cultures may be predisposed to violence, beyond the effects of recent conflicts. Generalizations about entire cultural groups tend to be controversial and might even be considered a form of racism, so there are not many empirical studies or findings on this topic. One of the few examples is Nisbett's study of differences in the reactions of males in the north and south of the United States to minor insults. Nisbett found what he describes as an "honor culture" in the south, with a cultural imperative to preserve honor by retaliating against even minor acts of aggression. The presence of honor cultures has been tied to levels of interpersonal conflicts and low-level feuds. There may be geographic and economic reasons for these norms, e.g., the need for herdsmen to establish a tough reputation to prevent theft, but the culture seems to persist even when the underlying circumstances change.[28] Similar dynamics of honor and retribution are common (but not ubiquitous) in the Middle East.[29]

DEMOGRAPHIC YOUTH BULGE

There has been recent attention to violence and youth bulges in a nation's demographics. Larger than normal proportions of young people may have contributed to historical conflicts including the European Revolutions of 1848, the rise of Nazism in Germany in the 1930s, and the American anti-war and civil rights protests (led by "baby boomers") in the 1960s. Henrik Urdal[30] examined a large historical dataset[31] and found a relationship between nations with a comparatively large percentage of young people (15–24 years old) and levels of smaller-scale political violence. Using a separate dataset[32] Urdal also found a relationship with levels of terrorism and rioting. This effect seemed to hold across different types of governance but seemed to be a greater risk factor in autocratic regimes.

There are several proposed reasons for this. Unemployment and lack of economic opportunity are usually considered the most important elements. Societies often cannot provide enough jobs for a sudden swell in young adults, so frustration and unemployment can contribute to a sense of grievance. Second, the presence of large numbers of military-age males (16–30 years old) provides a pool for existing protest or insurgent movements, and combined with high unemployment or underemployment, the risk is greater yet. A third factor may be that young people are more mobile and more likely to move to urban areas seeking employment,[33] where urban overcrowding and the concentration of young, restive people may create conditions for street protests and other more violent protests. A fourth factor may be more psychological; young people, especially those who are unattached, may be more likely than their elders to protest against grievances that are felt by all and may be more accepting of the risks associated with protest against the government, particularly autocratic regimes. The youth bulge hypothesis is relevant to current concern about Middle Eastern terrorism; areas of the Middle East including Saudi Arabia and large parts of Africa had large youth bulges approaching young adulthood coincident with these movements.

EXPLOITABLE PRIMARY COMMODITY RESOURCES

A provocative recent theory by Paul Collier and Anke Hoeffler[34] argues that violent rebellions are more likely when there is a "primary commodity resource" that can be used to finance a rebellion. Examples of insurgencies financed in this way would be the Fuerzas Armadas Revolucionarias de Colombia (Revolutionary Armed Forces of Colombia, or FARC), which is financed by cocaine trafficking, Nigeria's Movement for the Emancipation of the Niger Delta (MEND), which is financed with "bunkered" oil, and Angolan rebels, who are financed with "conflict diamonds."

This argument starts with the assumption that all countries have some groups with grievances and motivation to rebel but that most are incapable of sustaining themselves economically for long enough to pose a serious challenge. It is difficult for insurgencies to sustain themselves through legitimate businesses and voluntary contributions; therefore, many turn to more lucrative criminal enterprises. A highly valuable commodity that can be stolen and smuggled to finance a rebellion seems to make radical rebellion more likely. A second means of making money from valuable commodity exports is through extortion or "protection," which is a natural fit for insurgencies that already operate as mobile armed forces. However, it may be important that these resources be based in rural areas, where insurgents can operate more freely and the need for safe long-distance transportation facilitates extortion.

In this analysis, it seemed to be important that these valuable commodities were "primary" commodities, which means that they represented a large percentage of the country's gross domestic product. This could be simply because countries that have more diversified or well-developed economies are resistant to violence for other reasons, such as better governance or more stable government financing. There has also been considerable research specifically on oil-producing nations, which may be vulnerable to corruption, repressive governance, and political violence, sometimes referred to as the "oil curse." The theory here is that when governments can make a great deal of money by selling a resource that does not require either a highly educated populace or modern infrastructure to produce, they will tend to fall into patterns of corruption, patronage, and repression that are simpler to maintain than a modern economy.

A second economic risk factor identified by Collier and Hoeffler was the presence of a large diaspora living in more developed countries. Expatriates living in wealthier countries are a potential funding source for insurgents, as demonstrated by the substantial support provided by Tamils in North America for the insurgency in Sri Lanka; another prominent example is the Kurdish diaspora in Europe.[c]

TYPE OF TERRAIN

The terrain of a country may also be a risk factor in that some terrains provide more shelter and secrecy to insurgents than others. The Afghani Taliban (and many prior insurgent groups) benefit from mountainous terrain that makes surveillance and pursuit difficult. Mountainous terrain also tends to isolate villages, making them harder for the government to protect and easier for insurgents to influence or intimidate. At least two studies have found a correspondence between mountainous terrain and civil war.[35,36]

Dense forest might have a similar effect. Most experts would agree that the FARC in Colombia has benefitted from being able to hide and organize in the thick jungles of that country; the Viet Cong may have benefited similarly, despite American attempts to defoliate sections of the country. However, a general statistical relationship between forest cover and adverse political events has not been shown. It may also be that while difficult terrain does not necessarily encourage insurgencies to develop, it does tend to prolong them once they have developed by hindering government response.[37]

Of course, there are plenty of examples of insurgencies that survived without any particular geographic cover. The IRA relied on the "human terrain" of Catholic enclaves in larger cities for safe havens and bases of operations.

---

[c] Civil strife tends to encourage those with means to leave; so one could argue that it is the insurgencies causing the diasporas to form rather than the other way around. Likely, the causation goes both ways.

SUMMARY

Predicting where violence will break out remains an inexact science, but there is a great deal of useful research on underlying causes that may increase the probability of violence in a particular area. Economic deprivation contributes to instability, but the mechanisms are more complex than simply stating that poor people are more likely to rebel; the experience of "relative deprivation" related to either peer comparisons or internal expectations is an important contributor. Insurgencies tend to be led by the more affluent and educated members of poor countries but benefit from the ability to recruit unemployed or economically distressed members. The way a country is governed is another key predictor, and the failure to provide basic levels of justice, basic needs, and security undermine legitimacy and encourage rebellion. Marginalization of ethnic groups within a country is also a common catalyst to violence, especially if the state participates in discriminatory practices.

Countries that have a history of conflict internally or in nearby states are also at greater risk; violence begets violence via a number of pathways, including the availability of military-trained personnel and weapons and psychological damage from and desensitization to violence.

Countries experiencing youth bulge demographic shifts can become very vulnerable as these people reach young adulthood; failure to provide economic opportunity for large numbers of military-age adults can be a particularly dangerous precursor to violence.

Insurgencies need finances to survive; countries with an exploitable resource such as diamonds, oil, or cocaine can allow insurgencies to sustain themselves more easily than if they were dependent on external supporters or internal contributors. Insurgencies also need safe places to hide from superior government forces; national terrain that includes dense forest, rugged mountains, or other such features may increase an insurgency's chances of survival.

ENDNOTES

[1] Jack A. Goldstone, Robert H. Bates, David L. Epstein, Ted Robert Gurr, Michael B. Lustik, Monty G. Marshall, Jay Ulfelder, and Mark Woodward, "A Global Model for Forecasting Political Instability," *American Journal of Political Science* 54, no. 1 (January 2010): 190–208.

[2] Håvard Hegre, Tanja Ellingsen, Scott Gates, and Nils Petter Gleditsch, "Toward a Democratic Civil Peace? Democracy, Political Change, and Civil War, 1816–1992," *American Political Science Review* 95, no. 1 (March 2001): 33.

[3] Lars-Erik Cederman and Luc Girardin, "Beyond Fractionalization: Mapping Ethnicity Onto Nationalist Insurgencies," *American Political Science Review* 101, no. 1 (February 2007): 173.

[4] Goldstone et al., "Global Model," 190–208.

[5] Paul Collier and Anke Hoeffler, "Greed and Grievance in Civil War," *Oxford Economic Papers* 56, no. 4 (June 2004).

[6] Ibid.

[7] Theda Skocpol, *Social Revolutions in the Modern World* (New York: Cambridge University Press, 1994).

[8] Ibid., 36.

[9] Ted R. Gurr, *Why Men Rebel* (Princeton, NJ: Princeton University Press, 1970).

[10] Skocpol, *Social Revolutions in the Modern World*.

[11] Stephen G. Brush, "Dynamics of Theory Change in the Social Sciences: Relative Deprivation and Collective Violence," *Journal of Conflict Resolution* 40, no. 4 (December 1996): 523–545.

[12] Joan Neff Gurney and Kathleen J. Tierney, "Relative Deprivation and Social Movements: A Critical Look at Twenty Years of Theory and Research," *The Sociological Quarterly* 23, no. 1 (December 1981): 33–47.

[13] Monty G. Marshall and Keith Jaggers, *Polity IV Project: Political Regime Characteristics and Transitions, 1800–2007* (2007), http://www.systemicpeace.org/polity/polity4.htm.

[14] Hegre et al., "Toward a Democratic Civil Peace?" 33.

[15] Edward N. Muller, "Income Inequality, Regime Repressiveness, and Political Violence," *American Sociological Review* 50 (1985): 47–61.

[16] Hegre et al., "Toward a Democratic Civil Peace?" 33.

[17] Goldstone et al., "Global Model," 190–208.

[18] Philip G. Roeder, "Ethnolinguistic Fractionalization (ELF) Indices, 1961 and 1985," accessed April 16, 2011, http://weber.ucsd.edu/~proeder/elf.htm.

[19] Cederman and Girardin, *Beyond Fractionalization*.

[20] Collier and Hoeffler, "Greed and Grievance."

[21] Lindsay Heger and Idean Salehyan, "Ruthless Rulers: Coalition Size and the Severity of Civil Conflict," *International Studies Quarterly* 51, no. 2 (June 2007): 385–403.

[22] Goldstone et al., "Global Model," 190–208.

[23] Ted R. Gurr, "Why Minorities Rebel: A Global Analysis of Communal Mobilization and Conflict since 1945," *International Political Science Review* 14, no. 2 (January 1993): 161–201.

[24] Goldstone et al., "Global Model," 190–208.

[25] Dane Archer and Rosemary Gartner, "Violent Acts and Violent Times: A Comparative Approach to Postwar Homicide Rates," *American Sociological Review* 41, no. 6 (1976): 937–963.

[26] Craig A. Anderson, Akiko Shibuya, Nobuko Ihori, Edward L. Swing, Brad J. Bushman, Akira Sakamoto, Hannah R. Rothstein, and Muniba Saleem, "Violent Video Game Effects on Aggression, Empathy, and Prosocial Behavior in Eastern and Western Countries: A Meta-Analytic Review," *Psychological Bulletin* 136, no. 2 (March 2010): 151–173.

[27] Paramjit T. Joshi and Deborah A. O. Donnell, "Consequences of Child Exposure to War and Terrorism," *Clinical Child and Family Psychology Review* 6, no. 4 (2003).

[28] Dov Cohen, Richard E. Nisbett, Brian F. Bowdle, and Norbert Schwarz, "Insult, Aggression, and the Southern Culture of Honor: An Experimental Ethnography," *Journal of Personality and Social Psychology* 70, no. 5 (1996): 945.

[29] David Pryce-Jones, *The Closed Circle: An Interpretation of the Arabs* (London: Weidenfeld & Nicolson, 1989).

[30] Henrik Urdal, "A Clash of Generations? Youth Bulges and Political Violence," *International Studies Quarterly* 50, no. 3 (September 2006): 607–629.

[31] Nils Petter Gleditsch, Petter Wallensteen, Mikael Eriksson, Margareta Sollenberg, and Håvard Strand, "Armed Conflict 1946–2001: A New Dataset," *Journal of Peace Research* 39, no. 5 (2002): 615–637.

[32] Daniel C. Esty, Jack A. Goldstone, Ted Robert Gurr, Barbara Harff, Marc Levy, Geoffrey D. Dabelko, Pamela Surko, and Alan N. Unger, *State Failure Task Force Report: Phase II Findings* (McLean, VA: Science Applications International, 1998).

[33] Jack A. Goldstone, "Population and Security: How Demographic Change Can Lead to Violent Conflict," *Journal of International Affairs* 56, no. 1 (2002).

[34] Collier and Hoeffler, "Greed and Grievance."

[35] Ibid.

[36] James D. Fearon and David D. Laitin, "Ethnicity, Insurgency and Civil War," *American Political Science Review* 97, no. 1 (2003): 75–90.

[37] James D. Fearon and David D. Laitin, "Weak States, Rough Terrain, and Large-Scale Ethnic Violence Since 1945" (paper presented at the annual meeting of the American Political Science Association, Atlanta, GA, September 2–5, 1999).

PART I.

UNDERGROUNDS AS ORGANIZATIONS

CHAPTER 3.

ORGANIZATIONAL STRUCTURE AND FUNCTION

CHAPTER CONTENTS

Nathan Bos, Jason Spitaletta, and SORO authors

Despite their unique nature, resistance groups are human organizations that can be understood from the perspective of modern organizational studies. Academic research on industrial/organizational psychology, operations research, and social network analysis can each provide some insight into the functioning of insurgent organizations. Some core issues from these areas are already discussed in this book (*Chapter 4: Leadership*) and in the companion to this work, the second edition of *Undergrounds in Insurgent, Revolutionary, and Resistance Warfare (Chapter 1: Leadership and Organization)*.

COMPONENTS OF AN INSURGENCY

Insurgencies can be understood as having four components: the underground, auxiliary, armed component, and public component, as shown in Figure 3-1.

- Underground—A clandestine organization established to operate in areas denied to the armed or public components or conduct operations not suitable for the armed or public components.

- Armed component—The visible element of a revolutionary movement organized to perform overt armed military and paramilitary operations using guerrilla, asymmetric, or conventional tactics.

- Auxiliary—The support element of the irregular organization whose organization and operations are clandestine in nature and whose members do not openly indicate their sympathy or involvement with the irregular movement. Members of the auxiliary are more likely to be occasional participants of the insurgency with other full-time occupations.

- Public component—The overt political component of an insurgent or revolutionary movement. Some insurgencies pursue military and political strategies. At the termination of conflict, or occasionally during the conflict, the movement can transition to the sole legitimate government or form part of an existing government. Thus, the four spheres—armed component, underground, auxiliary, and public component—engage in a dynamic and evolving relationship that changes in response to internal and external drivers. The public component's overt position distinguishes it from the clandestine underground. However, it frequently overlaps with the underground in that the latter's functionality includes the management of propaganda and communications in general.

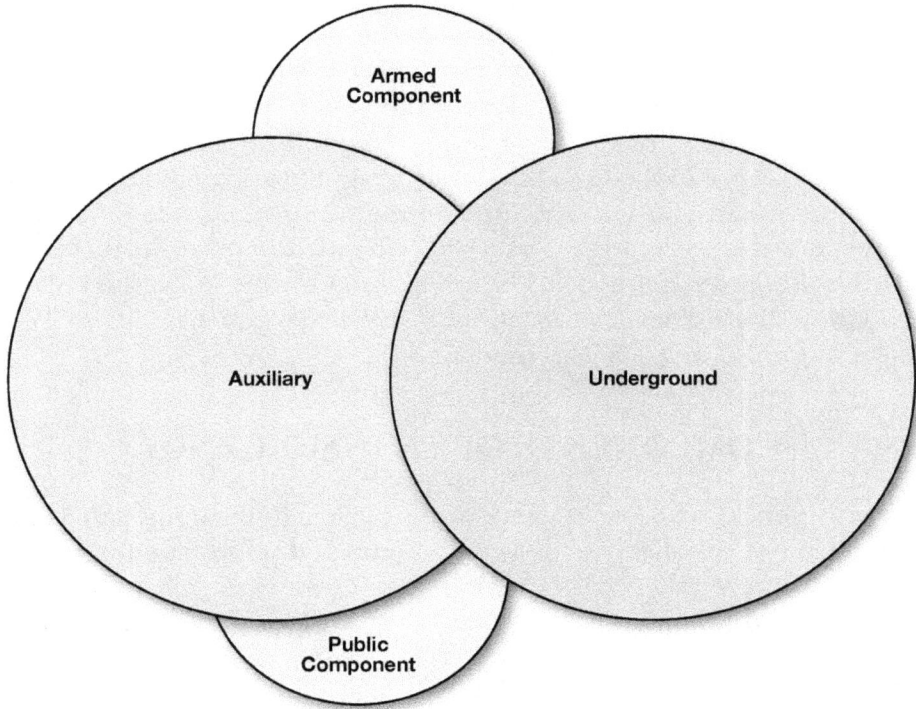

Figure 3-1. Components, phases, and functions of an insurgency.

Within this broad categorization of components, there are many different ways of organizing an insurgency. This chapter will examine three insurgencies in some detail—the Provisional Irish Republican Army (PIRA), the classical Communist model, and Al Qaeda. Each of these can be analyzed using the four-component model, but each instantiates the components within very different organizational structures and practices.

This chapter will not focus on the general model of insurgencies, which is treated elsewhere. Instead, we will focus on six cross-cutting issues, three of which are common to all organizations—command and control, aligning structure with strategy, and evolution and growth—and three of which tend to be unique concerns of resistance movements—secrecy and compartmentalization, underground and aboveground connections, and criminal connections.

COMMAND AND CONTROL

One way to characterize organizational structures is according to their hierarchical structure. A classic military unit has a very strict hierarchy, with clear lines of authority, often drawn as a pyramid. Authority is established by one's level and position in the hierarchy. Large corporations also usually have elements of hierarchical structures, often depicted in tree-like "organization charts." Over time, organizations have tendency to evolve more layers of hierarchy to deal with increasing size of the organization without any level of management being overwhelmed, and for other reasons, such as the need to prevent errors and create higher-status management positions. Hierarchies are often contrasted with network structures, which are organically evolved sets of relationships between individuals. Authority in a pure network comes from personal relationships, reputation, and to some extent network position, in that more-well-connected individuals may exert more influence. David Alberts and Richard Hayes popularized a third option, which they call an "edge" structure.[1] An edge has clearer vertical lines of authority than a network, and authority is at least partially established by position, but it is designed to be much flatter than hierarchies, intentionally allows many cross-cutting connections for information flow, distributes more decision-making authority to individuals, and is more flexible in allowing ad hoc teams to form and focus around specific problems.

The choice between hierarchy, network, and edge organizations involves trade-offs, with each type of system having different strengths. Modern organizational research has emphasized the importance of matching organizational forms to specific tasks and situations rather than identification of always-optimal strategies. Relevant questions related to choice of command structures are:

- How important is speed of response? Hierarchical command structures almost invariably involve delay as requests move up and down a chain of command, and bottlenecks when managers have to prioritize some decisions at the expense of others. Flatter systems typically gain speed but sacrifice reliability and quality control. The relative importance of speed is largely driven by the environment, e.g., the speed of competitors or enemy forces or time window of opportunities.

- How important is specialization? An insurgent organization at some point must begin to set up basic specialized functions such as media production or weapons training; if they do not, they risk wasting resources on redundant efforts.

- How costly are mistakes? In the corporate domain, a toymaker or radio station might be able to tolerate a great deal of variance in quality for the sake of novelty and freshness; likewise, a newly formed insurgency might highly value surprise at the cost of the occasional operational disaster. An established automobile manufacturer may have much less tolerance for manufacturing defects; likewise, a late-stage insurgency trying to establish legitimacy may have much less tolerance for accidental civilian deaths or public relations disasters.

- How important is unity of command? Effective hierarchical command structures maximize consistency. The archetype of this is a Napoleon-era army flawlessly executing a complex battle plan with preplanned, synchronized movements. In an insurgency, coordination of units, and coordination of military, political, and communications actions, can be crucial.

- How important is efficiency? Hierarchical structures can maximize efficiency by optimizing systems (manufacturing, training) and enforcing consistency across the organization. This optimization works best in low-margin business environments where small efficiencies can be large competitive advantages. This efficiency also presumes that the environment is sufficiently slow-moving to allow opportunity costs to be optimized. It also presumes that the cost of maintaining the hierarchy itself is not greater than intended gains.

- Are there cultural reasons to favor centralization? Hierarchies may also be preferable when an organization's membership strongly favors it. This could be the case in a resistance movement that recruits heavily from ex-military personnel, making military-style organization easy to implement. This might also be the case in national cultures that Hofstede[2] refers to as being high in "power distance"—where there are strong role expectations for leaders and followers.

None of these trade-offs are inviolable; examples can be found of fast-adapting hierarchies, efficient networks, etc., based on well-developed practices, competent leadership, or other factors. In general, effective decentralization requires these two factors:

- Communications technology. Information technology can allow faster communication and more widespread and efficient information dissemination, and successful organizations evolve to use this technology to flatten hierarchies and empower personnel.

- Highly trained and educated "edge" workers. When intelligent and capable personnel are available, it is more feasible to train members to adapt, coordinate, optimize processes, etc., without management directives. Research on self-organizing and self-optimizing systems is of great interest across many fields.

ALIGNING STRUCTURE WITH STRATEGY

A first principle of organizational design is that an organization's strategy should match its structure. If an insurgent group values media exposure, it should have an active and well-supported media arm. If it intends to affect the political process, it should have a strong public component with a political wing. Maoist organizations, as will be discussed, had the goal of undermining and then replacing existing political institutions. Consistent with this goal, they invested a great deal of effort into either infiltrating or creating nonmilitary civilian organizations of all types. Branches of Al Qaeda, in contrast, do not try to take or hold territory and generally have no ambitions of participating in political processes or directly governing nations, and this is reflected in their organizational structures as well.

Other types of organizations often must choose between regionally focused, product-focused, and functional divisions.[3] For an auto manufacturer, a regional division would focus on a manageable area (e.g., South America); a product division would focus on a product line (e.g., SUVs); and a functional division would focus on a technical area (e.g., transmissions or sales). Larger organizations will have all three types, with complex reporting relationships between them. Smaller organizations with less management and staff must choose which type of division will be primary and may need to manage difficult transitions between structures as strategy changes.

For insurgent groups, a key decision is often between regional and functional structures. Regional groups would have all functions located in an area and would be responsive to local conditions; functional groups would have specialized support groups located in a centralized command and would dispatch those specialists to regional groups as needs arise. Structure also strongly affects an organization's ability to learn and adapt. Regionally based structures are generally better at perceiving and adapting to local conditions. Functional structures, where specialists in an area (e.g., military strategy) mostly interact with other specialists, may attain a higher level of professionalism in that specialty and be able to take on more complex tasks.

It is also common for insurgents to have to decide between being primarily military or political organizations and between short-term and long-term structures; see Chapter 1 in *Undergrounds in Insurgent, Revolutionary, and Resistance Warfare* for discussion of these two issues.

A recent study of terrorist violence suggests these choices do make a difference in insurgent group functioning. This study[4] claims to show that violent groups that adopt a vertically integrated functional structure, and thus have specialized units dedicated to violence, commit more lethal attacks and sustain violence over longer periods of time. This is supported by a statistical analysis of attacks by groups listed in the MIPT Rand Terrorism Knowledge Base (TKB) dataset and illustrated by a case study of the Basque separatist group Euskadi Ta Askatasuna (Basque Homeland and Freedom, or ETA) in Spain. The ETA has reorganized itself between vertical and flat structures several times and has been more violent during periods of vertical organization. Specialization led to more competent and coordinated violence in this case; the trade-offs in slower responsiveness to local conditions and less compartmentalization were apparently not critical inhibitors.

Modern organizations also spend considerable resources on internal knowledge management. These initiatives may involve creating information systems to ease sharing of information across divisions and creating communities of practice that bring together organizationally separated specialists to share technical information and keep up with new developments. There is also a great deal of organizational research on use of social media, information markets, and other innovations to improve internal information flow with the goal of increasing learning and adaptability.

Insurgent groups operate under a different set of constraints, however, because of the need for secrecy and compartmentalization. Open, free flow of social contact and information is not a reasonable goal; cross-divisional communication must be more carefully managed or involve more anonymity. Nevertheless, new information technology and new media (see Chapter 8) are opening up new possibilities for facilitating internal information flow while managing risk.

SECRECY AND COMPARTMENTALIZATION

In resistance movements, the need for secrecy and security can trump all other concerns, particularly in the early stages or when operating in urban environments where governments maintain effective control. Compartmentalization thus becomes a critical design priority; as much as possible, forces are divided into cells where members have little knowledge or contact outside of the cell so that one compromised cell does not have cascading effects, endangering the larger organization. In a cell structure, usually only the leader has contact with the outside and relays instructions or requests through a single contact outside of the group. In some cases, even the leader may not have regular personal contact, and communication happens through coded messages or dead drops. Cell size is dictated by the demands of the type of work to be done (with the smallest capable group being best) traded off against the danger of penetration. Intelligence cells are most likely to use this level of secrecy. Cell size may be adjusted as conditions change; before World War II, the Communist Party in France had cells of fifteen to twenty or even thirty members. After the party was declared illegal in September 1939 and until the armistice in June 1940, cell size was reduced to three men to ensure maximum security. Later, cells were increased to five to eight and then reduced again under Nazi occupation.[5]

Communication and coordination must occur between cells, however, and the more complex the organization's goals, the more necessary this communication becomes.

Cells can be organized to accomplish different workflows. When possible, cells may be arranged in series, like an assembly line (Figure 3-2). In the Belgian Underground, six cells or sections were connected in a series to produce large-scale newspapers. One cell of reporters sent raw information to a cell of editors who composed stories that were handed off to other cells for printing and distribution.[6]

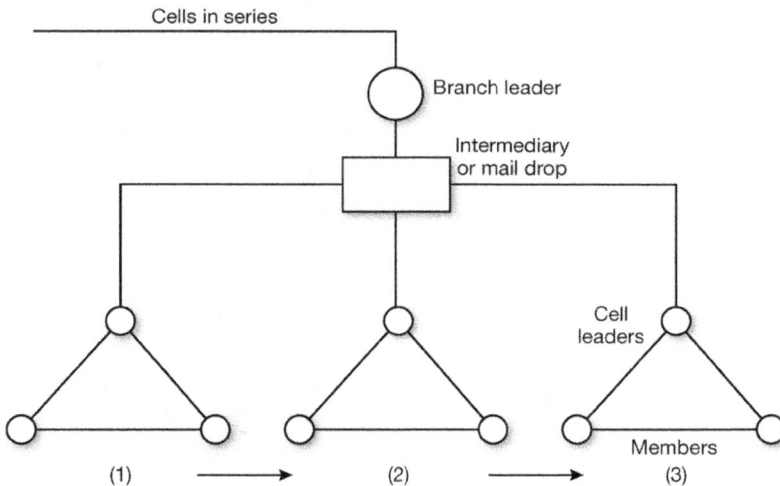

Figure 3-2. Cells in series.

Cells can also be set up in parallel (Figure 3-3). These can conduct work independently and report up a chain of command. Parallel cells are also sometimes set up to confirm or disconfirm information independently or set up as backups in case one cell is compromised.

41

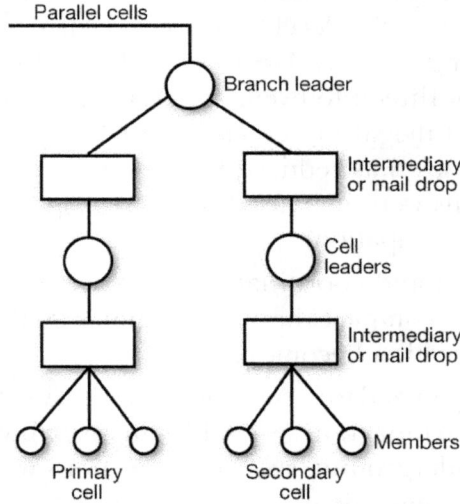

Figure 3-3. Parallel cells.

Task demands usually force some violations of strict compartmentalization. Erickson's[7] study of six secret societies documents some of the common reasons. For the White Lotus group, which was part of the 1813 Eight Trigrams rebellion in China,[8] the desire to spread new boxing skills led the group to employ traveling teachers who were not members of the upper hierarchy yet who gained knowledge of cell membership and function. The resistance cells within Auschwitz[9] were organized into a rigid hierarchical structure. Members of one five-man cell did not, in theory, know the identity of any of the members of any other cell, and cells only communicated through higher command. The Auschwitz group had difficulty maintaining compartmentalization because it was a relatively small community in very close quarters; members knew each other socially outside of the cell structures. Accidental discoveries also happened because cells had similar needs. Cells would find out about each other as each attempted to use the same couriers traveling outside of the camp or would recognize others' clandestine behaviors or convenient meeting locations simply because they were discovered in parallel. The main impediment to compartmentalization, according to Erickson, is the fact that groups such as organized crime families recruit from the same clusters of families, friends, and neighborhoods; the pre-existing social ties are too dense to allow strict separation. Strict cellular structures also have the same weaknesses as other types of hierarchies—slow response times and bottlenecks in the information flow between units.

EVOLUTION AND GROWTH OF ORGANIZATIONS

A key factor in an insurgency's success is the individual adaptability of its principal leaders and the strategic and ideological flexibility of the organization at large.[10] For example, recruiting tactics must change over the course of an insurgency. During the early stages, leaders seek to carefully select, investigate, and approach potential fellow insurgents. During the middle stages of an insurgency, leaders usually have to expand the recruiting effort in order to fill out the rank and file of forces—conventional, guerilla, terrorist, or other types of units and cells. It is during this transition period that some revolutionary movements fail while others succeed, and the question often comes down to the organization's ability to find new sources for recruitment and support. During the latter stages of an insurgency, recruiting is characterized by the momentum of the movement. A successful insurgency that is able to either take power (supplanting the former government) or achieve political, legal, or quasi-legal status will normally expand recruitment operations to include parallel efforts at political mobilization.

Ideology serves a dual function in an insurgency: it serves as the basis for recruitment, and it illuminates strategic direction. Ideology is a point of unification for insurgencies that have grown beyond the point where personal relationships are the major unifying force, and it provides an analytical framework from which new and peripheral members can interpret their environment. The subsequent worldview often serves as a cohesive meme that can further incentivize new members to join. An insurgency's ideology moves along a spectrum of exclusivity and inclusivity as leaders establish the movement's position on any number of social, political, and/or operational issues.[a] Exclusive ideologies aim at energizing the targeted sector of the population, helping them to define themselves in relation to the foes they oppose. Inclusive ideologies, conversely, seek to unify various groups and encourage them to coalesce around the insurgency's main goal.

The particular progression of an organization's ideology often coincides with changes in organizational structure and recruiting practices. As insurgencies evolve they tend to assume a greater degree of risk of exposure as their recruiting and communications operations become more overt. Moving beyond clandestine operations is necessary in order to build operational capacity; the span of control in a compartmentalized organization is such that information cannot be disseminated in a timely enough manner to make for effective command and control.

---

[a] See Chapter 1 of *Undergrounds in Insurgent, Revolutionary, and Resistance Warfare* for further discussion of the dilemma of inclusiveness.

While secrecy is often prioritized in deliberate planning phases of operations, it becomes impractical as the requirement to conduct surveillance and reconnaissance and/or tactical rehearsals becomes more tangible.

Organizational structures may also evolve in response to a changing security environment (as the PIRA did) or to reflect the increasing legitimacy of a movement (such as Hizbollah). The degree to which an insurgency recognizes these internal or external changes and adapts appropriately correlates with longevity. Although restructuring does not come without its costs (for example, the highly ideologically oriented may disapprove of a more inclusive recruiting strategy and thus may discontinue their association with the movement), overly rigid organizations are unlikely to enjoy success.

UNDERGROUND AND ABOVEGROUND CONNECTIONS

Nye suggests international relations exist on three separate but related levels: the world of traditional sovereign state power (e.g., military and diplomatic), economic power and influence, and the nongovernmental activities that affect the global system (e.g., human migration, criminal activity, black/gray markets, etc.).[11] This final level consists of both licit and illicit, light and dark, and overt and covert organizations and networks. Underground organizations must either interact with the legitimate networks or establish their own clandestine networks that parallel the legitimate ones.[12] The fuzzy boundary that is the light/dark nexus is the critical vulnerability of many underground organizations and thus is a priority for intelligence gathering. Modern insurgencies are microcosms of Nye's contention and are often required to operate effectively in both overt and covert domains.

Insurgencies exist in both overt and clandestine domains. Figure 3-4 shows some of the overt and covert functions of an underground. In the nascent period, the underground predominates. Often organized by charismatic individuals, the underground recognizes the underlying socioeconomic and/or political grievances and seeks to organize that dissatisfaction into a coherent narrative and to start forming a broader base of support. Much of the early activity involves disseminating information in order to generate internal and external support, shape perceptions, and set conditions for a broader mobilizations.

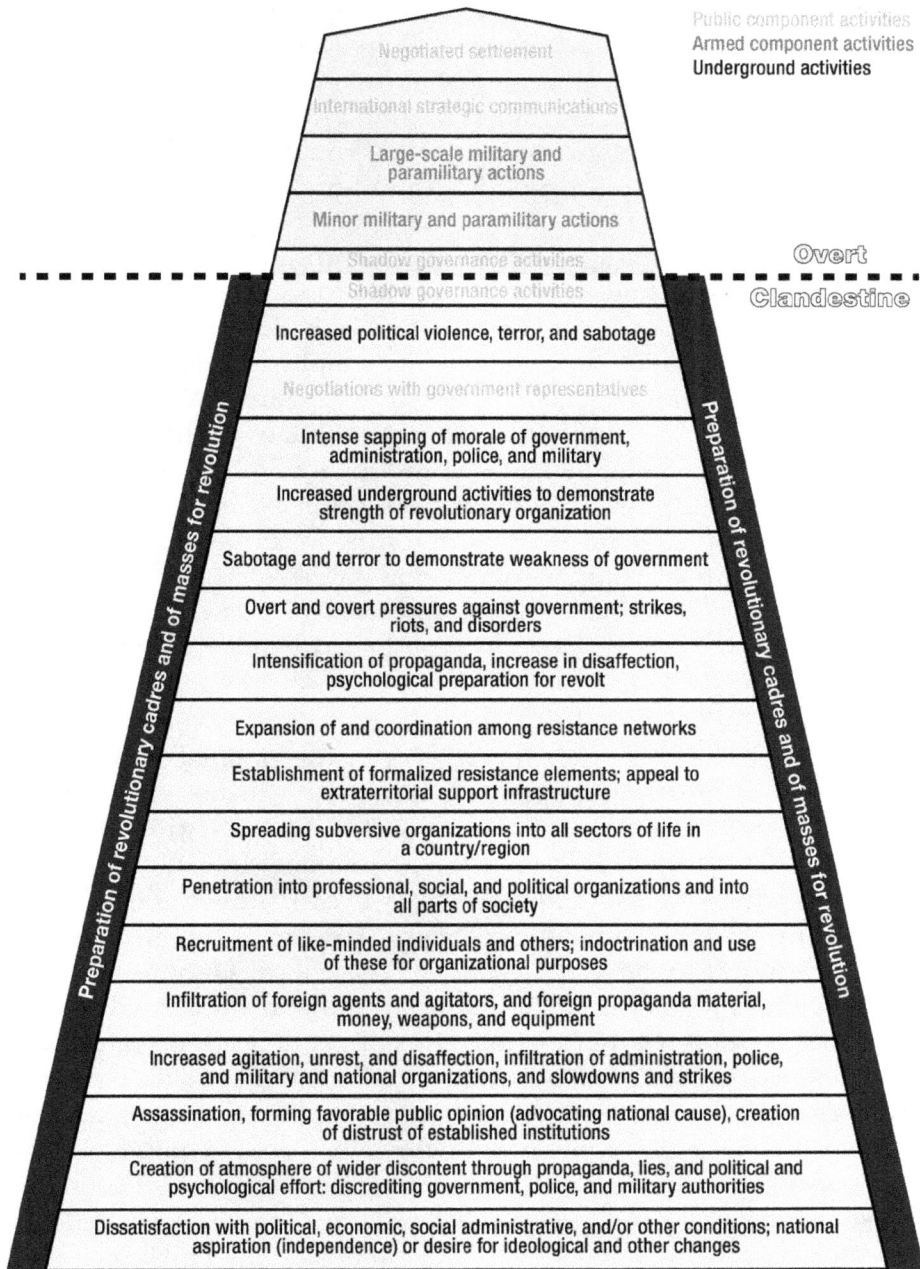

Figure 3-4. Covert and overt functions of an underground.

Undergrounds initiate recruiting, training, and infiltration, establish escape-and-evasion networks, raise funds, establish safe havens, and develop external support. The establishment of formal organizations and training programs and the coordinated penetration of government entities are historic functions of the underground, as is

45

intelligence and logistics support to these operations. Undergrounds may evolve to conduct subversive and psychological operations to undermine and delegitimize the government and cultivate popular support. They may ultimately perform command and control functions of the guerrilla force, establish shadow governments, and institute population control measures. An underground organization also requires financial resources to function: agents must be compensated and guerrilla forces armed; escape-and-evasion networks require money for extra food, for·safe houses, and to provide to escapees; psychological operations require funds for products and activities; and headquarters and administrative sections require office supplies. Undergrounds also often coordinate the engagement of diaspora communities for financial, logistic, and/or informational support. Many of these functions require overt movements and/or establishment of relationships outside the insurgent institution. These functions often exist in both the clandestine and overt domains and may be functions of either the underground or the public component. Shadow governance activities such as the provision (and withholding) of unofficial social welfare programs are types of activities that may be conducted by either overt and/or clandestine elements.

Perhaps the most vital of these functions is the establishment of political legitimacy not only in the eyes of the populace but also in the eyes of extraterritorial entities that inevitably influence a conflict. Today it is more likely that the successful insurgency will gain some level of legitimate, open political acknowledgment while simultaneously continuing quasi-legal and illegal activity (e.g., Irish Republican Army [IRA]/Sinn Féin, Umkhonto we Sizwe/African National Congress, Weather Underground Organization/Prairie Fire Organizing Committee). The integration of the underground and the armed components and (as a protective function) the obfuscation of the relationship between the two ensures both unity of effort and security. Although groups vary in their level of integration, from those that are tightly integrated (Hizbollah closely integrating the numerous components of the insurgency) to those that are loosely integrated (Al Qaeda continues to use numerous layers of individuals, organizations, and active denial measures to obscure their ties to any public components), success in modern insurgency requires coordination between each of the components of an organization (or organizations).

CRIMINAL CONNECTIONS

Most armed insurgencies at some point use criminal activities such as theft, smuggling, or extortion ("protection") as a means of fundraising. This often also brings the groups into contact with career criminals and organized crime elements, for better or worse. Criminal connections are a threat to the groups in more ways than one, threatening their ideological legitimacy and their support from the populace, and tempting members to become swept up in a different lifestyle and social network. Criminality may be viewed as distasteful by many of the more ideologically or politically motivated insurgents, particularly those drawn in for political or ideological reasons.[b] Mao cautioned against illegal activity lest the movement devolve into armed banditry for these reasons, but few groups manage to avoid criminal associations entirely. In some movements, the line between the strictly ideological insurgent and the strictly criminal drug lord becomes blurred. Pablo Escobar—the notorious Colombian drug lord killed in 1993—evolved from a criminal gang leader into a quasi-political figure who dabbled in social activism. Three cases of insurgency organizations will illustrate these points. The cases are as follows: the PIRA as an example of a regional insurgency; Communist insurgencies, using the blueprint described by Mao and elaborated with specific examples; and Al Qaeda, an example of a modern, decentralized insurgency. Each case will be used to illustrate the specific strengths of those organizations and treated in the order that makes for the clearest presentation.

Table 3-1. Organizational features illustrated by three cases.

	PIRA	**Communist**	**Al Qaeda**
Command and control	√	√	√
Aligning structure with strategy	√	√	√
Secrecy and compartmentalization	√		
Evolution and growth of the organization	√		√
Underground and aboveground connections	√	√	√
Criminal connections	√		√

THE PIRA AS A REGIONAL INSURGENCY

Figure 3-5 represents the circular evolution of the Provisional IRA and its complementary public component, Sinn Féin. After splintering

---

[b] This was particularly so for the PIRA, namely Eamon Collins, whose disillusionment with the petty criminality of his Active Service Unit (ASU) caused him to rethink his involvement and ultimately led to his disengagement and subsequent cooperation with the British.

from the IRA, the PIRA emphasized its small but prominent armed component with substantial militant auxiliary and international underground support. The relationship between the armed component and the public component began to equalize in the 1980s after the death of hunger striker Bobby Sands and the 1986 Sinn Féin convention ending abstentionism. Into the 1990s, Sinn Féin entered negotiations with the British government (without disestablishing the armed component or renouncing violence), eventually leading to the signing of the 1998 Good Friday Agreement calling for the decommissioning of the armed component and the entry of the public component into a devolved Northern Irish government.

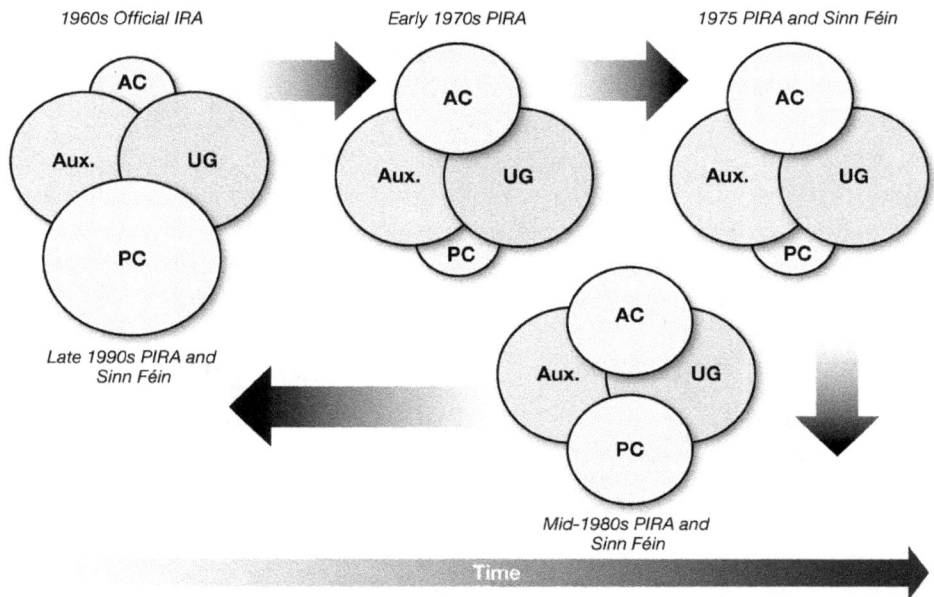

Figure 3-5. Cyclical evolution of the PIRA. AC, armed component; Aux., auxiliary; UG, underground; PC, public component. The above represents the evolution of the Provisional IRA and its complementary public component, Sinn Féin. The insurgency began with the moribund armed component of the Official IRA along with a robust public component that emphasized direct action. After the 1969 split, the Provisional IRA revived its military wing. Sinn Féin, the political party associated with the PIRA, gained importance and influence after establishing Advice Centers in 1975 to monitor the cease-fire terms of the same year. The relationship between the armed component and the public component began to equalize in the 1980s after the death of hunger striker Bobby Sands and the 1986 Sinn Féin convention ending abstentionism. Into the 1990s, Sinn Féin entered negotiations with the British government, eventually leading to the signing of the 1998 Good Friday Agreement, which called for the decommissioning of the armed component and the entry of the public component into a devolved Northern Irish government.

Evolution and Growth of the Organization

The IRA is an insurgent organization that dates back to the failed Easter Rising of 1916 against the British in Ireland. The early IRA, and its political wing, Sinn Féin, waged a successful guerilla campaign against a weakened Britain shortly after World War I in 1919–1921. Nearly all of Ireland gained independence from Britain, except the six counties of the traditional region of Ulster, now known as Northern Ireland. Catholic mobilization revived again during the worldwide political turbulence of the 1960s. During this decade, the IRA focused on direct action, influenced by socialist ideologies, at the expense of its armed wing. Waves of violence broke out in 1969 as the minority Catholic population marched in civil rights parades, calling for more equitable rights from the Protestant-dominated government. Mobs attacked each other across Belfast and Derry neighborhoods, and the British army was finally called in to regain control. As the violence continued to grow, a new offshoot of the IRA emerged to protect the Catholics from Protestant mobs and sectarian security forces and to drive British security forces back to England. This group called for a return to a focus on the armed component and a limited public component. The new Provisional IRA (PIRA), as it was called, just like the IRA from which it separated, intended to gain independence from Britain and unite with the Republic of Ireland through armed rebellion.

The Provisionals engaged in a three-decades-long campaign of urban guerilla warfare across Northern Ireland and England. The Provisionals underwent near collapses, rebuilding, and shifting tactics throughout the conflict until the 1998 cease-fire and the 2005 official end to its thirty-six-year armed campaign. In September 2005, the head of the PIRA's decommissioning, General John de Chastelain, confirmed that the weapons of the group had been destroyed and that the final event had been observed by the General, a Catholic priest, and a Methodist minister. No others were allowed at the event nor were any photographs taken. The original organization, the Official IRA, continued its socialist activities and officially ended its armed struggle in 2010, decommissioning its weapons in February 2010 under the same auspices as the PIRA.[13]

During the initial years of the fighting in the early 1970s, before the security forces refined their intelligence and counterinsurgent tactics, the Provisionals adopted the conventional military structure developed during the IRA's long history of struggle against the British. As the British's counterinsurgent tactics put more pressure on the Provisionals, the group re-aligned its strategy and structure. In the mid-1970s, the PIRA adopted a "long-war strategy," or a war of attrition. To complement this strategy, and to counteract British intelligence efforts, the PIRA also adopted a cell structure. Initially, the Provisionals relied on, and could expect, active community support. After the failed 1975 cease-fire and Provisional bombings that killed several innocent civilians, a more war-weary Catholic population was not as enthusiastic in their support of Provisional insurgents. However, because of the changing strategies of the PIRA—to a long-war strategy and a cell structure—the Provisionals did not require active community support but only passive support.[14] The secrecy required by insurgent groups often ensures their inefficiency and incompetence.[15]

Command and Control

The best estimate of the PIRA as an organization indicates that it had only a few hundred official Army members after it reached a steady-state campaign in the mid-1970s. After The Troubles and the formation of the Provisional movement, volunteers were very easy to come by, and membership peaked during 1972. As the British stepped up their countermeasures and narrowed the opportunities for action and movement, the PIRA implemented a smaller cellular structure for its active units and diminished the number of volunteers needed. The support network for the Army, however, stretched into the thousands in both Northern Ireland and the Republic.[16]

Active Service Units (ASU) contained the bulk of the membership of the PIRA and carried out its military operations. These units included both part-time volunteers and full-time members of the Army. Some men and women held normal jobs in the communities and participated in Army operations on the weekends or after work hours (the auxiliary). Far fewer members were paid by the PIRA with a weekly allowance to enable their full-time support to the Army. Members who had such a stipend often received additional support from the local community in the form of purchased clothes, donations, and food.[17] ASUs usually had four volunteers and one operations commander. If needed, an intelligence officer or an education officer could be made available for operational support or training, respectively. Each ASU was believed to be responsible for the bulk of their operational expenses, their local safe houses, and any required transportation. For security purposes, each operations commander most likely knew the identity of only one higher commander, the brigade adjutant. In turn, the brigade adjutant took orders from the higher geographical and/or central command.[18]

The PIRA was divided into two commands, the Northern and Southern commands[19]; the Northern Command covered the entirety of Northern Ireland plus the five Republic border counties of Louth, Cavan, Monaghan, Donegal, and Leitrim. This command was created in 1976 and was the predominate area of operations for the PIRA. Its leadership was tightly coupled with the commanders above it. The creation of the Northern Command coincided with a marked shift in the PIRA leadership from being dominated by members from the Republic to those from Belfast and Northern Ireland. Southern Command, created slightly later, encompassed the rest of the Republic and was far smaller in membership and importance. Southern Command functioned mainly to provide logistical support for the operations in Northern Command, including training, funding, safe houses, and the storage and movement of arms.

The upper authority echelons of the PIRA consisted of the General Army Convention, the Army Executive, the Army Council, and General Headquarters (GHQ). GHQ regulated and executed the daily tasks of the Army, ensuring the maintenance of the commands, and centralized the functions of the Army, including Finance, Security, Quartermaster, Operations, Foreign Operations, Training, Engineering, Intelligence, Education, and Publicity. GHQ consisted of approximately fifty to sixty people who were responsible for the overall conduct of the PIRA. It was officially located in Dublin, Ireland, although many staff members were based elsewhere. GHQ was headed by the Army Chief of Staff, who was selected by vote of the Army Council and was the key decision maker within the Army. The first Chief of Staff was Seán MacStiofáin, one of the original troika that formed the Provisionals. Traditionally, the Chief of Staff and other major positions were held by the trusted core members, with little turnover unless one was arrested. Members were required to give up their positions upon capture, and they were banned from partaking in decisions for a period of time after their release. There were nine Chiefs of Staff from the PIRA's inception to the peace accord in 2002.[20]

The Army Council was a seven-member panel, usually including the GHQ Chief of Staff, the Adjutant General, the Quartermaster, the Head of Intelligence, the Head of Publicity, and the Head of Finance. These men approved all major actions and established the majority of the strategy and policies of the PIRA, therefore serving as the main authority of the PIRA. The Council met at least once a month to review operations and to vote on major items, such as cease-fires or the expansion of operations to Britain.[21] The Army Executive was a board of twelve very senior and experienced PIRA veterans who met every six months or so to review the activities of the Army Council and to serve as the voting body that elected the seven members of the Army Council. Members were barred from sitting on both the Executive and the Council.[22] The Constitution of the PIRA placed the supreme authority of the movement within the General Army Convention. The Convention was a large body of delegates from the structures of the entire PIRA, and it was designed to meet to approve and vote on only the most important of issues, such as declarations of peace and the election of the Army Executive. The size of the Convention ranged from 100 to 200 members and included active volunteers, representatives for the imprisoned, staff from local brigades, the Commands, and GHQ, and usually all members of the Army Council. The constitution allowed for meetings to be held every two years, unless security deemed it prudent to delay the gathering of such a large group of members. The first convention instituted the new PIRA as an entity separate from the Official IRA, and others may have been held to debate major cease-fires, the end of abstentionism, and the final accords.[23]

Underground and Aboveground Connections

The relationship between the PIRA and its political arm, Sinn Féin, was often viewed as complementary and equal, but for the majority of the period of violence, the Army was the central organization and decision-making apparatus, while Sinn Féin was peripheral. The break of the Provisionals from the Official IRA in 1969 was due to perceptions that the Official IRA had emphasized a political strategy to the detriment of the armed rebelllion. Thus, the PIRA existed primarily as a fighting Army, seeing political solutions as undesirable and not worth pursuing until the British had been beaten into submission. Sinn Féin, therefore, operated mostly to explain the military operations to the public in the early days. It would take a generational change from within the Provisionals to make it possible for a political solution to take place in the 1990s. With the ascendancy of Gerry Adams to the top post of the political party in 1983, Sinn Féin became a powerhouse to the Republican cause, putting candidates up for office and acting as the proxy for the PIRA in the peace process. The PIRA also experienced a resurgence of support after the death of hunger striker Bobby Sands in 1981. His popularity, in part, derived from his election to the British Parliament while imprisoned. The success of this political strategy galvanized interest in the approach within the PIRA leadership. However, when the PIRA engaged in military actions, the British government would refuse Sinn Féin a place at the negotiating table until a cease-fire was called.

The PIRA's refusal to officially participate in any government changed in 1981 after a Sinn Féin *ard fheis*, or convention. They would still abstain from the London and Dublin governments, and any Belfast assembly, but they would pursue and occupy seats in the country's local elections. Sinn Féin first contested elections in 1985, winning 12 percent of the first preference votes.[24] Sinn Féin, however, was also up against its moderate counterpart, the Social Democratic and Labour Party (SDLP), which consistently did better at the polls than Sinn Féin. The SDLP offered a much milder brand of nationalism. In the 1982 election for the Northern Ireland Assembly, Sinn Féin received 64,191 votes to the SDLP's 118,891. Relations between the two parties were noticeably sour during most of the 1980s.[25]

Although Sinn Féin's candidates were often beaten, the threat of the political party, which openly supported armed struggle, worried both London and the unionists. In 1985, Britain and the Republic of Ireland struck an accord, the Anglo-Irish Agreement, which stated the status of Northern Ireland would not be changed except by vote of the Northern Ireland population and that the Republic would work with the United Kingdom on issues of security, human rights, and reconciliation. The agreement gave the Republic an "ongoing, consultative role" in matters relating to Northern Ireland. Prime Minister Margaret Thatcher engineered the agreement partly to boost the strength of the SDLP against Sinn Féin but primarily for Irish support in security matters, especially along their porous border with the North.[26] The agreement angered the unionists in Northern Ireland more than the PIRA, in part because the Republic and Britain had not requested any unionist presence or input during the negotiations. Gerry Adams argued that while, on one hand, the agreement legitimized the continued partition of Ireland, it also indicated that Republican activities were successful in gaining concessions from Britain.[27]

The PIRA continued its dual operations in both violence and politics, with the Sinn Féin now becoming as important as its military wing. Although Sinn Féin appeared to be gaining ascendancy over the latter, by no means did its members advocate a cessation of violence. The PIRA strategy gave equal measure to the ballot box and the ArmaLite. Both wings shared the same Republican aspirations and often even personnel. Released prisoners and other volunteers whose roles with the PIRA became public knowledge were often shuffled from the military wing to Sinn Féin.[28] Moreover, those who opposed the dual strategy were expelled from the movement.[c] Ironically, the shipment of Libyan arms received in 1985 also bolstered the Provisionals' emphasis on politics. Who could say that the volunteers advocating a political strategy were going soft on the armed struggle with an arms dump full of weapons?[30]

---

[c] A number of key Republicans disagreed with the Provisional Sinn Féin's position on abstentionism. Ruairi O'Bradaigh, president of Sinn Féin before Gerry Adams took the reins in 1983, formed a splinter Republican movement, Republican Sinn Féin, and its military wing, the Continuity IRA.[29]

Gerry Adams took charge of the political party in 1983.[d] By 1986, the Provisionals removed the ban on taking government seats in the Dublin parliament and also removed the official taboo on any volunteer openly discussing ending abstentionism.[31]

Criminal Connections

The desperation of the seemingly perpetual shortage of operational funds led the PIRA to use some creative fundraising tactics. These schemes included providing armed escorts for payments on international shipments (or riding shotgun); money-laundering schemes in the United States, Northern Ireland, and the Republic; gypsy taxi services; and counterfeit merchandise (including film piracy, unregulated alcohol production and distribution, designer clothing, etc.) as well as legitimate "front" businesses.[32] Social welfare fraud was a common exploitative tactic used by the PIRA; Irish-American fundraising bodies, such as NORAID (Northern Aid Committee) and FOSF (Friends of Sinn Féin), used local collections in pubs and clubs in Ireland and voluntary private donations as additional funding sources.[33] This further heightened the suspicion of more sophisticated forms of money laundering.[34] The PIRA turned to the United States for money and weapons as soon as they were organized enough to send agents abroad, and the Irish communities of Boston and New York proved very supportive. NORAID was founded by Irish expatriate Michael Flannery in New York City in 1970 to provide a steady stream of money to the IRA, mostly for the purchase of weapons. Republican sympathizers there cooperated at many stages of the weapons shipping routes to Dublin. For example, furniture might be filled with weapons at a warehouse, and customs in both New York and Dublin could be taken care of with a phone call. The arrival date would be given to someone in Dublin for pickup. Supply lines were established with Irish émigrés in the United States. One network alone shipped hundreds of light, powered, collapsible, concealable assault rifles during the 1970s. The security forces confiscated more than 700 weapons, 2 tons of explosives, and more than 150,000 rounds of ammunition in 1971 alone, most of which came from the United States.

The good fortunes of the PIRA hinged on the relationship they established with Colonel Muammar Gaddafi in 1972. Shipments began to flow to Dublin in 1973, often without success. The *Claudia* was boarded by Irish authorities, revealing that the PIRA was willing

---

[d] Adams, a northerner, replaced Ruairi O'Bradaigh, a southerner, in another instance in which control of the Republican movement was moving more firmly into northerner hands.

to accept large amounts of money and weapons from the state-sponsor. The relationship was a long one that was only reinforced with the later leadership of the PIRA under Adams. The *Eksund* was captured by French authorities on November 1, 1987, with more than 150 tons of armaments from Libya: 1,000 AK-47s, 1 million rounds of ammunition, 430 grenades, 12 rocket-propelled grenade launchers, 12 DHSK machine guns, more than 50 SAM-7 ground-to-air missiles, 2,000 electric detonators, 4,700 fuses, 106-mm cannons, anti-tank missiles, and 2 tons of Semtex.

The PIRA was allegedly involved in narcotics trafficking in the late 1970s and 1980s; however, condemnation from the Catholic Church and the subsequent decrease in popular support led to the abandonment of the tactic in favor of armed robbery.[35] This involvement was probably an individual ASU-level decision vice an institutionally approved method[e,36]—cellular structures often risk having decisions made at the cell level rather than at higher levels of authority.[f] The Real IRA, another offshoot of the movement, did not rely as extensively on popular support or the approval of the Catholic Church as the PIRA. The Real IRA continued its practice of fundraising through narcotics trafficking; however, they expended considerable effort in concealing the activity from the populace. In an effort to diminish the Real IRA and reassert its relevance, the PIRA began targeting drug dealers in Northern Ireland with a series of violent acts to not only intimate/dissuade the industry, but also to reassert the PIRA's role as the protectorate of the Irish Catholic community.[37]

---

[e] The social conservatism of many PIRA members often influenced their operations and policies. Seán MacStiofáin, first leader of the Provisional IRA, refused to allow contraceptives into Northern Ireland even though they had potential use in making acid fuses for bombs. His conservatism led one of his comrades to remark that MacStiofáin "would rather, it seemed, be caught with a Thompson [sub-machine gun] in his car boot than a packet of contraceptives in his pocket."

[f] Cellular structures, particularly blind cells, are not as susceptible to communications interdiction as hierarchal structures because there are fewer requirements to communicate in the interest of operational security. Decentralization, however, has its concomitant risk because the core leadership of an organization has less tactical control of a cell and thus the cell leader's freedom of movement may result in deviations from the group's strategy. The South Armagh Brigade of the PIRA had a notoriously aggressive ASU commander dubbed "The Butcher" for both his occupation and his propensity toward violence. Though his attacks on British soldiers were alienating the local populace, the PIRA leadership decided not to contact him lest the clandestine relationship be discovered. In this case, accepting local deviations from strategy was more palatable than the risk of comprising the political effort by exposing the network.

COMMUNIST INSURGENCIES AS ORGANIZATIONS

Mao, Lenin, and other leaders of global Communism paid a great deal of attention to organizational design and produced doctrinal statements of how Communist insurgencies should be organized. In his 1902 pamphlet "What Is To Be Done?" V. I. Lenin laid the organizational foundations of modern Communist insurgency. He formulated the notion that if revolutions are to be successful, they must be led by small, professional (i.e., Communist) elites. Later in his "Left-Wing Communism: An Infantile Disorder," written in 1920, he stressed the importance of political infiltration and the use of united fronts to disguise the Communist revolutionaries' purpose. He described the strategy of creating a covert parallel apparatus with interlocking leadership so that a small, highly disciplined elite could secretly direct and control a much larger revolutionary movement. Mao Tse-Tung formalized the strategy and tactics of a protracted guerrilla war among the rural peasantry as a means of extending international Communism into underdeveloped areas of Asia.

These models evolved over decades and through experience in multiple theaters. The outstanding features of Communist insurgencies as organizations were (1) the combining of aboveground and underground elements, especially through the infiltration and co-option of existing organizations, and (2) the matching of different divisions and committees to diverse functions.

Communist parties in Communist countries operated differently and are not a subject of this study. Legal Communist parties in non-Communist countries usually took part in legitimate political competition but very often also had underground and/or armed components; their goal was still to overthrow the system and replace it with a Communist-controlled one.

While the worldwide Communist movement appears to be well past its peak, and there may be no active insurgency that instantiates the classical model described here, there is much to be learned from these ingenious organizational designs. We will refer to the model in the present tense, as it was described in the first edition of this book.

Command and Control

Communist organizations are essentially hierarchical, although they are designed to appear democratic. Rank-and-file members participate in meetings and self-criticism sessions, and the organization's rhetoric is anti-elitist. But the power and decision making, by design, rest with a small number of committees reporting upward in a pyramid structure.

National central committees wield most of the power, although officially the supreme authority is the national congress. The national congress is composed of delegates elected by the various conferences and by the next lower level of the party. The congress meets every two or three years, when it is convened by the central committee, and is charged with four major responsibilities: (1) to determine the tactical line for the party on political issues; (2) to revise the official program and make new statutes; (3) to hear and approve the reports of the central committee; and (4) to elect the central committee. From time to time it is called upon to discipline top members of the leadership. In practice, most matters that are considered before the national congress have already been discussed. The central committee prepares and documents questions and problems, which are sent to the various party levels where they are discussed and agreed upon. The national congress usually approves what has already been decided. It also sanctions the decisions of the central committee. The national conference is called into special session by the central committee if urgent political matters arise in the period between party congresses. It is restricted in size to a small number of delegates. It is often used as a substitute for the national congress when the party, to minimize the chance of police detection, wishes to conduct clandestine meetings that can be quickly called and dispersed.

Between meetings of the national congress, the maximum authority of the party rests with the central committee, which is composed of top party leaders and varies in size from party to party. The members must have demonstrated competence in organizational ability. Their functions include carrying out the decisions of the national congress, supervising finances, enforcing programs and statutes, and controlling the party press and propaganda. The central committee sets up a finance commission for fundraising and a central control commission for carrying out party discipline and security. Its executive bureau, the political bureau (politbureau), is made up of ten to twelve members who are elected by the central committee and directs party activities between meetings.[38]

A secretary general and two aides are elected by the central committee to carry on the daily operations of the party. This secretariat transmits the decisions of the central committee and the party to the subordinate commands. The secretary general is the highest-ranking elected official and is responsible to the party congress. He makes decisions with the politburo and is responsible to the central committee. The party presidency, an honorary post, exists in some Communist parties.[39]

There are executive committees set up to discuss and resolve problems at the various echelons and pass issues to higher authorities for consideration. These committees supervise ideological instruction, the training of executive committees for finance and control committees, and the elections of delegates to the next higher level of the party. In addition, they are responsible for routing party business and directives through their area jurisdiction and executing the decisions of the party. All members of the party belong to a cell and have weekly or biweekly meetings.[40]

International Command and Control

National committees were intended to be part of a global hierarchy, with all national Communist organizations reporting to the Communist International (Comintern). In the 1930s, the Comintern was extended to every part of the world. In Western Europe and America, permanent bases were established. In many parts of the Western Hemisphere, seamen's and port workers' international organizations served as reception and reporting centers for agents. Agents could report to one of these groups and receive shelter, money, and further instructions. Agents within the maritime organizations made contacts for international functionaries, agents, and instructors passing through their districts and also provided cover addresses for covert communications. They received and handled international funds for local organizations. During the period from 1930 to World War II, through its Executive Committee (ECCI), the Comintern became a second arm of Soviet foreign policy. The role and power of the Comintern changed after the rise of China as a second-leading Communist power. Because of disharmony between the Soviet Union and China after World War II, the Comintern became less powerful than the Soviet and Chinese national committees, which exerted separate control over client state committees. International coordination still occurred, such as joint support for regional insurgencies, but these were negotiated rather than mandated by the Comintern.

Chain of Command Between Committee Levels

The Communist Party operates on principles of collective leadership and democratic centralism. Collective leadership means that executive and administrative decisions must be agreed upon by the majority of the officers at a given level of the party. Collective leadership, however, is an exception and practiced only during interparty conferences. In practice, the party functions in a highly centralized manner, with authority and command decisions flowing from the top to the lower echelons.[41]

Democratic centralism means that within the hierarchy of party organizations and committees, each lower body selects a representative to serve on a party committee; this committee in turn selects another representative to serve on a higher committee in the hierarchy. The principle of democratic centralism is followed throughout Communist organizations. Unit committees are elected by the membership or the delegates of the party organization. Each committee must report regularly on the activity of the party organization and must give an account of its work. These committees are responsible for carrying out the decisions of the higher party committees. All decisions of the higher committees are binding upon the lower body members.

In theory, each proposal is discussed at the lower levels of the party, and each committee member presents the opinion of the lower body to the next higher body until a decision is made at the central committee level. Once a decision is made, the entire party must carry it out. In general practice, a decision is determined at the central committee level and, although the lower echelons discuss it, the members are well aware that they must ultimately concur with it.[42]

Elections of committee members and their secretaries must be approved by the committee at the next higher level. This enables the leadership to exercise strict control over subordinates and to suppress any opposition from the outset.[43] Changes in leadership within the Communist Party are not frequent; elections become the equivalent of promotions. The leadership submits candidates and issues to the membership. In order to legitimize this authority, members are compelled to discuss these matters and overtly agree and vote on them.

This disposition of authority follows the party principle of reverse representation at all levels. The elected or designated leader of any organizational element of the party, regardless of the level at which he operates, represents among his associates the authority of the next highest party body. He is not the spokesman for his subordinates in higher party councils, but rather the latter's liaison with lower levels.

The Role of Self-Criticism Sessions

Institutionalized criticism or self-criticism serves two essential purposes in the Communist organization: (1) it increases the efficiency of the party by subjecting its operations to constant review and revision; and (2) it creates a norm of behavior in members and helps secure commitment and dedication to the party. The actual activities of criticism and self-criticism sessions consist of conferences, discussions, and meetings within the party in which attempts are made to determine and correct any weaknesses in the work of the party or party members. Criticism is practiced on all occasions and is an integral part of Communist life.

Theoretically, all decisions and basic policies of the party are open to institutional criticism and discussion in these sessions. But in actual practice, criticisms must never contradict the essential party line and are directed only toward improving the practice and implementation of existing revolutionary theory. A member is expected to analyze mistakes and shortcomings of the party operations only. Unless his criticism is constructive—that is, offers a concrete proposal for improvement in work or a method for correcting mistakes—it is not accepted and the individual making the criticism may find him or herself under attack. No criticism may be made of the central leadership, and no organized expressions of criticism or dissent are tolerated.[44]

Self-criticism sessions are a unique and effective feature of Communist organizations and are designed to develop absolute commitment and ideological dedication among members so that party orders are implemented, not mechanically but creatively. They attempt to make the individual member think in terms of a vanguard and how better to advance the current line and more effectively carry out revolutionary work. Members are compelled to report errors, mistakes, or weaknesses displayed by all party members no matter how small or trivial; they may also state and restate any change in policy.[45]

Every party member knows that if he does not make every effort to contribute seriously to criticism of his fellows, he will be obliged to confess this guilt in the subsequent comprehensive dissection of his own conduct. He also knows that participating fully in the identification of others' failings will not help him to escape his own eventual subjection to the same process. The thorough analysis of his conduct can proceed into the smallest details of his life, both private and public, both intimate and generally known. He must clearly acknowledge his faults before the group and promise to improve. He understands that an inadequate response in his own session can lead to a reduction of rank or even to expulsion from the party. Thus, the sessions instill in each member a need to demonstrate to his associates his unqualified responsiveness to the wishes of authority so that he can avoid undue attention by his cohorts and escape excessive criticism when his turn comes. In this fashion the Communist Party maintains a built-in, permanent uncertainty and apprehensiveness among the rank and file and can be assured of obedience from below.[46]

Underground and Aboveground Connections

Communist insurgencies, although past their peak of activity, may be unmatched in the sophistication with which they coordinate multiple organizations to achieve their ends. The Communist model laid a blueprint for parallel governance structures, infiltration of existing organizations, and coordination of multiple organizations in united insurgent fronts.

Parallel Structures

Article 12 of the Comintern Statutes called for Communists throughout the world to create secret illegal Communist organizations alongside legal organizations. Even in countries where the Communist Party is a legal political party, Communists typically maintained two major organizations, one open and one covert. The overt organization functions as an ordinary political party. However, the Communists everywhere organize their party into a system of cells and committees, regardless of the size or strength of the party or the degree of government opposition. Even in the legal party, the cellular structure serves to train members in conspiratorial behavior. Cell meetings are often held secretly so that members attending them can learn how to travel to and from them without arousing suspicion. Members are assigned minor intelligence gathering or sabotage missions, which in themselves have little or no practical use but which test and train members in clandestine behavior. A press (either open or clandestine) is usually set up in order to give members experience in writing, printing, and distributing material for the party.

In addition to the open legal party, a highly compartmentalized, clandestine organization is also created. Members of its cells are people who have potential value to the Communist Party in the event of an insurgency or coup d'état. Individuals recruited from government, vital communications centers, industry, or other organizations that the Communists seek to infiltrate may not be admitted directly to the party itself but may become members of clandestine cells. Thus a network of infiltrators and agents in important positions is in readiness. These cells may remain dormant for many years, being used only to collect selective intelligence, and are activated only in case of an insurgency.[47]

Communist doctrine also distinguishes between legal and illegal organizations for gathering intelligence and espionage internationally. While the party operates openly or through front groups, it also operates through embassies, foreign trade commissions, and news agency personnel. Those agencies that enjoy diplomatic immunity are termed the legal apparatus within a country. The term legal is used because the members of such agencies have diplomatic immunity and, if arrested for espionage activities, are not jailed but declared persona non grata and forced to leave the country.

The illegal apparatus is composed of espionage or intelligence agencies such as the Soviet GRU (Military Intelligence Directorate) and the KGB (Committee for State Security) and their agents and informers. If caught and arrested, members of these units can be legally tried for espionage. The GRU is in charge of military intelligence in foreign countries and the KGB units are responsible for nonmilitary espionage in foreign countries, operating parallel and often rival units to the GRU.[48]

Infiltration of Mass Organizations and Use of "Front" Organizations

A hallmark of Communist insurgencies is the infiltration, manipulation, and coordination of non-Communist groups, including trade unions, farmer's cooperatives, advocacy groups, and even government agencies.

Large groups that the Communists strive to infiltrate are called "mass organizations." Communist Party theory holds that a small group of highly disciplined individuals, operating through mass organizations, can rally the support required to win a revolution. V. I. Lenin recognized the vulnerability of mass organization to infiltration and manipulation. In turn, Joseph Stalin argued that Communist Party members must avoid the concept that efforts to build up the party should be directed solely toward recruiting new members. Instead, he suggested that the Communists systematically use mass organizations as "transmission belts" to the broad masses of nonparty workers. By working through mass organizations, Communist Party workers can reach and influence many thousands of workers "not yet prepared for Party membership." "[Through] these organizations, led by well-functioning fractions the Party must necessarily find its best training and recruiting ground. [Mass organizations] are the medium through which the Party . . . guides and directs the workers in their struggles and . . . keeps itself informed on the mood of the masses, the correctness of the Party slogans, etc."[49]

In addition to infiltrating existing organizations, Communists also organize "front" groups to use people who are sympathetic with causes that the Communist Party promotes but who are either unreliable or would not for personal reasons join the Communist Party. Front groups are organized around currently popular issues. While the party usually controls these organizations, they are kept separate. The Communists also attempt to gain control of governmental agencies through coalitions with other, non-Communist parties.

The use of front organizations in an insurgency can be illustrated by several cases. The Malayan Communist Party (MCP), early in its history, set about organizing a number of front groups, including the Proletarian Art League, Youth Corps, Racial Emancipation League, and General Labor Unions (GLU). The labor front was perhaps the most important. Using the demand for higher wages to match the rise in rubber and tea prices as a basis, the union movement organized a number of strikes and collective actions. The principle of organizing labor for collective action was new to Malaya in the thirties, and the Communists' efforts to develop labor unions were to pay off in the insurgency after World War II.

After the war, the MCP set up additional organizations. It organized a General Trade Union and a Youth League to attract Chinese students. Once the insurrection was underway, a Cultivation Corps, an Anti-British Alliance Society, a Students' Union, Women's Union, etc., as well as less overtly political organizations, such as youth and sporting groups, were organized. With employment hard to find, it was often necessary for a man to join a Communist union in order to get a job. The MCP also established its own schools and clubs so that it could approach the Chinese community to conduct political discussions and disseminate party literature.[50] During the insurgency in Greece, the Communists organized and controlled many front groups, such as the Seamen's Partisan Committee, the Communist Organization for Greek Macedonia, and the Democratic Women's Organization of Greece. In the rural areas, the Communist Party operated through the Greek Agrarian Party (AKE) and the United All Greece Youth Organization (EPON).[51]

In the Philippines, Communist Party officials spent much time before the war engaging in labor activities in Manila and other parts of Luzon. The printer's union was influenced by Mariana Balgos, and the League of the Sons of Labor was headed by Crisanto Evangelista, both noted Communist leaders. The League of Poor Laborers, the predecessor of the Confederation of Peasants, was among the mass support organizations that provided the base of support for the insurgency between 1946 and 1954. Most of the members of the Communist politbureau in Manila were officers in the unions affiliated with the Congress of Labor Organizations (CLO).[52]

Methods of Controlling Mass Organizations

Most voluntary, large-scale organizations are composed of a leadership (a small corps of individuals who represent the administration), a few faithful, full-time followers, and a large group of dues-paying members. The followers usually leave the operations and decisions to the leadership. Members may or may not agree with the leaders on all decisions and actions.[53] Members who are willing to work and accept responsibility are usually given the opportunity to do so and indeed, such willingness leads to a gradual promotion to leadership responsibilities.

When planning a takeover, the Communists first try to gain influence in the organization's membership office in order to control recruitment and to infiltrate Communist members. Once in, Communists are instructed to volunteer for all positions and for all work in the organization. They are instructed to be the first to arrive at and the last to leave meetings. They are taught how to harass non-Communist speakers and, through the tactics of attrition-through-tedium, win votes and offices within the organization. They seek the leadership of political and education committees and use these offices to identify people in the organization who might be sympathetic and those who are avowedly anti-Communist. Editorship of the organization's newspapers provides opportunities for expressing subversive ideas and gives access to printing materials that may be used to establish Communist distribution routes. Once the Communists have organized cells or fractions within the organization, they caucus and plan their organizational moves in advance.[54]

The Communist seeks leadership positions and represents himself as dedicated and loyal to the organization, taking the initiative in planning activities and volunteering for any job, no matter how time consuming or unpleasant. He is instructed to avoid the appearance of any subversive activity. Although his candidacy is supported by cell members in the rank and file, close ties between the candidate and the cell collaborators are hidden from the general membership so that the candidate's support appears spontaneous and unsolicited. Usually the most vocal members at a meeting pass resolutions and manipulate the apathetic majority. Therefore, a small, articulate group can readily influence the direction of the organization and eventually gain control.

In such organizations as labor unions, systems of rewards and punishments can be utilized to maintain the obedience of members. If a member is dropped from a union he may not be able to get employment. On the other hand, if the union leader improves the lot of the union members, they will more willingly go along with more purely political actions and obey strike calls. In addition, goon squads may be used to "persuade" uncooperative members. Having instruments of persuasion and coercion, the leadership can gain the compliance of a majority of members. Most members will comply with a strike decision, since higher wages may benefit them and failure to comply will only lead to punishment, loss of membership, or worse. When using front organizations, the Communists attempt to develop and maintain the loyalties of people who otherwise could not be persuaded to enter the Communist Party or who, even if willing, would not be sufficiently reliable. They also are able to mobilize many who are indifferent or even opposed to Communist ideology—uniting them instead behind causes such as "nationalism," "liberation," "pacifism," or other popular social issues within a particular society. The organization also attempts to gain support of those elements within the community, such as religious and fraternal organizations, that command the respect and loyalty of the workers.[55]

The cell attempts to evaluate the power structure of the group that it is trying to infiltrate. In professional groups such as industrialists, lawyers, or university presidents, the Communists seek to control executive staff functions because this is where the power resides in such organizations. They look upon the facade of distinguished citizens on the board of directors as an asset to the organizational infiltration. Hence, they do not seek positions of prestige but instead positions of control that affect the day-to-day operations of the organization.

In the Malayan Communist insurgency, for example, the MCP maintained its influence within the GLU through three separate control systems. The first system was made up of a president or secretary and two or three full-time organizers, who were part of the open membership of the labor union. Although they were party members, they avoided any connection with meetings or activities that might identify them with the party. They reported to and took orders from the GLU. They were told to operate within the law and to give the impression that their primary interest was the advancement and concern of trade unionism.

The second system of control was exercised through underground party members, who held no official office and were members of the open rank and file. They were activists who recruited new members for the union and for the MCP. They served to engender grassroots sentiment for policies favored by the party, enabling the leader to avoid the appearance of dictating to the union. This group also reported on the financial status of each member and provided information to the party on membership attitudes; the party then based organization policy on these reports. The underground members reported to the section of the party responsible for trade unionism, which was separate from the regular party. These members were more trusted than the leaders, who were considered expendable if discovered.

The third control system consisted of the regular party members who formed a fraction within the union membership. They held no official posts. They reported their activities to and received orders from the regular party.[56]

United Front Activities

After gaining control of existing mass organizations and creating front organizations, Communist strategy is to combine forces of these groups into united fronts while still maintaining the appearance of each group's independence. The groups in the united front need not agree with the objectives or goals of the Communist Party. However, if there is an aboveground party, they may offer support. In this manner the Communists maintain organizational integrity while becoming associated with other, legitimate organizations.[57] By drawing a number of legitimate groups into a united front, the Communists can gain the prestige of speaking for a large and diverse group of people. Once in the front, they may seek to discredit the leaders of the other organizations so as to gain control of their followings.[58]

In Venezuela, the creation of a united front was the first major step in initiating an insurgency and is characteristic of most Communist-dominated insurgencies. For example, the Venezuelan Communist Party (Partido Comunista Venezolano, or PCV) initially used its legal status to cover its illegal activities. Communists infiltrated the Acción Democrática (Democratic Action Party, or AD), and in 1960, under Domingo Alberto Rangel, the left wing of the party was expelled. The party formed a new group called the Movement of the Revolutionary Left (Movimiento de Izquierda Revolucionario, or MIR). In the mid-fifties, both the MIR and PCV were in militant opposition to President Rómulo Betancourt. In 1961, the MIR used its congressional immunity to carry out terrorism against the Betancourt regime. Finally, in 1962, both the PCV and MIR were ruled illegal by the Supreme Court of Venezuela. After this decision, a National Liberation Front (Frente de Liberación Nacional, or FLN) was formed to unite all left-wing elements against President Betancourt and initiate an insurgency. The FLN organized the Armed Forces of National Liberation (Fuerzas Armadas de Liberación Nacional, or FALN) to conduct urban terror and guerrilla warfare against the government.

In 1962, the Minister of Defense of Venezuela described the Communist plan as it appeared in captured FALN documents. The FALN proposed (1) agitation against the government; (2) demonstrations, disturbances, strikes, and terrorism; (3) sabotage and guerrilla actions throughout the country; and (4) insurrection culminating in violent takeover of power. The purpose was to create such chaos that the armed forces would take power through strong-arm methods; then the Communists would overthrow the army and gain control of the government. The Communists have gained support in the left wing of the Republican and Democratic Union (Unión República Democrática, or URD), which withdrew from the Betancourt coalition in protest against the government's anti-Castro action in 1960, as well as the AD opposition, which split from the AD in 1962. Both of these elements supported the terroristic campaign.[59]

Another example of a united front was the North Vietnamese opposition that united behind opposition to French rule. Communists were flexible enough to even temporarily disavow Communist affiliation to achieve longer-term gains. In May 1941, members of the Indochinese Communist Party (ICP) formed the Viet Nam Doc Lap Dong Minh Hoi (League for the Independence of Vietnam) or, as they were popularly known, the Viet Minh. The organization was a broad coalition of political parties, all of which wished to free Vietnam from French rule. But many nationalists did not join the coalition because it was Communist controlled, and Ho Chi Minh officially dissolved the ICP on November 11, 1945. In May 1946, in still another move to win nationalist support, he announced the establishment of the Mat Tran Lien Hiep Quoc Dan Viet Nam (Vietnamese Popular National Front), a broader front than the Viet Minh, whose goal was "independence and democracy."

In 1951, because the front received most of its aid from Communist China and the Communist bloc, the Communists felt that they had sufficient control over the movement and could reestablish a Communist Party. In addition, if some unforeseen event should occur in which they lost control of the front, they wanted to leave some official representation in the organization. The name, Dang Lao Dong (Workers' Party), was carefully selected. One party document describes how questions about the party's connection to Communism would be handled: "It should never be admitted outside Party circles that the Workers' Party is the Communist Party in its overt form for fear of frightening and alienating property owners and weakening national unity. To party members and sympathizers it can be admitted that the Workers' Party is the Communist Party, but to others it should neither be admitted nor denied."[60] In this way they avoided alienating people who for one reason or another could not accept Communism but at the same time won recognition from other Communist parties throughout the world.

Re-using the same tactics, in 1962, the Communists organized the National Front for the Liberation of South Vietnam (NFLSV). In order to control the movement, key members of the central committee of the Lao Dong Party went to the South to participate in the operations. Once they had firm control, a thinly disguised Communist Party (People's Revolutionary Party, or PRP) was formed that is ostensibly independent of the North but in fact is an extension of the Lao Dong Workers' Party. Except in the rare cases involving attempted coup d'état, the creation of a united front has preceded the initiation of Communist insurgency and guerrilla warfare.

When Radio Hanoi announced the formation of the PRP, it avoided the word "Communist" and described the party as "representatives of Marxist-Leninists in the South." A captured Viet Cong document that originated in North Vietnam and was sent to a provisional party committee in South Vietnam states that the formation of the PRP should be explained to party members as a tactical move to rebut accusations about the invasion of the South by the North and to permit the NFLSV to recruit new members and win sympathy and support from nonaligned nations. In July 1962, when the North Vietnamese signed the international agreements on Laos, a member of the delegation reported to foreign journalists that the list of members of the central committee of the Workers' Party was necessarily incomplete. Some names had been left off in order to protect the identity of men who were directing military operations in South Vietnam.[61,62]

Aligning of Structure to Strategy

Communist insurgencies, especially in the later stages of operations, employ a wide variety of specialized divisions, including military components, legal and illegal activists, and infiltrated external organizations brought into a united Communist front. Figure 3-6 shows the organizational structure of a mature insurgency.

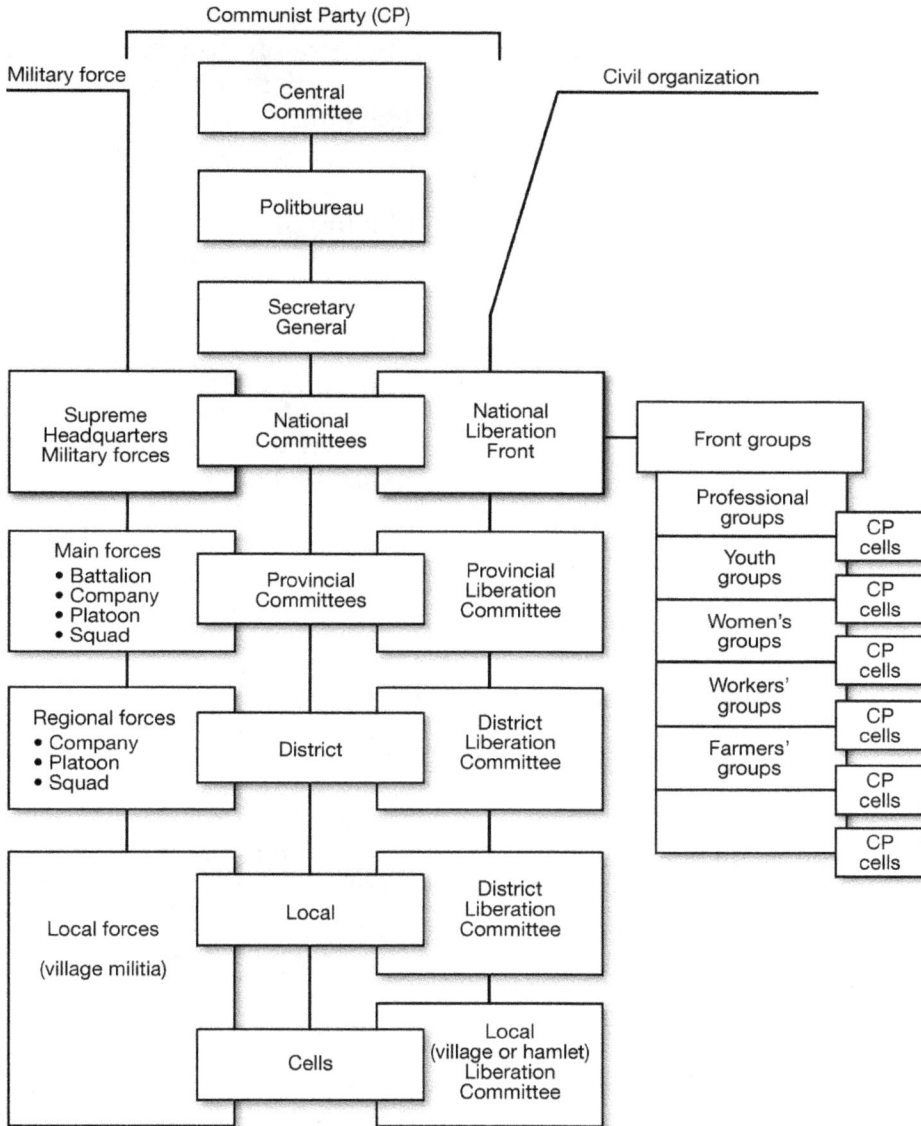

Figure 3-6. Organizational divisions and hierarchy of a mature Maoist insurgency.

Figure 3-6 shows that Communist organizations have some aspects of modern matrix designs, implementing both regional authority and also functional command.[63] Regionally, commands are hierarchically arranged from cells to local district, provincial, and national commands. But this does not imply that local leaders are independently directing local military actions; their involvement instead relates to recruiting, intelligence gathering, and interfacing with the population. There are parallel lines of authority enacting military strategies and

69

political strategies. The movement still retains paramilitary elements for special purposes even in the latter stages of an insurgency, when the overall emphasis has shifted from discrediting the government to capturing and holding territory. Paramilitary units have a parallel command structure and higher levels of secrecy and compartmentalization. There is also a parallel divisional line for control of associated front organizations, shown on the far right of Figure 3-6 and described earlier in this chapter.

The central committee maintains highest authority over all aspects of the movement, but the most important methods of exerting cross-control between specialized divisions are not necessarily through the formal lines of reporting. Another trademark of Communist organization design is the strategy of placing high-level functionaries on governing committees across different functional divisions. These elite members are not necessarily in the majority and do not necessarily have formal power (although they do have the backing of the official party); so in large measure, these party members have to rely on their well-practiced skills of both overt leadership and subversive influence to bring about coordinated action.

AL QAEDA AS A DECENTRALIZED NETWORK

The Al Qaeda "franchise"' represents a model of an insurgent organization that is very different from either the regional model of the PIRA or the global infiltration model of Maoist insurgencies, with different goals and a looser, more decentralized command structure. (This section on Al Qaeda is written for the most part in the past tense, because it is based on information that mostly predates Operation Enduring Freedom [OEF].)

Secrecy and Compartmentalization

Al Qaeda was conceptualized as a conspiratorial network that draws on leaders of all its regional nodes, and, when necessary, those leaders serve as an integral part of Al Qaeda's high command. Its structure is similar to that of a holding company, whose structure includes a core management group that controls partial or complete interests in other companies.[64] By some estimates, Al Qaeda had autonomous underground cells in some one hundred countries.[65] This apparent decentralization not only makes tracking affiliated militants and uncovering their operations more difficult, because there are no longer central "nodes" of individuals or institutions where significant activity is taking place, but it also increases the probability that cells will recruit on their own.[66]

Evolution and Growth of the Organization

Figure 3-7 depicts the evolution of Al Qaeda as will be described in this section. The organization started as an underground organization in Peshawar, Pakistan, providing support (in the form of Arab volunteers) to the Afghan mujahidin. When "Al Qaeda" was officially formed in the late 1980s, the organization was entirely underground and operated from Afghanistan, then from Sudan, then again from Afghanistan. Its armed component consisted of the 055 Brigade in Afghanistan as well as the training camp operations; beyond that, the elements were a loosely affiliated set of cells. In the wake of OEF, Al Qaeda has splintered and established regional underground franchise in Yemen (Al Qaeda in the Arabian Peninsula, or AQAP), North Africa (Al Qaeda in the Islamic Maghreb, or AQIM), and most recently Somalia (Al Shabaab).

Ideologically, Al Qaeda represents the extremist fringe of the global political Islamist movement[67] and the culmination of a religious struggle viewed in evolutionary terms. Al Qaeda's evolution corresponded roughly to the schools of thought prevalent among Islamic militants during different times. The first phase was associated with the charter of Egyptian Islamic Jihad, titled "The Neglected Duty" and written by Mohammad Abdel Salam Farraj. Early phases of this struggle were characterized by groups of dissident Muslims actively seeking to overthrow their own regimes, which they viewed as "apostate" governments. This "defensive jihad" was then expanded beyond the domestic arena to the entire globe. The second phase, epitomized by the writings of Abdullah Yusuf Azzam, was strongly influenced by the Soviet invasion of Afghanistan and the continued occupation of the Palestinian territories. Proponents of the radical doctrine argued that it was necessary for Muslims to fight alongside their co-religionists wherever apostate governments existed. This was exemplified by the participation of many Arab fighters in the Afghan struggle against the Soviets. Ultimately came the transition to an offensive jihad against the far enemy—Europe and the United States.[68] This third phase has long been present in the writings of political Islamists but became Al Qaeda's guiding principle when Azzam was killed and bin Laden secured singular control over the group's future direction.

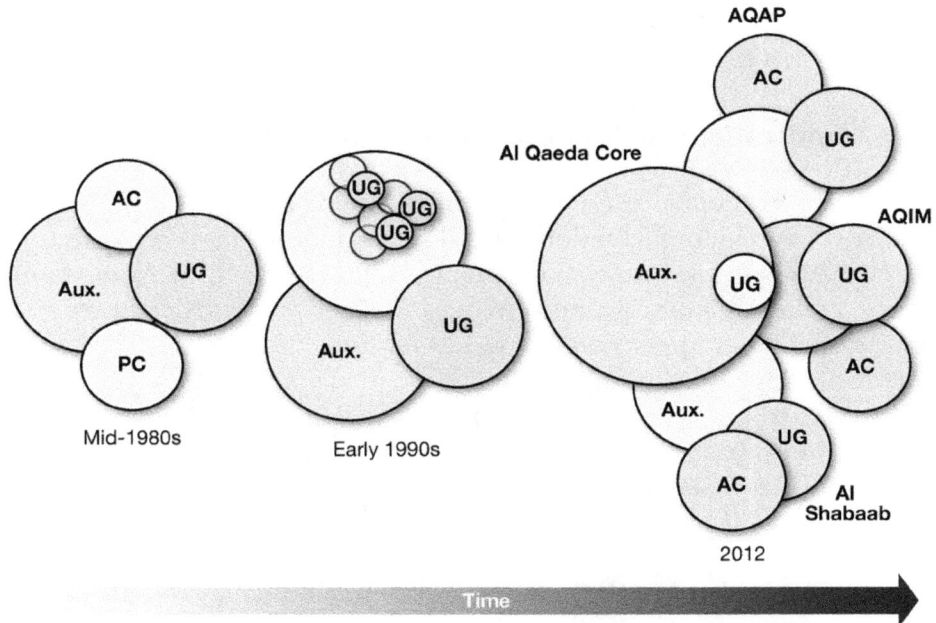

Figure 3-7. Evolution of Al Qaeda as a decentralized network, 1989–2012. AC, armed component; Aux., auxiliary; UG, underground; PC, public component. Al Qaeda, the loosely affiliated network at the center of the global Salafist jihad, started as primarily an underground organization in Peshawar, Pakistan, providing support (in the form of Arab volunteers) to the Afghan mujahidin. The MAK (Services Bureau) served as the overt element, although their role was more logical and financial than political. When Al Qaeda officially formed in the late 1980s, the organization was entirely underground and operated from Afghanistan, then from Sudan, then again from Afghanistan. Its armed component consisted of the 055 Brigade in Afghanistan as well as the training camp operations; beyond that the elements were a loosely affiliated set of cells. In the wake of OEF, Al Qaeda has splintered and established regional underground franchises in Yemen (AQAP), North Africa (AQIM), and most recently Somalia (Al Shabaab). After Bin Laden's death in May 2011, the leadership passed to Dr. Ayman Al-Zawahiri, whose core likely still operates from Pakistan.

Throughout this evolution, the structure of Al Qaeda was purpose-fully designed to be covert, secure, and ideologically (but not necessarily tactically) authoritative. The following excerpts from an Al Qaeda publication[g] (Figures 3-8 and 3-9) delineate the organization's principles, requirements, and operational priorities. Although just a snapshot, this translated component establishes a hierarchy between the military commander (a position also referred to as "emir") and his

---

[g] The "Manchester Manual" was discovered by the Manchester (U.K.) Metropolitan Police during a search of an Al Qaeda member's home. The manual was found in a computer file described as "the military series" related to the "Declaration of Jihad." The manual was translated into English and is available at the following website: http://www.justice.gov/ag/manualpart1_1.pdf.

advisory council and the individual members (or soldiers). This is a relatively simply replicated model that could be instituted by nascent (or franchised) cells. As those cells mature, specialization within the advisory council could emerge; however, the overarching strategy was retained by a new franchise and imbued in each new member.

UK/BM-12 TRANSLATION

Principles of Military Organization:

Military Organization has three main principles without which it cannot be established.

1.   Military Organization commander and advisory council
2.   The soldiers (individual members)
3.   A clearly defined strategy

Military Organization Requirements:

The Military Organization dictates a number of requirements to assist it in confrontation and endurance. These are:

1.   Forged documents and counterfeit currency
2.   Apartments and hiding places
3.   Communication means
4.   Transportation means
5.   Information
6.   Arms and ammunition
7.   Transport

Missions Required of the Military Organization:

The main mission for which the Military Organization is responsible is:

The overthrow of the godless regimes and their replacement with an Islamic regime. Other missions consist of the following:

1.   Gathering information about the enemy, the land, the installations, and the neighbors.
2.   Kidnaping enemy personnel, documents, secrets, and arms.
3.   Assassinating enemy personnel as well as foreign tourists.
4.   Freeing the brothers who are captured by the enemy.
5.   Spreading rumors and writing statements that instigate people against the enemy.
6.   Blasting and destroying the places of amusement, immorality, and sin; not a vital target.
7.   Blasting and destroying the embassies and attacking vital economic centers.
8.   Blasting and destroying bridges leading into and out of the cities.

Figure 3-8. Excerpt from page 13 of the "Manchester Manual."

UK/BM-13 **TRANSLATION**

Importance of the Military Organization:

1.   Removal of those personalities that block the call's path.
     [A different handwriting:] All types of military and
     civilian intellectuals and thinkers for the state.
2.   Proper utilization of the individuals' unused capabilities.
3.   Precision in performing tasks, and using collective views on
     completing a job from all aspects, not just one.
4.   Controlling the work and not fragmenting it or deviating
     from it.
5.   Achieving long-term goals such as the establishment of an
     Islamic state and short-term goals such as operations
     against enemy individuals and sectors.
6.   Establishing the conditions for possible confrontation with
     the regressive regimes and their persistence.
7.   Achieving discipline in secrecy and through tasks.

Figure 3-9. Excerpt from page 14 of the "Manchester Manual."

Organizationally, the roots of Al Qaeda lie in a number of different organizations, including the Maktab Al Khidamat (MAK) or Services Office, a clearinghouse established to facilitate the recruitment, transportation, organization, training, and equipping of Arabs to support the Afghan resistance. Established by Abdullah Yusuf Azzam, a Palestinian scholar of Islamic law, and Osama bin Laden in Peshawar, Pakistan, in 1984, the MAK consisted of a network of international recruiting offices, bank accounts, and safe houses and was also responsible for the construction of paramilitary camps for the training of militants and the fortifications used by Arab fighters.[h] The network of international recruiting offices (including the Al Kifah Refugee Center at the Farouq Mosque in Brooklyn, New York) and bank accounts collected and distributed nearly $600 million annually from a variety of Islamic charities and state sponsors. The MAK was responsible for training approximately 12,000–15,000 of those fighters,[i] with approximately 4,000 remaining connected through either chain of command or ideological affinity after the conflict. The MAK also played a role, albeit minor, in the U.S. Central Intelligence Agency's (CIA) Operation Cyclone, which helped provide funding and weapons to the Afghan mujahidin through Pakistan's Interservice Intelligence Directorate (ISI). Al Qaeda as an organization can be traced to a meeting held on August 11, 1988, between Azzam, bin Laden, and Zawahiri among others that addressed what to do with financial resources and a cadre of trained operatives once the Soviet withdrawal from Afghanistan was

---

[h] Azzam originally established the MAK and later persuaded bin Laden to join. Bin Laden used his family's relationship with the Saudi Royal Family to support the effort overtly—through a strategic communications plan—and covertly, eventually matching U.S. financial contributions to the resistance.

[i] Between 1982 and 1992, estimates report approximately 35,000 foreign fighters contributed to the Afghan effort, although there were probably never more than 2,000 in Afghanistan at any one time.

complete. The 1989 Soviet withdrawal from Afghanistan emboldened the Al Qaeda leadership, who believed their success in Afghanistan could be replicated in other Muslim territories. Some fighters remained in Afghanistan, but many traveled to other conflict sites, not only to support the global religious struggle but also because they were unwelcome in their home countries.[j]

To grow their nascent organization, Al Qaeda targeted mosques, schools, and boardinghouses throughout the world to serve as recruiting stations.[69] Although many fighters came from middle- and upper-class backgrounds, many also drew monthly paychecks from the organization.[70] During the 1980s, many Muslim nations supported the efforts of the Muslim Brotherhood and the more militant Islamic groups to recruit, encourage, and, to a lesser degree, facilitate the movement of young Muslim males from their native nations to Pakistan. Upon arrival, they were received, housed, and trained by Azzam and bin Laden's MAK and inserted into Afghanistan to support the Afghan mujahidin resistance of the Soviet occupation.[71] Al Qaeda does not necessarily use a centralized recruiting approach, often allowing subordinate leaders to determine the most appropriate course of recruitment for a given operational need,[72] but Al Qaeda does use a variety of models and approaches to recruiting. Some (referred to as The Net,[k] The Funnel,[l]

---

[j] The existence of so many conflicts, spread across a wide geographic space, facilitated the entry of many individuals into the militant apparatus. When enhanced security in one region made movement difficult (as was the case on the Afghanistan–Pakistan border after the 1998 bombings in East Africa), Al Qaeda's leadership reached out to movements located elsewhere that could house and train militants (in this case the Philippines offered an alternate location).

[k] In The Net, target populations may be engaged equitably. Some members will respond positively and others negatively, but in general the whole population is viewed as primed for recruitment. The target audience is viewed as homogeneous and receptive (and therefore does not require a differentiated message). This approach is used when serious opposition to the group (or cause) is absent in the audience's environment. It was initially used during by MAK during the anti-Soviet recruiting campaign throughout the Middle East and is often endorsed by various regimes.

[l] In The Funnel, the recruiter uses a phased approach when the target population is ripe for recruitment yet may require significant transformation in identity and/or motivation. With Al Qaeda, progress is measured "by validation of commitment to Salafist principles through the recruits' demonstrated knowledge of al-Qaeda's interpretation of Islam and both the willingness and ability to employ violence to achieve its goals." This approach can backfire (as it has in certain cases of Al Qaeda recruits) because individuals may not fully accept the ideology or may become turned off by the justifications for and types of violence perpetrated on behalf of Islam.

The Infection,[m] and The Seed Crystal[n]), but not all, are detailed in the footnote below.[73] Within the aforementioned models, Gerwehr and Daly also describe four recruiting approaches that are interchangeably used depending on the environment: public and proximate, public and mediated, private and proximate, and private and mediated.[74] In the first approach, public and proximate, recruiters commingle with the target population and make appeals individually or in small groups irrespective of any opposition or observation by authorities.[75] Islamic groups have used this approach for generations in prisons and concentration camps throughout the Muslim world with great success. The public and private mediated approaches are often broader and exploit mass media and marketing tactics, techniques, and procedures and modern information technology to convey their message.[76] The Muslim Brotherhood has historically used print media in various forms to further their social outreach message, while others have used the tactic to a lesser degree, often preferring Friday sermons at the mosques or public gatherings as better venues. Al Qaeda's use of the print media is best described as propaganda or psychological warfare.[77] During the 1990s, Zawahiri's writings served to elevate his personal status in the eyes of Egyptian Islamic Jihad (EIJ) members and legitimize the message to potential recruits.[78] By the end of the decade, the message became less of an espousal of his individual ideology and more of that of Al Qaeda itself and its recognized emir. Use of the Internet as a recruiting medium by, with, and through which Al Qaeda conveyed its message and targeted recruits did not reach maturity or operational effectiveness until after the 2001 merger with EIJ. Through its emergence as a distinct federation of interested parties after the Soviet withdrawal from Afghanistan, Al Qaeda's preferred recruiting approach was private and proximate.

---

[m] The Infection model is typically used when a target population is insular or a sub-component of a larger group that would actively deny terrorist group recruitment. Often, a trusted agent is inserted into the target population to rally potential recruits through direct, personal appeals. This method leverages individual charisma and persuasive strength of source credibility, social comparison and validation, and appeal specificity and was a favored strategy of Zawahiri as a clandestine cell leader. Egyptian Islamic Jihad (EIJ) under Zawahiri's guidance used this method extensively; EIJ veterans of the Afghan resistance would be dispatched to specific locales in Egypt to recruit quality fighters with military and/or engineering backgrounds to contribute both technically and operationally to the mujahidin resistance. This approach is often used when targeted recruits are in the existing security apparatus, such as police and military forces.

[n] The Seed Crystal is used when a target audience is too remote or inaccessible to penetrate and/or insert an agent. Recruiters here would provide the context and often the tools for individuals to seek out the groups, a pattern that tends to evolve into that of the Infection model. This approach was used by a variety of groups who were affiliated with the MAK during the 1990s when much of their leadership, now veterans of the Afghan resistance, were operating outside their native countries but still required access, which was actively denied by the authorities, to the younger generation of young men to execute terrorist operations domestically.

In this context, pitches were made in intimate settings with the explicit intent of avoiding observation. The technique leverages the influential power of conformity and relies heavily on personal appeals tailored specifically for a targeted individual, often using peers and/or relatives in making the pitch.[79]

After the Iraqi invasion of Kuwait in August 1990, bin Laden offered the services of his mujahidin to King Fahd to protect Saudi Arabia;[o] however, the king demurred, opting to allow U.S. and allied forces to deploy troops into Saudi territory. The deployment angered and offended bin Laden, because he believed the presence of foreign troops in the "land of the two mosques" (Mecca and Medina) profaned sacred soil. In 1992, bin Laden travelled from Jeddah, Saudi Arabia, to Kabul, Afghanistan, to help stop the intertribal militia conflict; his trip ultimately terminated in Khartoum, Sudan, where the Al Qaeda inner circle established a base of operations. From 1992 to 1996, Al Qaeda and bin Laden based themselves in Sudan at the invitation of Islamist theoretician Hassan al Turabi. During this time, bin Laden assisted the Sudanese government, bought or established various business enterprises, and established camps where insurgents trained.[80]

A key turning point for Al Qaeda occurred in 1993 when Saudi Arabia gave support for the Oslo Accords, which set a path for peace between Israel and Palestinians. On April 9, 1994, bin Laden's Saudi citizenship was revoked and his family publicly disowned him (although controversy persists over whether and to what extent he continued to garner support from members of his family and/or the Saudi government).[81]

The debate over targeting the near (Middle Eastern) versus the far (e.g., the United States) enemy seems to have shifted Al Qaeda's operational focus. The rebuke by the Saudi monarchy denied bin Laden the opportunity to surpass his own father's contributions to the land of the two holy places. Zawahiri, who had recently had a falling out with Sayyed Imam al-Sharif (aka Dr. Fadl) over the rationalization of terrorist acts against Muslim civilians, joined bin Laden in Sudan (though he is said to have preferred Yemen as a training ground for EIJ). At this point, Zawahiri had essentially integrated the senior leadership of EIJ into Al Qaeda, be it for pragmatic or strategic purposes.

During the early 1990s, alumni of MAK camps were participating in jihad (in support of Islamic causes) worldwide; however, by 1994, the

---

[o] In August 1990, Saddam Hussein's Iraq invaded neighboring Kuwait (over a petroleum resource dispute that was left unresolved by the United Nations [UN]). The Iraqi invasion placed the Saudi kingdom, its oil fields, and its ruling House of Saud at risk. In the face of a massive Iraqi military presence (at the time, Iraq possessed the fourth-largest standing army in the world), Saudi Arabia's own forces were well armed but outnumbered.

groups in Bosnia saw a decline in their aggressiveness and effectiveness.[82] Al Qaeda recognized this and decided to intervene by assuming operational control of many of the terrorist cells in Bosnia. The Al Qaeda leadership instructed its recruiters to search for like-minded Muslims (those who believed in a globalized Islamic insurgent movement) with academic (mostly science, mathematics, and engineering) and religious (Sunni Muslims with Salafist beliefs) backgrounds to increase the capability and reach of the still (publicly) unnamed organization. From 1991 to 1996,[p] Al Qaeda operated from the Northwest Frontier Province (NWFP) of Pakistan along the Afghan border or inside Pakistani cities, though it has autonomous underground cells in some one hundred countries, including the United States.[83]

As of this writing, it is unclear what effect bin Laden's death[q] will have on Al Qaeda. Since the initiation of Operation Enduring Freedom, Al Qaeda has grown increasingly decentralized, with the most active franchises in Yemen (AQAP), Northern Africa (AQIM), and Somalia (Al Shabaab).

Command and Control

The Al Qaeda strategist Abu Musab al-Suri has said: "Al Qaeda is not an organization, it is not a group, nor do we want it to be. It is a call, a reference, a methodology." Though the exact structure of Al Qaeda is still unknown, information acquired from former members provided U.S. authorities with a rough picture of how the group was organized.[84] Al Qaeda was administered by a *Shura* (consultative) council that discussed and approved major actions, including terrorist operations. Its structure is similar to that of a holding company, with a core management group controlling partial or complete interests in other companies (Al Qaeda affiliates). The core of the organization comprised a dozen or so militants, primarily Egyptians who had previously been held as political prisoners and traveled to Afghanistan to fight against the Soviets. This inner cadre was surrounded by an outer layer of one hundred or so highly motivated and well-trained loyalists from throughout the Muslim world. U.S. and coalition counterterrorism operations have been reasonably effective at capturing or killing many in this layer, and thus, as of early 2012, while inexact, this number is estimated to be somewhat less.

---

[p] During this period, the Taliban began to emerge in Afghanistan as the most powerful of the groups competing for power.

[q] Osama bin Laden was killed by U.S. Special Operations Forces in a raid on his Abbotabad, Pakistan, compound on May 1, 2011.

Osama bin Laden was, until his death, the group's undisputed leader—to whom the newly admitted swore an oath of allegiance (bayat).[85] Bin Laden's personal credibility was among the highest of any modern Muslim leader. His ascetic lifestyle contrasted sharply with those of the state-sponsored clergy, as well as those of many previous movement leaders, who amassed significant personal fortunes. Although the textual and visual narrative of bin Laden as the warrior-scholar has been carefully crafted, it does rest on a body of proven exploits. Although he lacked many of the serious religious, academic, and military credentials of other movement leaders, his image as a billionaire's son who forsook wealth and comfort for the austerity and deprivation of the life of a militant was admirable to many.

Throughout the 1990s, Al Qaeda used satellite phones and computers to organize and maintain plans and faxed copies of religious rulings issued by bin Laden throughout the Muslim world and Europe where they were picked up by Arabic-language media outlets.[86] The group also exploited informal, traditional forms of communications, such as pre-existing social ties, to transfer information.[87] Ultimately, their communications infrastructure and operations made extensive use of electronic media for mobilization, communication, fundraising, and planning attacks.

Subsequent to the 9/11 operation and the initiation of Operation Enduring Freedom, Al Qaeda's command and control apparatus was degraded to the point that there was very limited communication between the core leadership and its followers.[88] This degradation, however, did not necessarily limit the operational effectiveness of Al Qaeda as a social movement, as attacks such as the July 7, 2005, bombing of the London Underground indicate. The ideology that serves to unify Al Qaeda's narrative with its tactics does not require technologically sophisticated information systems to endure.

Aligning Structure to Strategy

The Al Qaeda command core is augmented by a set of specialized committees, each with a tailored set of goals, missions, and budgets. The Military Committee is responsible for training operatives, acquiring weapons, and planning attacks.[89] The most operationally viable component of the military committee (and Al Qaeda writ large) was, at one time, the 055 Brigade. At its peak in 2001, the 055 Brigade had an estimated 2,000 soldiers and officers comprising Arabs, Central Asians, and South Asians, as well as Chechens, Bosnians, and Uighurs from Western China.[90] The 055 Brigade, a guerrilla organization, was made up of veterans of the war against the Soviets, many of whom remained in Afghanistan after that conflict, and a second generation of younger, better-educated recruits shunned by their native countries and very loyal to bin Laden. The 055 Brigade fought with the Taliban (as their shock troops) against U.S. and Northern Alliance forces in 2002 and suffered significant losses.[91] Bin Laden ordered what remained of this force to retreat to the mountainous region along the Afghanistan–Pakistan border to wage a "protracted campaign" and to fight another day. The 055 Brigade, as an organization, did not possess a capability to operate outside of the region. Individual members of the brigade may join terrorist cells if selected to do so.[92]

The Finance and Business Committee, comprising professional bankers, accountants, and financiers, funded the recruitment and training of operatives, often through the *hawala*[r] banking system, providing airline tickets and false passports, issuing paychecks, and overseeing the group's vast network of businesses, as well as dealing with large organizational issues, such as developing financial resources to meet Al Qaeda's payroll and fund its operations and those of its various affiliates.[93] The Al Qaeda Finance and Business Committee managed the group's resources across four continents. The Foreign Purchases Committee was responsible for acquiring weapons and other technical equipment.[94] Regional money managers also appear to have exercised considerable influence and control over the network's financial decisions. The Sharia Committee determined whether particular courses of action conformed to established Islamic law. The Fatwa (formal legal opinion) Committee issued religious edicts, while the Media Committee supplied video and audio materials. The Media Committee was an essential component to the organization because it served as the mechanism though which bin Laden and/or Zawahiri could articulate their message to the faithful. In 2005, Al Qaeda formed Al-Sahab, a media production house, to supply its video and audio materials; this greatly professionalized the products and increased the receiving audience as the dissemination mechanisms evolved with technological advances in information technology.[95] Al Qaeda's strength and appeal did not lay solely in its sophisticated theological discourse, but also lay in its ability to comprehend, co-opt, and exploit modern grievances. Al Qaeda's leadership was not composed of highly trained religious scholars, and

---

[r] *Hawala* is an alternative or parallel remittance system outside of, or parallel to, traditional banking or financial channels. It was developed in India, before the introduction of Western banking practices, and is currently a major remittance system used around the world. The components of *hawala* that distinguish it from other remittance systems are trust and the extensive use of connections, such as family relationships or regional affiliations. Unlike traditional banking, *hawala* makes minimal use of any sort of negotiable instrument. Transfers of money take place based on communications between members of a network of *hawaladars*, or *hawala* dealers.

their religious rhetoric was far from complex or nuanced. Al-Sahab, in turn, transformed these edicts into accessible memes that resonated across geographic boundaries and at almost any position along the socioeconomic spectrum. Over the past decade, Al Qaeda has increasingly relied on groups external to this structure through its affiliates or franchises in Yemen, North Africa, and East Africa.[96]

Unlike the Communist model, nowhere in Al Qaeda "doctrine" is there explicit instruction as to how a cell must be formed; rather, a loose set of requirements for vertical (strategic-operational-tactical) integration is established. The proliferation of consistently themed Internet chat rooms and secure blogs have further socialized different "best practices" or "lessons learned" from aspiring cell members and leaders; however, those ideas (while often valid) are not necessarily sanctioned by Al Qaeda's core leadership.

The post-9/11 Al Qaeda has been categorized as somewhere between a centralized hierarchy and a flat network comprising a core (its central leadership based in Pakistan), the movement, and the periphery. This classification is not necessarily categorical but rather reflects the relative strength of the ties between an individual (or group) and the original membership of Al Qaeda. The core represents the strongest ties to bin Laden and his ideology; the franchises such as Al Shabaab in Somalia, AQAP in Yemen, and AQAM in Northern Africa would be considered part of the movement; and the self-radicalized or part-time supporters would be enmeshed in the periphery, or auxiliary.[97]

Underground and Aboveground Connectivity

Al Qaeda deviates from the four-component model of an insurgency because there is no explicit "public component" of the group that engages in legal or quasi-legal activity in support of political objectives. The combination of a transnational area of operation and Islamic fundamentalist worldview distinguished Al Qaeda from other insurgencies. Al Qaeda did, however, attempt to enhance its capabilities by acquiring and integrating existing militant organizations. Al Qaeda was a supporter of EIJ for years prior to June 2001, when the two organizations merged. Thereafter, all activities and operations of former EIJ members were carried out under bin Laden's banner. Al Qaeda may also have formed a strategic partnership with other militant groups, including possibly Hizbollah, from whom it may have received technical assistance, training, and intelligence. Israeli intelligence also believes that Al Qaeda infiltrated the Palestinian Occupied Territories with the support of Ḥarakat al-Muqāwamah al-'Islāmiyyah (Islamic Resistance Movement, or HAMAS).

Al Qaeda was, for a time, aligned with the de facto government of Afghanistan during the reign of the Taliban. The Taliban were a group of mujahidin veterans and orphans of the conflict, many of whom had been educated in the expanding network of Islamic schools (madrassas) either in Kandahar or in the refugee camps on the Afghanistan–Pakistan border. Bin Laden's organization and/or affiliates thereof laundered much of the financial support to these institutions in the form of donations from an array of Islamic charities (some were simply front organizations while others were more legitimate). The continuing internecine strife between various mujahidin factions in Afghanistan and the concomitant lawlessness subsequent to Soviet withdrawal set the conditions for the principled and well-disciplined Taliban, under the leadership of Mullah Mohamed Omar, to expand their control over the territory they called the Islamic Emirate of Afghanistan. The Taliban captured the regional center of Kandahar in 1994 and the national capital of Kabul in September 1996. Taliban-controlled Afghanistan provided a suitable location for Al Qaeda to establish its headquarters. Al Qaeda enjoyed the Taliban's protection and a measure of legitimacy as a component of their Ministry of Defense, although only Pakistan, Saudi Arabia, and the United Arab Emirates recognized the Taliban as the legitimate government of Afghanistan. Bin Laden quickly established a close relationship with the Taliban (now the de facto government of Afghanistan) leader, Mullah Mohammed Omar, whom he acknowledged as the "Amir Ul-Momineen" (Commander of the Faithful), a title once reserved for the head of the Islamic caliphate.

Criminal Connections

Among the reasons for Al Qaeda's success has been their ability to move operational funds through overt donations, quasi-legal gray markets, and patently illegal black markets. While some of these activities are considered illegal, Al Qaeda's public religiosity has somewhat insulated them against the "criminal network" label within much of the Middle East as well as in central and southeast Asia. From the MAK's operations in Peshawar to the modern affiliates throughout the world, Al Qaeda has relied upon and integrated criminal entities out of necessity. However, they have done so with sufficient discipline (for the most part) to accomplish their required missions without the concomitant excesses that tend to draw attention from law enforcement and/or other security forces.

The origins of many Islamic economic and financial regulatory organizations date back to the Muslim Brotherhood's development of political, economic, and financial infrastructures that enabled fulfillment of their religious obligations. The founder, Hassan Al-Banna, viewed finance as a critical weapon in undermining the infidels and reestablishing the Islamic caliphate. To do so, he believed Muslims would need to create an independent Islamic financial system that would parallel and later supersede the Western economy.[98] Al-Banna's successors set his theories and practices into motion, developing uniquely Islamic terminology and mechanisms to advance the Brotherhood's system of faith, as well as their unique financial apparatuses. In 1962, the Brotherhood convinced the king to launch a global financial joint venture that established numerous charitable foundations across the globe, and these foundations became the cornerstone and spread Islam (and later funded terrorist operations) worldwide. The first of these charitable foundations were the Muslim World League and Rabitta al-Alam al-Islami, which united Islamic radicals from more than twenty nations. In 1978, the kingdom backed another Brotherhood initiative, the International Islamic Relief Organization (IIRO), an entity that has been implicated in funding organizations such as Al Qaeda and HAMAS.

Al Qaeda used an extremely sophisticated, complex, and resilient money-generating and -transferring network.[99] The organization's highly resilient financial infrastructure spanned the globe with fund-raising operations, various types of accounts, and financiers on every continent in approximately one hundred countries.[100] The versatility of Al Qaeda's financial infrastructure was primarily due to its compartmentalized structure. Sources of funding were kept separate from the cells receiving the money, and high priority was assigned to financial training and management, as well as to the sustained generation and investment of funds.

To move funds clandestinely from source to recipient, Al Qaeda's financial network disguised the identities of both parties and established several legitimate institutions, including state and privately owned charities, banks, and companies, through which to funnel funds.[101] Revenues went directly to central headquarters. With the exception of some operational expenses, cells were typically expected to be self-sufficient. Operational cells could deploy without much information regarding the larger organization's underlying financial network.[102]

Al Qaeda favored charities as a primary source of income.[s] Charitable contributions are a sizeable source of financing for humanitarian, educational, and foreign aid activities in Muslim countries. Donations were largely cyclical; they peaked during the months of Ramadan, suggesting funds came from *zakat* and other obligatory charitable activities, often from unsuspecting religious institutions.[103] Al Qaeda utilized two different approaches in redirecting funds from charities to its own organization. The first involved creating, subsuming, or collaborating with a cooperative charitable organization. Donors to these charities may or may not have known the true nature of the organization to which they were contributing. Once donations were made to the charity, money was then sent to the network's headquarters. The charities would print false documentation for the benefit of donors, typically showing the money had been spent on humanitarian causes. The second, less common and riskier approach involved Al Qaeda operatives infiltrating unwitting charities with the intent of having those charities send funds to support the group's efforts in various parts of the globe.[104] Most donors to these types of benevolent organizations probably did not know that the organization had been infiltrated. However, some donors may intentionally have provided funds to such charities, providing an additional layer of plausible deniability to hide the true intentions of their funding activities.

Al Qaeda was adept at exploiting the environment of limited financial regulation prevalent in the Middle East and many Asian states but was also able to make use of formal financial institutions (including wire transfers and bank accounts). Like other components of their logistical and operational networks, redundancies were incorporated to avoid interruption, and nodes were replaced frequently to avoid detection and surveillance.[105] A presence was established in most countries using indigenous or migrant Muslim communities. Although Al Qaeda had no single, central financial repository, it could never operate in isolation because mounting a terrorist operation required financial and technical logistical support that often had to be in place years in advance. In the Middle East, especially the Gulf states, Al Qaeda had a great deal of covert support among the public and received practical

---

[s] Most Muslim nations collect mandatory Islamic charity (*zakat* is the third pillar of Islam, an annual wealth tax for charitable purposes) of about 2.5 percent from Muslim institutions and companies. *Zakat* is proscribed to go to those less fortunate. However, the Brotherhood has determined that those engaged in jihad against the enemies of Islam are entitled to benefit from the charitable offering. The interpretation that modern jihad is a serious, purposefully organized work intended to rebuild Islamic society and state and to implement the Islamic way of life in the political, cultural, and economic domains is widely accepted among Muslims and thus those engaged in jihad are viewed as legitimate recipients of *zakat*.

help from Islamic philanthropists and foundations, particularly from the United Arab Emirates and Saudi Arabia.[106]

The exploitation of existing commercial, financial, and transportation systems and institutions is common practice in organized crime and was implemented adeptly by Al Qaeda. The *hawala* system, an informal network of money brokers originally established to facilitate long-distance trade beginning around eighth century, was one such system.[107] Partly because of the weakness and undependability of the existing financial and banking system, MAK relied heavily on the *hawala* system to move money from its satellite entities (recruiters, arms dealers, logisticians) to the training camps during the Afghan resistance.[108] *Hawaladers* used formal financial institutions and couriers to transfer funds, but the process was not subject to substantial government oversight or record keeping, which was written in shorthand and kept only for brief periods. This enabled operatives to access funds without opening an account. Couriers came from inside the network and were chosen for certain characteristics that facilitated their movement (ethnicity, documentation, language skills), but they were rarely privy to operational details.[109]

ENDNOTES

[1] David S. Alberts and Richard E. Hayes, *Power to the Edge* (CCRP Publication Series, 2003).

[2] Geert Hofstede, "National Cultures in Four Dimensions: A Research-based Theory of Cultural Differences Among Nations," *International Studies of Management & Organization* 13, no. 2 (1983): 46–74.

[3] Jay R. Galbraith, *Designing Organizations: An Executive Briefing on Strategy, Structure, and Process* (San Francisco, CA: Jossey-Bass, 1995).

[4] Lindsay Heger, Danielle Jung, and Wendy Wong, "Organizing for Resistance: How Group Structure Impacts the Character of Violence" (49th Annual Conference of the International Studies Association, San Francisco, CA, 2008).

[5] Angelo Rossi, *A Communist Party in Action* (New Haven, CT: Yale University Press, 1949).

[6] George K. Tanham, "The Belgian Underground Movement 1940–1944" (unpublished Ph.D. diss., Stanford University, Stanford, CA, 1951).

[7] Bonnie H. Erickson, "Secret Societies and Social Structure," *Social Forces* 60, no. 1 (1981): 188.

[8] Susan Naquin, *Millenarian Rebellion in China* (New Haven, CT: Yale University Press, 1976).

[9] Jozef Garlinski, *Fighting Auschwitz: The Resistance Movement in the Concentration Camp* (London: Friedman, 1976).

[10] Frans P. B. Osinga, *Science, Strategy, and War: The Strategic Theory of John Boyd* (New York: Routledge, 2006).

[11] Joseph S. Nye, *The Paradox of American Power: Why the World's Only Superpower Can't Go It Alone* (Oxford: Oxford University Press, 2002).

12 John R. Dodson, "Man-hunting, Nexus Topography, Dark Networks and Small Worlds," *IO Sphere* (Winter 2006): 7–10.

13 Chuck Crossett and Summer Newton, "The Provisional Irish Republican Army: 1969–2001," in *Casebook on Insurgency and Revolutionary Warfare, Volume II: 1962–2009*, ed. Chuck Crossett (Alexandria, VA: U.S. Army Publications Directorate, in press).

14 Ibid.

15 Summer Newton, *The Provisional Irish Republican Army: 1969–1998* (Alexandria, VA: U.S. Army Publications Directorate, in press).

16 Crossett and Newton, "The Provisional Irish Republican Army: 1969–2001."

17 Ibid.

18 Ibid.

19 John Horgan and Max Taylor, "The Provisional Irish Republican Army: Command and Function Structure," *Terrorism and Political Violence* 9, no. 1 (1997): 1–32.

20 Ibid.

21 Ibid.

22 Ibid.

23 Ibid.

24 Newton, *The Provisional Irish Republican Army: 1969–1998*.

25 Ibid.

26 Ibid.

27 Ibid.

28 Ibid.

29 Ibid.

30 Ibid.

31 Ibid.

32 John Horgan and Max Taylor, "Playing the 'Green Card' - Financing the Provisional IRA: Part 1," *Terrorism and Political Violence* 11, no. 2 (1999): 1–38.

33 Newton, *The Provisional Irish Republican Army: 1969–1998*.

34 Horgan and Taylor, "Playing the 'Green Card.'"

35 Vanda Felbab-Brown, *Shooting Up: Counterinsurgency and the War on Drugs* (Washington, DC: Brookings Institution Press, 2010), 165.

36 Crossett and Newton, "The Provisional Irish Republican Army: 1969–2001."

37 Felab-Brown, *Shooting Up*.

38 Rollie E. Poppino, *International Communism in Latin America* (Glencoe, IL: The Free Press, 1964).

39 Ibid.

40 Ibid.

41 Ibid.

42 J. Peters, *The Communist Party: A Manual on Organization* (New York: Workers Library Publishers, 1935).

43 Werner T. Angress, *Stillborn Revolution: The Communist Bid for Power in Germany, 1921–1923* (Princeton, NJ: Princeton University Press, 1963).

44 Peters, *The Communist Party: A Manual on Organization*.

45 Louis Francis Budenz, *The Techniques of Communism* (Chicago, IL: Henry R Finery Co., 1954).

46 Ibid.

47 Arthur Koestler, "The Initiates," in *The God That Failed*, ed. Richard Crossman (New York: Harper and Brothers, 1949).

[48] Andrew R. Molnar, *Undergrounds in Insurgent, Revolutionary, and Resistance Warfare* (Washington, DC: Special Operations Research Office, 1963).

[49] Philip Selznik, *The Organizational Weapon: A Study of Bolshevik Strategy and Tactics* (Glencoe, IL: The Free Press, 1960), 118.

[50] Lucian W. Pye, *Guerrilla Communism in Malaya* (Princeton, NJ: Princeton University Press, 1956).

[51] Edward R. Wainhouse, "Guerrilla War in Greece 1946–1949: A Case Study," *Military Review* 37, no. 3 (1957): 17–25.

[52] Molnar, *Undergrounds in Insurgent, Revolutionary, and Resistance Warfare.*

[53] Selznik, *The Organizational Weapon.*

[54] Budenz, *The Techniques of Communism.*

[55] Selznik, *The Organizational Weapon.*

[56] Pye, *Guerrilla Communism in Malaya.*

[57] Selznik, *The Organizational Weapon.*

[58] Vladimir Ilyich Lenin, "What Is to Be Done?" in *Collected Works* (New York: International Publishers, 1943).

[59] Special Operations Research Office, *Area Handbook for Venezuela* (Washington, DC: U.S. Government Printing Office, 1964).

[60] P. J. Honey, "North Vietnam's Workers' Party and South Vietnam's People's Revolutionary Party," *Pacific Affairs* 35, no. 4 (1962): 375–383.

[61] Ibid.

[62] Van Chi Hoang, *From Colonialism to Communism* (London: Pall Mall Press, 1964).

[63] Peters, *The Communist Party: A Manual on Organization.*

[64] Peter L. Bergen, *Holy War, Inc.: Inside the Secret World of Osama bin Laden* (New York: The Free Press, 2001), 32.

[65] Bruce O. Reidel, *The Search for al-Qaeda: Its Leadership, Ideology, and Future* (New York: Reed Elsevier Inc., 2009).

[66] Ibid.

[67] Jason Spitaletta and Shana Marshall, "Al Qaeda," in *Casebook on Insurgency and Revolutionary Warfare, Volume II: 1962–2009*, ed. Chuck Crossett (Alexandria, VA: U.S. Army Publications Directorate, in press).

[68] Ibid.

[69] Ibid.

[70] Bergen, *Holy War, Inc.*, 30.

[71] Fawaz A. Gerges, *The Far Enemy: Why Jihad Went Global* (New York: Cambridge University Press, 2005), 60.

[72] Spitaletta and Marshall, "Al Qaeda."

[73] Scott Gerwehr and Sara Daly, "Al-Qaida: Terrorist Selection and Recruitment," in *The McGraw-Hill Homeland Security Handbook: The Definitive Guide for Law Enforcement, EMT, and All Other Security Professionals*, ed. David G. Kamien and Michael Kraft (New York: McGraw Hill, 2005), 80.

[74] Ibid., 80.

[75] Ibid., 82.

[76] Ibid.

[77] Youssef H. Aboul-Enein, *Ayman Al-Zawahiri: The Ideologue of Modern Islamic Militancy*, Counterproliferation Papers, Future Warfare Series, no. 21 (Maxwell Air Force Base, AL: USAF Counterproliferation Center, 2004), 8.

[78] Ibid.

[79] Gerwehr and Daly, "Al-Qaida: Terrorist Selection and Recruitment," 82.

[80] Spitaletta and Marshall, "Al Qaeda."

[81] Ibid.

[82] Ibid.

[83] Reidel, *The Search for al-Qaeda.*

[84] Spitaletta and Marshall, "Al Qaeda."

[85] Ibid.

[86] Ibid.

[87] Ibid.

[88] Marc Sageman, *Leaderless Jihad: Terror Networks in the Twenty-First Century* (Philadelphia: University of Pennsylvania Press, 2008).

[89] Spitaletta and Marshall, "Al Qaeda."

[90] Ibid.

[91] Ibid.

[92] Ibid.

[93] Steve Kiser, *Financing Terror: An Analysis and Simulation for Affecting Al-Qaeda's Financial Infrastructure* (Santa Monica, CA: RAND Corporation, 2005), 67.

[94] Spitaletta and Marshall, "Al Qaeda."

[95] Ibid.

[96] Scott Helfstein and Dominick Wright, "Success, Lethality, and Cell Structure Across the Dimensions of Al Qaeda," *Studies in Conflict & Terrorism* 34, no. 5 (2011): 367–382.

[97] Ibid.

[98] Rachel Ehrenfeld, "The Muslim Brotherhood New International Economic Order," *The Terror Finance Blog* (blog), October 13, 2007, http://www.terrorfinance.org/the_terror_finance_blog/2007/10/the-muslim-brot-1.html.

[99] Kiser, *Financing Terror,* 67.

[100] Jason Spitaletta, "Egyptian Islamic Jihad," in *Casebook on Insurgency and Revolutionary Warfare, Volume II: 1962–2009,* ed. Chuck Crossett (Alexandria, VA: U.S. Army Publications Directorate, in press).

[101] Spitaletta and Marshall, "Al Qaeda."

[102] Kiser, *Financing Terror,* 67.

[103] Spitaletta and Marshall, "Al Qaeda."

[104] Ibid.

[105] Ibid.

[106] Ibid.

[107] Ibid.

[108] Ibid.

[109] Ibid.

CHAPTER 4.

LEADERSHIP

CHAPTER CONTENTS

Jason Spitaletta and Nathan Bos

Leadership over human beings is exercised when a person mobilizes institutional, political, financial, psychological, and other resources to arouse, engage, and satisfy the motives of followers.[1] The core leadership of underground organizations develops ideology, finds a common grievance to garner popular support, and develops a strategy and organizational pattern based on the physical, human, and security environments. Leadership is one of the most observed, but least understood, phenomena in the social sciences, yet the study of leaders is integral to a thorough comprehension of revolutionary and insurgent warfare. There has been a great deal of research within the U.S. government on leadership analysis and related methodologies. A comprehensive discussion of theoretical and analytical methods used to assess leaders is beyond the scope of this chapter; what follows is an attempt to introduce some of the central concepts and their applications.

TRANSACTIONAL AND TRANSFORMATIONAL LEADERSHIP

Modern studies of leadership in the corporate domain distinguish between transactional and transformational leadership styles.[2] Transactional leadership, which focuses on the monitoring of employees and explicit reward and punishment contracts, has been shown to be effective but counterproductive when overemphasized. Transformational leadership, which focuses on inspirational leadership that communicates common purpose, has been shown to have a more direct impact on productivity than transactional strategies and a much greater impact on trust and group identity.[3] Charismatic political leaders (described in the next section) generally fit the description of transformational leadership.

Leadership also includes a broad range of activities, usually more than can be performed by a single person. Yukl's synthesis of leadership literature[4] (mostly from studies in the corporate domain) summarizes the tasks of leaders into the list below. Necessary behaviors fall into three categories: task focused, relationship focused, and change focused.

Task Behaviors
- Plan short-term activities
- Clarify task objectives
- Monitor operations and performance

Relationship Behaviors
- Provide support and encouragement
- Provide recognition for achievement and contributions

- Develop member skill and confidence
- Consult with members when making decisions
- Empower members to take initiative in problem solving

Change Behavior

- Monitor the external environment
- Propose an innovative strategy or vision
- Encourage innovative thinking
- Take risks to promote necessary changes

As organizations expand, this range of leadership functions must be accomplished by teams, not individuals, not only because of the amount and complexity of the work to be done, but also because different leadership styles have different advantages and disadvantages. Effective leaders surround themselves with a team that both accentuates their strengths and complements them where they are weakest.

CHARISMATIC LEADERSHIP IN UNDERGROUNDS

Rare or extraordinary personal factors such as charisma are often necessary to begin underground movements. Nascent movements are extremely vulnerable, and many if not most are crushed before they have a chance to grow beyond small cells; it often takes an extraordinary leader (or leaders) to motivate the initial cadre and keep it together through inevitable setbacks. The leadership cadre of an insurgency later becomes the core of the larger organization and consists of persons who devote their full time to the cause.

Charismatic leadership is the "quality of an individual personality by virtue of which he is set apart from ordinary men and treated as endowed with supernatural, superhuman, or at least specifically exceptional powers or qualities."[5] These leaders are seen as visionaries who demonstrate some combination of emotionality; activity; sensitivity to the sociopolitical landscape; intense interest in and empathy toward their followers; superior rhetorical and persuasive skills; and exemplary behavior in the form of sacrificing their personal ambitions to those of the movement.[6] Abimael Guzman[a,8] (Sendero Luminoso), Velupillai

---

[a] Abimael Guzman was the undisputed leader of Sendero Luminoso. This physically unremarkable man, rarely seen by Sendero members apart from the highest leadership, employed an extraordinary capacity for persuasion and organization to create a cult-like organization whose members literally revered him as a god in many cases. For the Sendero members, Guzman was shrouded in mystery—a charismatic, almost hypnotic leader who held the one true vision of the future and the means to achieve it. In their minds, he was almost superhuman, and his commands were obeyed without question or hesitation. He demanded and received absolute devotion. Indeed, Guzman saw himself as a

Prabhakaran[b] (Liberation Tigers of Tamil Eelam [LTTE]), Osama bin Laden[c] (Al Qaeda), and Mohandas Karamchand Gandhi have all been considered charismatic leaders of their respective groups. Gandhi, of course, espoused nonviolent tactics and would not be considered an insurgent leader; but he did lead people to endure hardship and violence, and his charismatic leadership style is relevant.

Gandhi's leadership of India's independence struggle verged on the mystical. Thousands of villagers from rural India, who perhaps could not be touched or aroused by any modern means of communication or organized population pressure, were stimulated into action by Gandhi's fasts and his religious mystique. Hundreds of thousands of peasants gathered to meet Gandhi, although they often did not understand his language and could barely see him. It is difficult to estimate the role of Gandhi's mystique in coalescing public opinion against the British, but it is clear that resistance to colonial rule had never appeared on such a large scale before Gandhi. Movements headed by charismatic leaders are prone to a high degree of volatility. Whether it is because of the lack of institutional parameters (such as legislated checks and balances) or the unwillingness of the true believers to challenge the leader, groups with unconstituted leaders tend to be comparatively less stable.

Charismatic leadership is difficult to maintain, especially as movements grow larger. Charismatic leaders must accomplish these four functions: (1) maintain the public persona of the leader; (2) moderate the effects of the psychological identification of followers with the leader; (3) negotiate the routinization of charisma; and (4) achieve frequent new successes.[9] To address the first and second, charismatic

---

"revolutionary Moses who will lead his followers across a river of blood into the Maoist promise land of communism." Sendero believed that Peru was the epicenter of a world revolution and that ultimate victory depended on absolute obedience to Guzman, the leader of the world revolution. In one sense, the near deification of Guzman ensured unparalleled organizational unity and clarity of vision.[7]

[b] Velupillai Prabhakaran's veneration by Tiger members led some observers to liken the LTTE to a religious cult rather than an ethno-nationalist movement, although Prabhakaran himself was believed to be an atheist. There is disagreement as to his background— some analysts claim he was the son of a smuggler, while others report that his father was a tax commissioner and his grandfather a postal worker. This latter account would indicate that his family had access to more educational resources than the average rural Tamil family.

[c] Osama bin Laden was the group's undisputed leader to whom the newly admitted swore an oath of allegiance. Although few of the Afghan Arabs or later Al Qaeda recruits had direct contact with bin Laden, he was able to build a popular following throughout the Islamic world and continues to be regarded as the supreme symbol of resistance to U.S. imperialism. Although bin Laden was principally a financier, logistician, and facilitator, he also retains some distinction as a military commander. In 1987, Soviet Special Forces attacked a training camp in Afghanistan. Bin Laden and a small group of fighters repelled the attack, earning bin Laden a reputation for tactical prowess.

leaders must be seen and heard from on a regular basis through both staged public displays and small appearances before regular members of the group. However, the leader must balance this exposure with an aura of mystery and in some cases a sense of supernatural power (to address the third issue, above), and this requires occasional segregation or isolation from his or her followers. Such isolation has the potential to stem negative feedback from group members; it can also lead to future decisions being made without the consideration of all necessary information and a subsequent failure and internal fracture over direction.[10] The smaller the group, the greater the impact a leader's personality[d] can have on the group. Of course, public exposure entails great risk when groups are still in clandestine stages.

Another limitation of charismatic leadership can be demonstrated by the LTTE's Prabhakaran, whose force of personality also meant he had little capability for political compromise. This probably perpetuated the kinetic component of the conflict while decreasing the bargaining room for the Sri Lankan government, thus limiting the Tamil minority's options for achieving political objectives without further bloodshed and ultimately resulting in the destruction of Prabhakaran's movement.

George Jackson, founder of the Black Guerrilla Family (BGF), is an example of a charismatic source of inspiration[11] whose ideology did not seem to survive him (though his organization has). Jackson's writings highlight a unique manifestation of Black Power[e] inside prison walls—political mobilization of a disenfranchised demographic was transformed into riot and ultimately revolution.[12] In 1966, Jackson formed the BGF in San Quentin State Prison with the goal of restoring the inherent dignity to the black prisoner and eventually overthrowing

---

[d] Personality is the outward manifestation of how individuals tend to experience, use ideation, modulate affect, manage stress, view him or herself, and interact with others.

[e] Black Power is a political slogan and an umbrella concept for various associated ideologies from black separatism to emancipation from economic neo-slavery. The movement came to prominence in the late 1960s and early 1970s, coincident with the U.S. civil rights movement. Black Power expresses a range of political goals, from defense against racial oppression to the establishment of separate social institutions and a self-sufficient economy. Not only did the Black Power movement encourage separatism, but it also helped usher in black radical thoughts and action against what was considered to be an elusive, yet visible, oppressive power. Black Power stood in dramatic contrast to the civil rights movement led by the Reverend Dr. Martin Luther King, Jr., who advocated inclusion and stressed the similarities among all peoples. The more radical elements disagreed with the use of a nonviolent approach, with criticism ranging from claims of its ineffectiveness to claims that it represented appeasement. Black Power gained notoriety and prominence through its advocacy by the Black Panther Party for Self-Defense founded by Huey Newton and Bobby Seale in Oakland, California, in 1966. The Black Panthers achieved international renown through their deep involvement in the Black Power movement and in the U.S. politics of the 1960s and 1970s, and the intense antiracism of the time is today considered one of the most significant social, political, and cultural currents in twentieth-century U.S. history.

the white-run government in America and the prison system. Often sequestered in solitary confinement, Jackson studied political economy and radical theory and wrote two books, *Soledad Brother* and *Blood in My Eye*, which brought him worldwide attention during the zenith of American prison literature's popularity.[13] Jackson's thesis can be considered a fusion of Marxist class struggle and Guevarist revolutionary fervor.[14] Jackson's prison experience and its influence on his psychological state cannot be overstated. While many insurgent leaders consider periods of political imprisonment transformative, Jackson's was particularly so. It was while imprisoned that he saw extremes of racial hostility. Prison was certainly not the first time Jackson experienced racism; however, for him, like many African American males of his generation,[15] it reinforced his preexisting negative identity as nothing more than an emasculated black man whose incarceration was inevitable.[16] Prison is also where Jackson's propensity toward violence was exacerbated; his subjective experience in the low socioeconomic status neighborhoods of Chicago and Los Angeles and later in the California penal system reinforced the pro-social components of violence.[17] He saw the social value of violence as protective, deterrent, and admirable and prioritized those positive ramifications over the acknowledged negative consequences. His letters indicate a passing hint of remorse for resorting to violence, but he is able to rationalize violence as necessary to keeping his word and protecting those of his in-group, actions he seemingly prioritizes over his personal freedom.[18] George Jackson's ideological heritage includes the BGF,[f] The George Jackson Brigade,[g, 19] and his insightful writings on the Marxist–Leninist–Maoist–Fanonist Revolutionary movement in late-1960s America. While his ideal was never quite realized by his organization and his political objectives were not met, his legacy survives as motivation for both the modern

---

[f] BGF at its founding was the most politically oriented of the major prison gangs; Jackson's goals were to form a revolutionary organization with specific objectives to eradicate racism, struggle to maintain dignity in prison, and overthrow the U.S. government. Some original BGF members were formerly associated with the Black Liberation Army, the Symbionese Liberation Army, and the Weatherman Underground organization. Today, the BGF is said to have upward of 50,000 members in both state and federal prisons, with the majority of them concentrated on the East and West coasts. Prospective members must be black and nominated by an existing member; if selected, they must pledge a life of loyalty to the group. BGF is aligned with some African American and Hispanic prison gangs and against many of the white prison gangs. A feature unique to the BGF is that members must be incarcerated in order to advance in rank within the organization.

[g] A Seattle-based revolutionary group, The George Jackson Brigade was involved in violent acts and claimed to use force to overthrow the U.S. government or the government of the State of Washington. The group justified their actions by claiming to further the ends of a revolution of the masses to overthrow the present governmental and international business structures and to establish a system of Communism. In various communiqués, the group claimed credit for bank robberies, bombings, and attacks against custom houses, court houses, grocery stores, public utilities, and correction facilities.

revolutionary as well as those activists who continue to fight for social justice for the disenfranchised.

Charismatic leaders needn't be violent or in positions of authority to be effective. Throughout his twenty-seven-year incarceration in South Africa's prison system, Nelson Mandela was not only able to lead African National Congress's guerilla component, Umkhonto we Sizwe, but also become the international symbol for the antiapartheid movement. Mandela's experience as an enlightened political prisoner and later the first black president of South Africa continues to serve as an archetype for both dignified nonviolent resistance and racial equality.

PROFILING LEADERS

The first remote psychological profile of a foreign leader was led by the Office of Strategic Services' Walter C. Langer in collaboration with Henry A. Murray (Harvard Psychological Clinic), Ernst Kris (New School for Social Research), and Bertram D. Lawin (New York Psychoanalytic Institute). Their target was Adolph Hitler, and their report, *A Psychological Profile of Adolph Hitler: His Life and Legend*, was a vital piece of intelligence that supported Allied psychological warfare efforts and military deception operations.[20] Since World War II, numerous academic and operational methods of assessing leaders have been devised and implemented. Among the more prevalent methods are a psycholinguistic approach developed by Margaret Hermann and a psychodynamic approach postulated by Jerrold Post. The following section provides a brief overview of some of those methods and an abbreviated case study that includes elements of each of them.

Trait/Motivational Approaches

Early psycholinguistic approaches to personality analysis date back to Walter Wientraub's work with clinical populations in the 1960s. Wientraub employed content analytic methods to identify syntactic structure in patient responses to ambiguous stimuli (given during projective assessments such as the Thematic Apperception Test or TAT) and clinical interviews. David Winter's motivational analysis of political behavior applied similar content analysis to code for need for achievement, power, and affiliation to determine a leader's interpersonal behavioral preferences.[21] In Margaret Hermann's profiling method (trait analysis of leadership style), each trait is assessed through content analysis of the leader's public statements as well as other secondary sources of information. Although both prepared speeches and statements from interviews are considered, the latter is given preference because of its spontaneity. The data are collected from interviews and analyzed or content coded, and then a profile is developed. The profile is then compared with the baseline scores developed for the database of leader scores. One is considered to have high score on a trait if he or she is one standard deviation above the average score for all leaders.

Hermann's work exploits the stable patterns or personality traits vice the psychopathology model.[22] Hermann's method assesses these aspects of leadership style:

- Belief that one can influence or control what happens (self-efficacy)
- Need for power and influence
- Conceptual complexity, or the ability to differentiate among things and people in one's environment (related but not identical to intelligence)
- Self-confidence or self-esteem
- The intensity with which a person holds an in-group bias
- General distrust of others
- Task versus relationship focus[h]

No single aspect dominates behavior; Hermann analyzes how combinations of these aspects lead to certain observable patterns of behavior. The combination of beliefs and need for power and influence often determines whether the leader will challenge or respect constraints. The combination of conceptual complexity and self-confidence determines how open a leader will be to new information and new ideas. The extent of their in-group bias and general distrust of others provides evidence concerning a leader's motivation, particularly whether the leader may coordinate or form coalitions with other groups. The leader's outlook about the world and its problems largely determines the confrontational attitude of the organization and may help predict whether groups will espouse violence.

Jerrold Post has a method of profiling leaders that has a great deal of overlap with Hermann's, as well as some points of differentiation. His method is based less on public statements and more on biographical information. It also takes the political and cultural context into account in understanding a leader's development; the extended analysis of Al Qaeda leader Dr. Ayman Al-Zawahiri is adapted from a longer case study using this approach.

---

[h] Hermann expresses the two distinct leadership functions as a continuum between moving the group toward completion of a task (solving problems) and maintaining group spirit and morale (building relationships).

Cognitive Approaches

Operational code analysis has transformed from a manual process to near-automated (with human in the loop) processes. Regardless of the means, cognitive approaches assume perception and beliefs are more easily inferred than personality given the availability of data sources (often transcripts of speeches, letters, or the personal communications of a particular leader). The integrative complexity approach to political personality assessment is an extension of operational code analysis in that it is more rooted in cognitive psychology and social cognition than in personality psychology.[23] Content analytical measures of integrative complexity can be applied to verbal (or written) materials to assess the extent to which the individual can differentiate and integrate multiple perspectives on a particular issue. Low integrative complexity correlates with cognitive rigidity, where the individual is either unable or unwilling to consider varying frames of reference.[24]

Psychodynamic Approaches

From Langer's team's initial work through today, remote psychodynamic assessments have been employed to determine "what makes a leader tick."[25] Psychoanalysis[i] is based on the proposition that much of mental life is unconscious, that a person's psychological development is important for understanding their current state, and that psychological distress derives from unresolved intrapsychic conflict. Psychoanalytic studies of narcissism and paranoia may shed light on the personalities of some insurgent leaders.[26]

Egocentricity is a normal component in infantile development; however, as a child develops into adolescence, he or she is supposed to become less self-absorbed and more cognizant of others. Narcissism is a psychoanalytic theory that holds that primary narcissism (or self-love) in the form of the grandiose self does not diminish as the individual

---

[i] The dynamics of this theory were derived from nineteenth-century concepts of physics (as interpreted by Sigmund Freud), in which the flow of mental and libidinal energy is deterministically expressed, repressed, or discharged. The theory has variants, but they share the notions that (1) parenting (as opposed to intrinsic temperament) determines psychological temperament and health; (2) active, unconscious forces exclude unpleasant thoughts from the consciousness; and (3) relationships with others, object relations, are controlled by unconscious forces such as projection—the theory that one irrationally attributes one's own attitude to others. Strengths of psychoanalytic interpretations of terrorism are their acknowledgment that individual developmental factors beginning in early childhood probably influence adult behavioral proclivities, their recognition of the enormous power of the unconscious to influence conscious thought, and their observation that covert psychodynamic forces of groups may subsume individuality. The weakness of psychoanalytic interpretations is their lack of falsifiability.

develops and expands his or her social network. If this fails to occur, regardless of reason, the resultant grandiose self-image can result in individuals who are sociopathic, arrogant, and devoid of compassion for others. Some leaders demonstrate a marked desire for admiration and attention, a hallmark of narcissism.[27] Their chosen methods of violence are often spectacular and attention grabbing, suggesting a more narcissistic clinical presentation. There are also those who exhibit a narcissistic leadership style although they probably do not meet the clinical criteria for an Axis II disorder.[28] In fact, this leadership style is heavily represented in the military, industry, and academia. Characteristics of the narcissistic leadership style include a vulnerability to biased information processing that results in an overestimation of their own strength and an underestimation of their adversary's, a grandiose and self-serving disposition, a lack of tolerance for competition, difficulty relying on experts, and a desire for sycophantic subordinates.[j] Often displaying superficial arrogance over profound personal insecurity, they actively seek admiration, are vulnerable to insults, slights, and attacks, and are prone to rage.[29] Key observables that indicate this style are the leader's sensitivity to criticism, surrounding themselves with sycophants, and an overvaluation of his chances of success and an underestimation of the strength of an opponent.[30]

Paranoia theory states that violent radicalism is the result of a particular personality trait that predisposes one to mistrust others and display aggressive behavior. Paranoia theory states that the salient feature of terrorist psychology is projection, an infantile defense that assigns intolerable internal feelings to an external object. This occurs when an individual who has grown up with a damaged self-concept idealizes the good self and splits out the bad self. This projection is proposed to be the root of an adult persistence of the infantile phase called the "paranoid schizoid position."[31] Two personality traits evident in some individuals drawn to radical groups are the psychological mechanisms of externalization and splitting.[32] Splitting is an extreme tendency to see the world in black and white, at the extremes of attraction and rejection.[33] Problematic feelings are not accepted as part of the self but are instead ascribed to something external. Socially acceptable feelings are located within an idealized "good self," whereas bad feelings are split

---

[j] Another manifestation of the theory is the malignant narcissist style of leadership: a combination of narcissism, paranoia, and sociopathy. These individuals exhibit grandiosity and suffer from poor underlying self-esteem with attendant sensitivity to slights, insults, or threats. They suspect and blame others, have no compunction regarding the use of violence, and lack empathy or concern for the impact of their actions on others. Observable characteristics of the malignant narcissist leader are displays of extreme grandiosity, paranoia, and other antisocial traits, the lack of inhibition in the use of violence, dreams of glory, lack of empathy regarding the impact of his or her acts on others, and the target of anger (subsequent to personal or group setbacks) being an external entity.

99

out and projected onto something external. Individuals with such traits are hypothesized to be likely to strongly idealize the group to which they belong (the in-group) while demonizing outsiders.[34] The paranoid position inflames the terrorist with suspicions that justify bloody acts of "self-defense" against his victims.[35] For leaders with rigid and extreme ideological belief systems centered on themes of oppression and persecution, paranoid personality disorder would seem a logical fit.[36] Paranoia regarding the survival of the in-group creates the psychological foundation that permits terrorists to kill random civilians who, seen from the outside, do not directly threaten the terrorist group.[37]

There are also those who exhibit a paranoid leadership style and again probably do not meet the clinical criteria for the Axis II disorder.[38] Characteristics of the paranoid leadership style include the view that one is surrounded by enemies, suspicion and distance, preoccupation with unjustified doubts about the loyalty and/or trustworthiness of those around them, hypervigilance and invulnerability to influence, and the deeply rooted need to externalize blame for their own.[39] Observable characteristics of the paranoid leadership style are the blaming and demonization of adversaries, the stockpiling of weapons to defend against imminent attack, and the frequent purging of the inner advisory group.[40] Experts tend to agree that paranoid personality disorder may be more common among leaders than among nonleaders of terrorist organizations.[41] Some degree of paranoia can be a useful trait in the highly vulnerable position of an insurgent leader, especially in early stages where infiltration and security lapses can lead to rapid annihilation. A combination of the narcissistic and paranoid styles may be somewhat common among underground leaders.

Personological Approaches

Another approach to remote leadership analysis is Theodore Millon's personological model. A clinical criticism of the aforementioned approaches is that they do not adequately incorporate disciplines of personality theory and psychodiagnostics.[42] Millon's theory of personality was the intellectual driver behind the DSM-IV Axis II personality disorder diagnostic criteria as well as the Millon Clinical Multiaxial Inventory (MCMI), a widely used personality assessment.

Millon divides the personality into functional and structural attributes. Functional attributes are the processing and modulating features of the personality such as behavior, cognitions and perceptions, and intrapsychic regulatory mechanisms. There are four functional attributes: expressive acts (observable behavior), interpersonal conduct (how individuals interact with others), cognitive style (perceptions, attention, information processing, organization of thoughts), and regulatory mechanism (unconscious processes to resolve needs, protect the ego, and mediate conflicts).[43]

Structural attributes are the enduring components of the personality that help define one's worldview. There are four structural attributes: self-image (perception of oneself/identity), object representations (interaction with memory and its influence on information processing), morphologic organization (structural strength and internal cohesion of the mental system), and mood temperament (how one interacts with and processes emotion).[44]

A CASE STUDY: DR. AYMAN AL-ZAWAHIRI

An illustrative example of the evolution of underground leadership is the case of Zawahiri, the Egyptian Islamic scholar, surgeon, and radicalized Islamist whose contributions to the global Salafist jihad are the extension of those of his hero and ideological forefather, Sayyid Qutb, and his Egyptian contemporary, Salam Faraj. His synergistic codependent relationship with Osama bin Laden has changed the international security landscape as well as political and military strategies of the twenty-first-century world. Both Zawahiri's original organization, Egyptian Islamic Jihad (EIJ),[k] and his current organization, Al Qaeda, can credit undergrounds with a large component of their operational successes. During the mid-1980s (after most had been released from prison for their actual or suspected role in the Sadat assassination), many EIJ and Egyptian Islamic Group (EIG) members relocated to Peshawar, Pakistan, to escape the Egyptian government and to participate in the Afghan resistance to the Soviet Union's occupation. Although some EIJ members recognized the still-imprisoned Abud al-Zumur as emir, others began to consider

---

k EIJ was formed in 1979 when Faraj united four to six clandestine jihadi cells under his leadership, including one that was still underground and not yet operational led by Zawahiri and two army officers. During a Shura in late spring of 1981, leaders of the two branches (Karam Zuhdi of EIG and Faraj of EIJ) agreed to merge their groups under the leadership of Sheik Umar Abd-al Rahman, who agreed to be the *mufti*. Rahman, an alumnus and former professor at Al-Azhar University in Cairo, became one of the most outspoken clerics to denounce Egypt's secularism. Although respected by both groups, he did not have a unanimous mandate to serve as the group's emir. Zawahiri's cell, and later EIJ as a whole, employed a blind-cell structure like that of the Leninist Communist Party where members in one cell did not know the identities or activities of those in another, so that if one member was captured or compromised, he would not be able to endanger more than a few people.[45]

101

Sayyed Imam al-Sharif,[l] also known as "Dr. Fadl," EIJ's leader. At some point during the mid-1980s, Sharif's academic, professional, and ideologically like-minded colleague, Zawahiri,[m] became the recognized leader of EIJ.[n,47] The opportunity to select, train, and indoctrinate operatives from the pool of Egyptians seeking to martyr themselves in Afghanistan enabled Zawahiri to slowly unite the still-disparate cells of EIJ and build capability.[48] It also brought Zawahiri into contact with Osama bin Laden, who, along with Abdullah Yusuf Azzam, a Palestinian scholar of Islamic law, established the MAK or Services Office, a clearinghouse to facilitate the recruitment, transportation, organization, training, and equipping of Arabs to support the Afghan resistance.[o] Osama bin Laden was the group's undisputed leader, to whom the newly admitted swore an oath of allegiance.[49] Although few of the Afghan Arabs or later Al Qaeda recruits had direct contact with bin Laden,[50] he was able to build a popular following throughout the Islamic world and continues to be regarded as the supreme symbol of resistance to U.S. imperialism.[51] Although bin Laden was principally a financier, logistician, and facilitator, he also retains some distinction as a military commander. Bin Laden's principal ideological advisor, and

---

[l] Sharif, a physician who was administering to the refugees from Afghanistan under the Red Crescent, emphasized the importance of the Qutbist ideology and the condemnation of those who deviated from it. Sharif had established a series of guesthouses and training camps in Peshawar to receive young Muslims from the Middle East and to stage and prepare them to move and integrate into the Afghan resistance. His was a parallel effort to that of Osama bin Laden and his spiritual advisor (a role Zawahiri would eventually assume), Abdullah Azzam, who had established Maktab Al Khidamat (MAK), or Services Office, to address administrative problems for foreign Muslim fighters.

[m] Both Zawahiri and Imam al-Sharif were pious, high minded, prideful, and rigid in their views. They evaluated matters of religion as a series of immutable rules established by God, a mindset representative of the well-educated engineers and technocrats who occupied the higher ranks of Salafists.

[n] In 1992, as the conflict between EIG and EIJ reached the stage of mutual accusations of apostasy and individual acts of assassination attempts, Zawahiri emerged as the winner, largely because of bin Laden's support and because of the murder of Abdullah Azzam, the spiritual leader of bin Laden, which some attribute to EIJ.[46]

[o] There is a long history of Arab migration into Afghanistan, much of it following the Russian Revolution of 1917 when Arabs living throughout central Asia fled encroaching Soviet control for the religious freedom afforded them in Afghanistan. Later, under Stalin, many Muslims of Arab and non-Arab descent living in the Caucasus were deported en masse to Siberia and Central Asia. The current Islamist resistance in Chechnya owes much to these earlier Soviet policies. Although this resistance was originally led by indigenous fighters, it has received substantial financial and human support from Muslims throughout the Middle East, where many deportees settled and integrated into the local communities. Therefore, to suggest that the Arabs who supported the anti-Soviet resistance are religious mercenaries focused only on global jihad and without personal ties to Muslims living outside the Middle East is misleading. More likely, the intense migration and intermarriage between Muslims of Arab and non-Arab descent throughout Central Asia, the Caucasus, and the Middle East created strong personal ties that supported the narrative of religious jihad and resistance to foreign occupation.

the organization's deputy chief of operations, was Zawahiri,[p] who provided the scriptural and juridical substance for bin Laden's more general political ideals.[53] Al Qaeda's senior leadership (top aides, media representatives, military advisors, etc.) were more likely to have technical and professional backgrounds—in business, public administration, law, engineering, or medicine—than religious ones.[q]

Bin Laden (particularly after September 11, 2001) was the charismatic leader of the movement until his death in 2011; his name recognition and affiliation connote credibility and legitimacy.[54] To many Islamists, bin Laden represents the personification of strong oppositional leadership: resolutely defying those who attempt to exploit Muslims (regardless of nationality) and/or fail to live and govern by the principles set forth in the Koran and Hadith. Bin Laden's legacy will probably extend beyond his death regardless of whether Al Qaeda in its current structure and function continues to be a viable organization/movement.

Bin Laden's proclivity toward conformity (the opposite pole for the personality trait of openness), conscientiousness, and introversion (the combination that results in deference to religious authority)[55] rendered his personality an ideal fit with that of the abrasively negativist Zawahiri. Both individual personality types, despite their differences, resulted in a shared religious extremism and strict reliance on a literal interpretation of the Koran. Their ability to rationalize actions (such as intentionally targeting civilians or killing innocent women and children of all faiths) as justifiable in the eyes of God (although debatable) is considerate, careful, and rational (hallmarks of conscientiousness)

---

[p] Zawahiri, a trained surgeon, was born into a pious, middle-class Egyptian family. A key figure in EIJ, he spent time in prison in Egypt on suspicion of being involved in the assassination of Egyptian President Anwar Sadat in 1981. Zawahiri is now an older man whose long beard has turned to gray and white and whose large glasses cover eyes set among the wrinkles of a man who has lived a life of secrecy that has been under siege for more than two decades. His most prominent physical feature is probably a large callous in the middle of his forehead that indicates hours spent bowing prostrate while reciting the Koran or praying. His obsessive-compulsive tendencies were probably a benefit in his academic pursuit of medicine, which requires not only an expansive memory but also high general fluid intelligence in the application of the learned material in a clinical setting (particularly as a surgeon). It is said that he does not have many hobbies (or vices) and that his chosen activities are often solitary and intellectual (praying, studying, or reading); both his academic accomplishment and his theological knowledge attest to his attempts at mastery. Most individuals that have come in to contact with Zawahiri recall his vast intellect; a gift evident at an early age not only by objective measures but also though subjective interactions. Given the competitiveness and rigidity with which the socialist-leaning Nassertist regime's education system functioned, Zawahiri was subjected to (and performed well on) numerous standardized tests throughout his formative years, qualifying for admittance into Cairo University (from which he later graduated with distinction).[52]

[q] This is as much a product of changes in the region's educational system as it is indicative of psychological links between certain occupations and support for terrorist tactics.

and not necessary antisocial. The actions ascribed to bin Laden's direction seem indicative of a charismatic destructive narcissist[56]; however, accounts of his personal behavior present the image of someone introverted by nature who seems to prefer isolation, drawing energy from being quiet, contemplative, and focused on his religion. Among the personality traits that remain uncertain are his agreeableness and/or neuroticism (particularly his proneness toward worrying). Zawahiri, who has exhibited both highly disagreeable and paranoid tendencies, would probably be able to, through the strength of his personality, compel even a stable bin Laden to worry. Again, this confluence of personality styles may be a contributor to their collective (and repeated) warnings about the influence of the United States in the Middle East.

Zawahiri, in some respects, can be considered a representative example of an underground strategist[57] in that he lacks the interpersonal charisma or inspirational aptitude of bin Laden and prefers to work behind the scenes. Despite his tendency toward and preference for secrecy, Zawahiri was often selected to represent various groups to which he belonged, whether as an EIJ member in Tula prison or as bin Laden's deputy in Al Qaeda, because of his intellect[r] and articulate yet demonstrative manner of speaking.[59] Zawahiri is not described as a man whose affective reactions compromise his ability for rational thought; however, he does speak with an impassioned tone and often waves an extended right index finger with his palm toward the camera or interviewer in a demonstrative, often condescending manner.[s]

---

[r] Zawahiri's judgment (particularly in operation matters), however, has been questioned. In the late 1980s, while preoccupied with the mujahidin in Afghanistan, his financial mismanagement of EIJ necessitated a closer alliance with Al Qaeda that ultimately detracted from their initially stated objective (driving some members away). The string of failed operations in the 1990s at the hands of the Egyptian authorities were also attributed to Zawahiri, who was still considered a valuable strategic thinker but who lacked the operational experience to effectively plan and execute terrorist operations.[58]

[s] Zawahiri's public statements reveal not only some of the aforementioned personality traits but also a degree of Machiavellian leadership evidenced by his willingness to subordinate certain components of his nuanced Islamic beliefs to the strategic objective. Zawahiri's words (both written and spoken) are the extension of Qutb's, calling for militant opposition not only to Christian and Jew infidel crusaders but also to Muslim apostates who deviate from Salafist practice. Zawahiri's writings blend his militant ideology with pragmatic recommendation for vanguard Islamist movements, tactical employment of diverse weapon systems, and underground tradecraft recommendations that render his writings must-reads for the budding jihadist. His acknowledgment of the populace as the center of gravity of the global Salafist jihad (for both the insurgents and counterinsurgents) indicates an appreciation for both those people and an understanding of the U.S. and coalition strategic communication and/or psychological operations themes, messages, and targeting priorities. In advising Abu Musab Al-Zarqawi (the Jordanian leader of Al Qaeda in Iraq) not to highlight his Wahhabist ideals (as neither the majority Shia nor the minority Sunni population in Iraq subscribes to Wahhabism), he acknowledges the limitation of proselytizing by force as well as the risk of imposing one's religion. In a further divergence from his Qutbist revolutionary core, he urges Al-Zarqawi against branding

It is unclear whether this gesture is Zawahiri's way of paying homage to Qutb (who exhibited a similar demonstrative gesticulation) or whether it is indicative of his inability to regulate the anger he seems to experience when speaking about the plight of the *umma* at the hands of the apostate regimes or in response to rumors of his death or his organization's reduced operational capabilities. The value of the underground strategist is thought leadership and planning acumen; his role is akin to operational art in that he translates the intent of the inspirational leader into a proscribed set of tasks for subordinates to execute. Their role is vital yet often unrecognized and a potential source of personal and organizational tension.

Zawahiri has exhibited narcissistic and paranoid personality traits, but none have been exhibited to the degree that would meet the clinical criteria for a disorder. He exhibits contentious/oppositional and dominant/controlling features, with secondary features of the dauntless/dissenting and ambitious/self-serving pattern.[60] The amalgam of contentious (negativistic or passive aggressive) and dominant (aggressive or sadistic) patterns in Zawahiri's profile suggests the presence of abrasive negativist syndrome, a set of characteristics whereby seemingly minor frictions are perceived as significant confrontations.[61] Zawahiri's self-concept is that of the Islamic scholar carrying the banner of his heroes (Qutb[t] most prominent among them) who lives life in anticipation of martyrdom because that is perceived to be the noblest path to paradise. Both his reluctance to participate in contact sports as youth and what some saw as his hesitance to act in his early years as a cell leader indicate a lack of physical courage that he attempts to conceal behind his academic demeanor. Zawahiri also may be haunted by the memory of betraying some of his friends to Egyptian authorities under torture. Both the shame of this confession and the lack of physical courage manifested though his thoughtful, contemplative,

---

himself a cruel terrorist by conducting and broadcasting hostage executions because it is inconsistent with the strategy of Al Qaeda and the loosely federated (and successful) resistance movement in Iraq.

[t] The most defining relationship of Zawahiri's personality is that with the ideal of Sayyid Qutb. In adolescence, accounts of Qutb's character, binary worldview, and steadfast devotion to Islamic principles caused the ambitious Zawahiri to identify Qutb as a surrogate father figure to emulate and to which he continually aspires. Zawahiri has exhibited at times narcissistic, obsessive-compulsive, and paranoid leadership styles but seems to favor a mixture of obsessive-compulsive and paranoid. His contentious/oppositional and dominant/controlling personality traits, along with his hypersensitivity and introversion, cause him to prefer serving in a Rasputin-like advisory capacity to being the individual who makes the decisions (and thus experiences the ramifications and criticisms of those decisions). He is often dismissive, hypercritical, and condescending when threatened or challenged, causing some Western analysts to dismiss his comments holistically as propaganda. However, his choice of tone is deliberately calculated to appeal to his target audience and should not be overlooked.[62]

and sensitive private self conspire to create a sense of vulnerability, weakness, and shame that he aggressively protects with his most prolific gift: his intellect. The operationalizing of his intellect as an ego defense mechanism is emboldened by the respect his academic accomplishments have afforded him as a young man and his willingness to live an austere life of sacrifice under constant threat in pursuit of his beliefs as an older man.[u] Abrasively negativistic individuals are often eager to identify inconsistencies in others' actions or ethical standards and are adept at constructing arguments that amplify observed contradictions.[63] Zawahiri has been known to act in such a manner, but it is unclear whether it was to achieve a particular goal in Machiavellian fashion or because he felt threatened and needed to defend himself or his position. Zawahiri has and will probably continue to take the moral high ground, dogmatically and contemptuously exposing his antagonists' perceived hypocrisy and contemptuously, derisively, and scornfully betraying those who cross his path.[64]

The decision to strike the United States, although not exclusively attributed to Zawahiri, has come under criticism by many Islamist groups throughout the Middle East who saw their freedom of movement (and subsequent opportunity to strike) diminished by the post-9/11 United States-led global war on terrorism; these groups felt that the Al Qaeda leadership traded a tactical victory for strategic defeat.[65] These criticisms, although significant, are insufficient to make the assertion that he lacks judgment, for clearly someone who has been at or near the top of multiple organizations with rigid vetting processes could not have attained or retained positions of leadership lacking some capacity for judgment. This particular debate (if one did occur) is an important component of the underground decision-making process. Understanding not only the personality of the leader but also how he receives and processes information are vital in the dynamics of the underground. This is particularly so during crises where one's leadership is stressed, challenged, or removed. These crises can be assassination attempts on a leader, successful counternetwork operations by an adversary, or even planned political events such as the guerilla conferences held by the Revolutionary Armed Forces of Colombia (Fuerzas Armadas Revolucionarias de Colombia, or FARC).[v]

---

[u] Intellectualization is also a defense mechanism that satiates both a practical and aesthetic need, as one of the quotes attributed to the Prophet Mohamed in the Hadiths is (to paraphrase): "the ink of the scholar is more holy than the blood of the martyr."

[v] The FARC hold periodic conferences to assess progress and plan for the future. The 1st Guerrilla Conference was in late 1965 in Río Chiquito. Several disparate groups joined together, forming what was called the "South Bloc" and what would eventually become the FARC.

Zawahiri's leadership stands in stark contrast to that of the stabilizing and politically flexible Hassan Nasrallah, the leader of Hizbollah since 1992. Nasrallah's subtle charisma has endeared him to strict Islamists and secular Lebanese alike: his face appears on billboards, key chains, and screensavers and excerpts of his speeches are even used as cell phone ringtones.[66] Although highly intelligent and deeply devout, Nasrallah (unlike Zawahiri) is not as ideologically rigid nor is he passive aggressive. In fact, his willingness to deviate from Hizbollah's initial manifesto has led to the increased political prominence of Hizbollah. His recognition of the social landscape and political acumen, coupled with his religious credentials, have continued to provide him a degree of legitimacy unrivaled in insurgent groups. While bin Laden's visage is as popular in the Middle East as it is condemned in the West, he lacked both the political authority and the legitimacy of Nasrallah. Nasrallah earned broad support by cultivating a social welfare network that provided schools, clinics, and housing in the predominantly Shiite parts of Lebanon and received considerable credit for Israel's withdrawal from Lebanon. Although he cannot claim full credit for the military operations that Hizbollah waged, he was largely responsible for the information campaign that secured Hizbollah broad Shiite support and decreased Israeli public opinion (and ultimately political will) toward the occupation of Lebanon.

TARGETING LEADERS

Leadership analysis can facilitate the targeting process, particularly when the goal is not necessarily removal but manipulation or marginalization. Influencing a group's membership and/or civilian populace to discredit an effective leader or sow dissension can be far more effective when based on psychological characteristics in additional to other sources of intelligence. Conversely, a counterinsurgent force or rival group may want to strengthen a relatively ineffective leader to lessen the organization's effectiveness.

Kinetically targeting leaders is a common counterterror and counterinsurgency tactic; however, decapitation strikes have not been proven to be a cost-effective strategy.[67] The underlying logic behind targeting leaders lies in eliminating the impact of charismatic leadership, which was discussed earlier. Viewing leaders as moral centers of gravity presupposes that the movement centered on them will cease to exist if they are removed. In cases such as Guzman and Sendero Luminoso, this was the case; however, Guzman's public recanting was likely more catastrophic than his death. An organization that decides to execute a leader runs the risk of martyring that leader and thus further

107

incentivizing resistance; however, execution is still a viable option in many campaigns. Qutb's refusal to renounce his beliefs, temper his rhetoric, or plead for his life has made him an admirable martyr in the minds of many devout Islamic insurgents.

In a rigorous analysis of the 298 cases of leadership decapitation from 1945 to 2004, Jordan concluded that younger and smaller organizations are more vulnerable to such strikes than larger and/or more-developed institutions. Furthermore, whereas religious organizations are resilient to decapitation, nonreligious ideological organizations are more susceptible to collapse after decapitation.[68] Other approaches may include discrediting a key leader or using his relative isolation as an opportunity to manipulate/modify/disrupt his communication with his extended network of subordinates and associates as a means of sowing friction within the group.

SUMMARY

Leadership is a critical factor for understanding underground movements. Charismatic leaders are often at the center of such movements, perhaps by necessity. These leaders exhibit the characteristics of transformational leadership and can help nascent organizations form against opposition and survive their most vulnerable stages. Charismatic leaders such as George Jackson also form ideologies that can unite movements and propel them forward. Charismatic leadership has limitations as well, however. Maintenance of such personas can be costly, these leaders may be resistant to compromise, and they may lack other organizational skills necessary as underground organizations grow.

Leadership profiling methods, such as those of Margaret Hermann and Jerrold Post, can shed some light on both the personalities and likely behaviors of leaders. Psychoanalytic theory, particularly studies of narcissism and paranoia, may also lend some insights. There is, however, no single approach that is either universally applicable or resistant to criticism. The most appropriate approach is understanding the tools and methods available and selecting the appropriate technique given the existing data. The Millon-based approach is the most consistent with current (as of 2012) diagnostic criteria; however, it (much like the others) requires the expertise of a clinician and/or behavioral scientist to ensure the necessary analytical rigor and context. Leadership analysis entails concomitant risks such as the fundamental attribution error and/or deception. To mitigate some of those risks, it is vital to ensure individual actors (and their actions) are placed in the appropriate sociocultural context.[69] This is compounded by the fact that insurgent group leaders do not make themselves readily available for analysis.

Access to credible primary source materials (family members, former associates, written documentation, audio/video recordings, interviews, intelligence materials, etc.) that can be used to conduct a leadership analysis may be the single most important limiting factor in determining which methodology can be used.

ENDNOTES

[1] Jerrold M. Post, *The Psychological Assessment of Political Leaders: with Profiles of Saddam Hussein and Bill Clinton* (Ann Arbor, MI: University of Michigan Press, 2003).

[2] Bernard M. Bass, "Two Decades of Research and Development in Transformational Leadership," *European Journal of Work and Organizational Psychology* 8, no. 1 (1999): 9–32.

[3] Philip M. Podsakoff, Scott B. Mackenzie, Robert H. Moorman, and Richard Fetter, "Transformational Leader Behaviors and Their Effects on Followers' Trust in Leader, Satisfaction, and Organizational Citizenship Behaviors," *Leadership Quarterly* 1, no. 2 (1990): 107–142.

[4] Gary Yukl, Angela Gordon, and Tom Taber, "A Hierarchical Taxonomy of Leadership Behavior: Integrating a Half Century of Behavior Research," *Journal of Leadership & Organizational Studies* 9, no. 1 (2002): 15–32.

[5] Jay A. Conger and Raabindra N. Kanungo, "Toward a Behavioral Theory of Charismatic Leadership in Organizational Settings," *The Academy of Management Review* 12, no. 4 (1987): 637–647.

[6] Ibid.

[7] Ron Buikema and Matt Burger, "Sendero Luminoso," in *Casebook on Insurgency and Revolutionary Warfare, Volume II: 1962–2009,* ed. Charles Crossett (Laurel, MD: The Johns Hopkins University Applied Physics Laboratory, 2009), 58–59.

[8] Ibid.

[9] Post, *Psychological Assessment.*

[10] Conger and Kanungo, "Charismatic Leadership."

[11] Fathali M. Moghaddam, *From the Terrorist's Point of View: What They Experience and Why They Come to Destroy* (Westport, CT: Praeger Security International, 2006), 102.

[12] John Pallas and Robert Barber, "From Riot to Revolution," in *The Politics of Punishment,* ed. Erik Olin Wright (New York: HarperCollins, 1973), 237.

[13] H. Bruce Franklin, "The Inside Stories of the Global American Prison," *Texas Studies in Literature and Language* 50, no. 3 (2008): 235–242.

[14] George L. Jackson, *Soledad Brother – The Prison Letters of George Jackson* (New York: Bantam Books, 1970).

[15] James B. Jacobs, "Race Relations and the Prisoner Subculture," *Crime and Justice* 1 (1979): 1–27.

[16] Dennis D. Dorin and Robert Johnson, "The Premature Dragon: George Jackson as a Model for the New Militant Inmate," *Contemporary Crises* 3, no. 3 (1979): 295–315.

[17] Ibid.

[18] George L. Jackson, *Blood in My Eye* (New York: Bantam Books, 1972).

[19] Daniel Burton-Rose, *Guerrilla USA: The George Jackson Brigade and the Anticapitalist Underground of the 1970s* (Berkeley, CA: University of California Press, 2010).

[20] Walter C. Langer, *The Mind of Adolf Hitler: The Secret Wartime Report* (New York: Basic Books, 1972).

[21] Audrey Immelman, "Political Psychology and Personality," in *Handbook of Personology and Psychopathology*, ed. Stephen Strack (Hoboken, NJ: John Wiley & Sons, Inc., 2005).

[22] Margaret G. Hermann, "Explaining Foreign Policy Behavior Using the Personal Characteristics of Political Leaders," *International Studies Quarterly* 24, no. 1 (1980): 7–46.

[23] Immelman, "Political Psychology and Personality."

[24] Dean K. Simonton, "Presidential IQ, Openness, Intellectual Brilliance, and Leadership: Estimates and Correlations for 42 U.S. Chief Executives," *Political Psychology* 27, no. 4 (2006): 511–526.

[25] Langer, *The Mind of Adolf Hitler.*

[26] Jeff Victoroff, "The Mind of the Terrorist: A Review and Critique of Psychological Approaches," *Journal of Conflict Resolution* 49, no. 1 (2005): 3–42.

[27] Robert D. Hare, *Without Conscience: The Disturbing World of the Psychopaths Among Us* (New York: Guilford Publications Inc., 1993).

[28] Jerrold M. Post, Keven G. Ruby, and Eric D. Shaw, "The Radical Group in Context 1: An Integrated Framework for the Analysis of Group Risk for Terrorism," *Studies in Conflict and Terrorism* 25, no. 2 (2002): 73–100.

[29] Ibid.

[30] Ibid.

[31] Victoroff, "The Mind of the Terrorist."

[32] Jerrold M. Post, *The Mind of the Terrorist: The Psychology of Terrorism from the IRA to Al Qaeda* (New York: Palgrave McMillan, 2007).

[33] Rex A. Hudson, *The Sociology and Psychology of Terrorism: Who Becomes a Terrorist and Why* (Washington, DC: Library of Congress Federal Research Division, 1999).

[34] Anja Dalgaard-Nielsen, "Studying Violent Radicalization in Europe I: The Potential Contribution of Social Movement Theory" (DIIS Working Paper no. 2008/2, Denmark, Danish Institute for International Studies, 2008).

[35] Victoroff, "The Mind of the Terrorist."

[36] Ronald N. Turco, "Psychological Profiling," *International Journal of Offender Therapy and Comparative Criminology* 34, no. 2 (1990), 147–154.

[37] Dalgaard-Nielsen, "Studying Violent Radicalization in Europe."

[38] Post et al., "The Radical Group in Context 1."

[39] Ibid.

[40] Jerrold M. Post, Keven G. Ruby, and Eric D. Shaw, "The Radical Group in Context 2: Identification of Critical Elements in the Analysis of Risk for Terrorism by Radical Group Type," *Studies in Conflict and Terrorism* 25, no. 2 (2002): 101–126.

[41] Jordan Maile, Tali K. Walters, J. Martin Ramirez, and Daniel Antonius, "Aggression in Terrorism," in *Interdisciplinary Analyses of Terrorism and Political Aggression*, eds. Daniel Antonius, A. D. Brown, Tali K. Walters, J. Martin Ramirez, and S. J. Sinclair (Cambridge: Cambridge Scholars Publishing, 2010).

[42] Audrey Immelman, "The Assessment of Political Personality: A Psychodiagnostically Relevant Conceptualization and Methodology," *Political Psychology* 14, no. 4 (1993): 725–741.

[43] Ibid.

[44] Ibid.

[45] Jason Spitaletta, "Egyptian Islamic Jihad," in *Casebook on Insurgency and Revolutionary Warfare, Volume II: 1962–2009*, ed. Chuck Crossett (Alexandria, VA: U.S. Army Publications Directorate, in press).

[46] Jason Spitaletta and Shana Marshall, "Al Qaeda," in *Casebook on Insurgency and Revolutionary Warfare, Volume II: 1962–2009*, ed. Chuck Crossett (Alexandria, VA: U.S. Army Publications Directorate, in press).

47 Montasser al-Zayyat, *The Road to Al-Qaeda: The Story of Bin Laden's Right-Hand Man* (Sterling, VA: Pluto Press, 2004).

48 Fawaz A. Gerges, *The Far Enemy: Why Jihad Went Global* (New York: Cambridge University Press, 2005).

49 Rohan Gunaratna, *Inside Al Qaeda: Global Network of Terror* (New York: Berkley Books, 2002).

50 Peter L. Bergen, Holy War, Inc.: *Inside the Secret World of Osama Bin Laden* (New York: The Free Press, 2002).

51 Gunaratna, *Inside Al Qaeda.*

52 Spitaletta, "Egyptian Islamic Jihad."

53 Lawrence Wright, *The Looming Tower: Al-Qaeda and the Road to 9/11* (New York: Vintage Books, 2007).

54 Moghaddam, *From the Terrorist's Point of View,* 103.

55 Post, *The Mind of the Terrorist.*

56 Ibid.

57 Moghaddam, *From the Terrorist's Point of View.*

58 Chuck Crossett and Jason Spitaletta, *Radicalization: Relevant Psychological and Sociological Concepts* (Ft. Meade, MD: Asymmetric Warfare Group, 2010).

59 al-Zayyat, *The Road to Al-Qaeda.*

60 Audrey Immelman and Kathryn Kuhlmann, "Bin Laden's Brain: The Abrasively Negativistic Personality of Dr. Ayman Al-Zawahiri" (paper presented at the Twenty-Sixth Annual Scientific Meeting of the International Society of Political Psychology, 2003).

61 Ibid.

62 Spitaletta, "Egyptian Islamic Jihad."

63 Immelman and Kuhlmann, "Bin Laden's Brain."

64 Ibid.

65 Gerges, *The Far Enemy.*

66 "Profile: Hassan Nasrallah," Council on Foreign Relations, accessed May 9, 2011, http://www.cfr.org/publication/11132/profile.html.

67 Jenna Jordan, "When Heads Roll: Assessing the Effectiveness of Leadership Decapitation," *Security Studies* 18 (2009): 719–755.

68 Ibid.

69 Jennie L. Johnson and Matthew T. Berrett, "Cultural Topography: A New Research Tool for Intelligence Analysis," *Studies in Intelligence* 55, no. 2 (2011): 1–22.

PART II.

MOTIVATION

CHAPTER 5.

JOINING, STAYING IN, AND LEAVING THE UNDERGROUND

CHAPTER CONTENTS

Jason Spitaletta, Nathan Bos, and SORO authors

This chapter is about the process by which individuals come to join underground and insurgent groups, why they stay, and why some leave, including notes on demobilization (or so-called "deradicalization") programs. This chapter includes a substantial amount of material related to southeast Asian insurgencies—in particular the Viet Minh and Philippine Huk insurgency—that appeared in the original version of this book. This chapter will not delve as deeply as subsequent chapters into the psychological forces, group dynamics, and influence methods behind these decisions. Rather, we will give an overview of these behaviors and the most apparent causes. Because these processes differ greatly between settings and groups, where possible we will give multiple diverse examples of each process.

REASONS FOR JOINING

Recruitment

As discussed in greater detail in Chapter 2 of the companion to this work, the second edition of *Undergrounds in Insurgent, Revolutionary, and Resistance Warfare,* recruiting and retention of personnel is a *sine qua non* of any sustained movement. Insurgent recruitment is typically very different at different insurgency stages, as insurgencies move from small conspiracies to larger organizations with specialized roles. An insurgency that is in its nascent stages might take an extended period of time to investigate and cultivate a relationship with a potential front leader in an area of a country targeted for expansion of the movement. As an insurgency expands, the underground organization must recruit and develop middle-level leaders, such as front or provincial leaders, influential agents within a university or government agency, and military leaders. Recruiting key figures that already wield great influence in a society is a boon for the insurgent underground because it subtracts this individual's influence and reach from the government and adds significant new capability to the movement. During the middle stages of an insurgency, leaders usually have to expand the recruiting effort in order to meet growing operational and functional requirements as well as to replace members lost to attrition. During the latter stages of an insurgency, the momentum of the movement characterizes recruiting. A successful insurgency that is able to either take power (supplanting the former government) or achieve political, legal, or quasi-legal status will normally expand recruitment operations to include parallel efforts at political mobilization.

The type of individual an insurgency attracts and the nature of their motivations for joining also changes as an insurgency develops. In the early stages of an underground movement, recruitment is selective and restricted as much as possible to known, trusted associates of current members. Strong ideological sympathies are most important at this point, when joining entails greater risk and thus requires stronger convictions. If they are successful and survive to later stages of expansion and militarization, undergrounds move into mass recruitment. The motivations for joining an underground movement during later phases are more complex, with no single motivating reason being dominant.

Three contrasting examples will illustrate the range of experiences and recruitment tactics that have been observed. The examples are as follows: (1) "classic" rural insurgencies, illustrated with the Philippine Huks and the Viet Minh; (2) Provisional Irish Republican Army (PIRA) recruitment, a largely urban insurgency that drew from the large pool of sympathizers born into multigenerational Republican families, and (3) campus recruitment efforts of the Egyptian Islamic Group (EIG) in Egypt.

Recruitment in Rural Insurgencies: the Huks and the Viet Minh

Rural insurgencies are typical of Communist movements in the developing world, where the class struggle of poor farmers versus wealthy landowners is a central theme, and the tactics and motivations that characterize rural Communist insurgencies are typical of non-Communist rural insurgencies as well. A study of the Philippine insurgency suggests that individuals usually join as the result of a combination of factors—most often reflecting immediate needs and situational constraints.[1] A chance to obtain personal advantage—ownership of land, leadership, or position of authority—was frequently cited. For some individuals, situational problems, such as family discord, violations of minor laws, and so on, also influenced decisions to join because the insurgency offered a convenient way to disappear. First contact with the movement usually came through chance. An individual joined if the movement fulfilled a personal need or served as an escape, or if social pressure or actual force was applied. Once he was in the movement, indoctrination and other organizational processes helped him to rationalize his commitments.

A chain of interlocking acts eventually led to full-fledged membership. While not invariable, this process was typical of most of the ex-Huks interviewed. The story of one former Huk exemplifies the interplay of motives. This man was recognized as a leader in his barrio and became the local contact for the Huks. He got involved in the arrangement of a rally. When the local people then decided to form a Huk unit, they chose him as its leader. He could not refuse without antagonizing the Huks nor could he leave the town and move elsewhere. The Huks sent him to a "Stalin University," where he was exposed to Communist thinking and propaganda, and the propaganda points that particularly impressed him were the promises to eliminate usury and government corruption and to distribute land to the poor.[2]

The role of ideology varied between the Vietnamese and Huk example. A survey of a group of captured Viet Minh showed that 38 percent of the prisoners expressed a belief in the Viet Minh cause; only 17 percent of a Huk sample of 400 prisoners expressed sympathy with the political objectives of the Communist Party.[3]

Promises and propaganda appear to have been involved in a number of cases, although their actual effects are difficult to determine. One source has noted than among ex-Huks, a majority had joined the movement without any noticeable propaganda influence; most had been primarily concerned with issues like land distribution and lower interest rates. Less than 15 percent gave their only reason for joining the Huks as propaganda or verbal persuasion, although 27 percent reported persuasion or propaganda as a contributory factor. Thirty-eight percent of the Huks became involved through personal friends. It was probably later, after being exposed to propaganda and indoctrination, that motives for joining were related to specific grievances.[4]

A number of prisoners claimed that they were coerced into joining the movement, but because they were prisoners, their claims may have been exaggerated. Huk leaders realized that recruiting could not wait for the slow process of persuasion and free decision. A simple example of a coercion technique was related by a farmer's young son who, while working in the fields, was asked by the Huks to help carry supplies to their mountain hideout. Since they were armed, he complied. After they got to the mountains, the Huks told him he had better stay or else they would report him to the constabulary, who would punish him for helping them. In one study it was found that 20 percent of ex-Huks had been forced to join at the point of a gun or because of threats of violence against their families; for another 13 percent, violence was one important factor among several others.[5]

During the Indochinese War, some cases were reported of young men who had been forced to join the Viet Minh by direct physical coercion. Other men entered the movement because of indirect pressure on their families or on village leaders to provide recruits. The Communists in general combined strong-arm and other pressure techniques with propaganda appeals stressing independence. They generally avoided open appeals for Communism. At any rate, in one study 25 percent of the Viet Minh prisoners stated that they had been forced to join against their wishes and had resented being coerced. Another 23 percent also claimed to have been forced to join but did not appear to resent the fact.

Coercion alone did not seem to be a large factor (20–23 percent) in either the Huks or Viet Minh. Coercion combined with other positive incentives related to personal and situational factors, however, accounted for a larger proportion of joiners (33–48 percent). Another important factor was the action of government troops. Of the 95 ex-Huks interviewed, 19 percent said they joined the Huks because of persecution or terrorization by government forces. The effort of the army to suppress the revolt apparently was a factor that led many to join the movement.[6]

PIRA Recruitment of Republican Sympathizers

The IRA's recruitment methods were quite different from those of the rural, ideology-based, and relatively new insurgencies in Southeast Asia. In many ways, the Provisional IRA's recruitment strategies were easier than those of most other insurgencies because the resistance had been continuing for so long at the point when this group became most active. Many potential recruits were from "Republican" families, and these sympathies sometimes went back generations. When "The Troubles" started in August 1969, approximately one hundred active PIRA members were in Belfast, with another 300–500 supporters providing cover, hiding, or communications. Through 1971 and 1972, the Provisionals had smaller membership numbers than the Official IRA, but the perception that the Provisionals were protecting the Catholic population against the brutality of the British Army and Protestant mobs gained them a following and support that began to drain the popularity and support from the older, politically oriented Official IRA.[7]

The sectarian violence and actions of the British Army during The Troubles were sufficient catalysts to drive young men to the Provisionals. Many volunteers came from Republican (not necessarily IRA) backgrounds and had already experienced a violent confrontation with mobs or the British Army, and retribution more than politics was the proximal factor that motivated volunteers to seek out PIRA affiliation. Nevertheless, the PIRA leadership carefully weeded out those who were motivated purely by revenge and violence. Prospective members were required to endure tedious lectures and reading assignments about Republicanism in order to (1) fully indoctrinate the individual into the cause of a united Ireland and (2) discourage those seeking solely prestige or violence.[8] The volunteers who were selected were typically in their late teens or earlier twenties, mostly middle class, and usually adequately educated.[9]

As the PIRA grew, it focused on recruiting a professional army and were thus interested in sound, stable members, not academics, thrill seekers, the well connected, or the rich, as illustrated by this section from the IRA's *Green Book*:

> The Army as an organisation claims and expects your total allegiance without reservation. It enters into every aspect of your life. It invades the privacy of your home life, it fragments your family and friends, in other words claims your total allegiance. All potential volunteers must realise that the threat of capture and of long jail sentences are a very real danger and a shadow which hangs over every volunteer . . . potential volunteers should think about their ability to obey orders from a superior officer . . . whether they like the particular officer or not . . . He should not join the Army because of emotionalism, sensationalism, or adventurism. He should examine fully his own motives, knowing the dangers involved and knowing that he will find no romance within the Movement.

Most PIRA members had day jobs and took part in operations during the weekends or evenings. Their regular jobs often facilitated their support of the PIRA. If they worked in a government office, then collecting official paperwork, forms, or intelligence was part of their PIRA duty. Those volunteers who had training or experience in politics or in organizational planning were put to work in Sinn Féin, while those who wished to serve without exposure to violence were asked to be part of the auxiliary, providing safe-houses, transportation, donations, and communications. The support and cover that the PIRA received from the auxiliary and from the broader Catholic community allowed them to operate somewhat overtly, establish weapons caches, and quickly disappear into sympathizer's houses after an ambush. While this deep connection to the community afforded freedom of movement, it also made the PIRA reliant upon popular opinion.[10]

Campus Recruitment by EIG: Filling a Void

Communist groups were notorious for their campus recruitment: in Europe, Latin America, and the United States, it was not uncommon for groups to have an overt presence in many universities. Islamic groups adopted similar tactics where the environment was suitably permissive (predominantly in the Middle East and North Africa). As groups like Egyptian Islamic Group (EIG) and Egyptian Islamic Jihad

(EIJ) began to distinguish themselves from the Muslim Brotherhood in the 1960s and 1970s, they were often led by individuals with academic backgrounds in science and engineering (a demographic overrepresented among Islamist movements in the Muslim world). Therefore, a crowded (in some cases, overcrowded) university setting made for a logical recruiting ground. Many of these young men were intelligent but not yet worldly and knew little outside of their upbringing. While educational opportunities were available (if one were to do well enough on his exams), the supporting institutions such as housing and transportation were unable to meet the demands of the influx of students each year. Groups like the Muslim Brotherhood and EIG stepped in, not necessarily to recruit but rather to provide necessary services to their perceived in-group.

Early in its campus recruiting and social outreach programs during the 1970s, EIG subjected their recruiting pool to a single, undifferentiated message to which some, but not all, were expected to respond positively.[11] The Muslim Brotherhood historically used print media in various forms to further their message, while EIG employed the tactic to a lesser degree, often preferring Friday sermons at the mosques or public gatherings as better venues. This served to provide a social component to the ideological indoctrination process.[12] In its role as a provider of social services (housing, transportation, tutoring, mentoring, etc.) on college campuses, EIG served as a conduit for young Muslim men seeking both intellectual and spiritual guidance to local mosques and Muslim organizations, including both the Muslim Brotherhood and EIG. EIG favored a public and proximate approach whereby prominent members, such as civic leaders or clerics, would serve as recruiters, openly espousing the benefits of affiliation.[13] This occurred in full view and often with the tacit support, if not overt consent, of the Sadat regime. EIG activists were often prominent university students or recent graduates, including rural-to-urban migrants and urban middle-class youth whose fathers were middle-level government employees or professionals. Their fields of study—medicine, engineering, military science, and pharmacy—were among the most highly competitive and prestigious disciplines in the university system. The rank-and-file members of EIG came from the middle class, the lower-middle class, and the urban working class.

EIG's domestic recruiting efforts fared better than those of EIJ during the 1980s (particularly since the Egyptian regime, with Saudi Arabian financial backing, provided tacit support to the Islamic groups contributing personnel to aid the anti-Soviet mujahidin in Afghanistan) and early 1990s, and their membership, though always significantly larger than that of EIJ—an estimated tens of thousands compared with hundreds—remained strong until the Egyptian law enforcement and counterintelligence apparatus began to systematically dismantle the networks in the mid-1990s.[14]

Ideology

While an organization rarely survives on ideology alone, "men who are participating in a great social movement always picture their coming action as a battle in which their cause is certain to triumph."[15] Successful movements often rely on a narrative for recruitment, legitimacy, and support that resonates with a deep cultural, ethnic, or historical myth/memory within the population.

Common to most underground movements is an ideology, a set of inter-related beliefs, values, and norms. Ideologies are usually highly abstract and complex and are more than a group of rationalizations and myths that justify the existence of a group; they can be used to manipulate and influence the behavior of the individuals within the group. In every society, ideas, knowledge, lore, superstitions, myths, and legends are shared by its members. These are cultural beliefs. Associated with each belief are values—the "right" or "wrong" judgments that guide individual actions. This value code is reinforced through a system of rewards and punishments dispensed to members within the group. In this way, approved patterns of behavior, or "norms," are established.

Because beliefs and values are only distantly related to concrete action in daily life, an interpretive process is essential to derive specific rules of behavior. Commonly agreed-upon historical truths are used to justify the norms, values, and beliefs of the group. Significant events that occurred in distant times are given symbolic meanings, and a reinterpretation or "reification" of these events in the form of myths or legends supporting the group's purpose is developed. In doing this, the group may select certain concepts and adapt or distort them to justify specific forms of behavior; where existing concepts conflict with current activities, the group may deny that a particular concept is relevant in a particular case.

Within organizations, certain rules specify desirable behavior and the consequences of not conforming. The rules are enforced by organized rewards and punishments that are relevant to the objectives of the group. Normative standards are also enforced by surveillance of members. In established groups, many beliefs are based on authority; that is, since the leaders of the group voice the beliefs, they are accepted as true. When a leader controls the dissemination of information to the members of an organization, he censors and approves various types of information. As a result, the group receives a restricted range of information, and group members tend to develop a set of common beliefs. Thus, in some cases, members need not be persuaded by argument, induced by reward, compelled by pressure, guided by past beliefs, or influenced by the opinions of other people; the restricted range of information to which they have access is sufficient to determine their beliefs.

Human beings dislike ambiguity and uncertainty in their social and physical environment. Through generalized beliefs, individuals seek to give meaning and organization to unexplained events. Common agreement on certain beliefs also enables individuals to operate collectively toward a desired goal. Leaders can interpret situations in terms of the group's beliefs or ideology, translating abstract, ideological beliefs into specific, concrete situations in which actions are to be taken.

Uncertainty reduction theory holds that most people do not tend to categorize themselves into groups unless there is a motive to alleviate uncertainty. This uncertainty arises from instability and the inability to obtain confirmation of one's beliefs and attitudes from objective criteria[16] and is a source of stress that has cognitive, affective, and behavioral consequences[17] that can inhibit decision making. Intolerance of uncertainty (IU) is the tendency to adversely react to (or in anticipation of) uncertain situations.[18] IU is a cognitive bias that affects individual observation, *orientation, decision, and action* during uncertain situations[19] and is an indicator of susceptibility to anxiety.[20] IU can be categorized as the desire for predictability and/or the predilection towards paralysis[21] when confronting a task where the effect of one's input on a system is indeterminate.[22] High IU individuals prefer either very easy or very difficult tasks because of the low risk of an unexpected outcome resulting in cognitive disorientation, unsettling experience, or (at worst) lowered self-esteem.[23] Intermediate challenges moderately push the limits of one's knowledge, skills, and/or abilities and thus do not provide predictable estimates; such challenges are therefore avoided by people with high IU.[24] Similarly, people with high IU may not effectively interact with strangers or tolerate being exposed to situations in close interpersonal relationships whose consequences are unpredictable.[25] Intolerance of uncertainty as a characterological trait does present a cognitive risk factor for anxiety;[26] however, there are no empirical data to support it as a risk factor for either joining or remaining in an insurgent organization. Nevertheless, employing an ideology (often in the form of a narrative) provides an individual a conceptual framework through which to interpret the existing world and process novel stimuli.

Qutbism: Ideology of the Modern Global Salafist Jihad

The modern global Salafist jihad, as proposed by Marc Sageman, consists of three phases. The first was the largely clandestine Islamist movement in Egypt, with Egyptian Islamic Jihad (EIJ) and Egyptian Islamic Group (EIG) being the most prominent of the groups that emerged from the Muslim Brotherhood. The second phase was the mass mobilization of Muslim youth to participate in the defensive jihad in support of the Afghan mujahidin's resistance to Soviet occupation. The final (and ongoing) phase is the Al Qaeda social movement, a loosely affiliated group of networks linked by a common ideology.[27]

No modern theorist influenced Al Qaeda ideology more than Sayyid Qutb, whose unique Salafist interpretation of Islamic theory provides the ideological foundation of the late-twentieth-century global Salafist insurgency. The concept of Salafism has been employed by Sunni theologians since at least the fifth Muslim generation to differentiate the creed of the first three generations (the *Sahabah*, or Companions of the Islamic prophet Muhammad, and the two succeeding generations, the *Tabi'un* and the *Tabi' al-Tabi'in*) from subsequent variations in the Muslim belief system and method of interpretation. Salafists view those generations as an eternal model for all succeeding Muslim generations, especially in their beliefs and methodology of understanding the texts, and also in their method of worship, mannerisms, morality, piety, and conduct. Salafists place a particular emphasis on monotheism (*tawhid*); they condemn religious and political practices that are insufficiently deferential to God. Salafist ideology strictly prohibits Muslim practices such as venerating the graves of Islamic prophets and saints and in general opposes both Sufi and Shia doctrines. Salafists also reject Islamic speculative theology (*kalam*), which involves the use of discourse and debate in the development of the Islamic creed. Salafists maintain that innovation in the Islamic creed (*bid'ah*) or actions of worship are totally without sanction, and, based on scriptural evidence, that they are unacceptable to God. Salafists idealize an uncorrupted, pure Islamic religious community. They believe that Islam's decline after the early generations results from religious innovations (*bid'ah*) and from an abandoning of pure Islamic teachings; they believe that an Islamic revival will only result through the emulation of the three early generations of Muslims and the purging of foreign influences from the religion. Salafists, as do strict adherents of other Islamic denominations, place great emphasis on ritual, from their five-times-daily prayers to activities such as eating, sitting, and manner of dress so as to follow the example of Muhammad and his companions as they endeavor to make religion part of every activity in life.[28]

A well-educated literary critic from Cairo, Qutb returned to Egypt in 1950 after spending two years at the Colorado State College of Education under the sponsorship of the Egyptian Ministry of Public Education. Qutb left Egypt a secular nationalist opposed to British colonial occupation of Egypt and returned to his homeland a radical Islamist. His writings stoked the revolutionary fire within disillusioned young Muslims who sought a more active role in returning Egypt to the center of the Islamic world. Throughout the 1950s and 1960s, political Islam served as an intellectual and ideological counterweight to the uniquely Egyptian blend of Arab nationalism and socialism espoused by the Nasser government. The ideological war over Egypt's future reached a climax on the night of October 26, 1954, when a Muslim Brother, Abdul Munim Abdul Rauf, attempted to assassinate Nasser as he spoke before a crowd in Alexandria; six conspirators were quickly executed and more than 1,000 Muslim Brothers, including Qutb, were imprisoned. Qutb's imprisonment, and the accompanying physical and psychological torture he endured, influenced the articulation of his revolutionary Salafist doctrine. His most famous work, translated as *Milestones* or *Signposts Along the Road*, published in Cairo in 1964, was quickly banned, and anyone caught with a copy was charged with sedition. Qutb was released from prison the same year.

While a detailed analysis of Qutb's worldview and its development, much less a historiography, is beyond the scope of this section, it is important to acknowledge the effects of historical influences and his subjective experience on his worldview. Humans process information through their subjective experience and acquired biases, and Qutb was no different. Qutb believed that the world was meant to serve man if it was understood properly; that understanding was to be realized through Islam. One's faith in Islam should be approached as a means to change society, to remove man from the enslavement of other men to the servitude of God. He viewed Islam as a system of morality, justice, and governance, whose laws and principles (*Shari'a*) should be the sole basis of political, public, and personal life. After the 1952 coup by the Free Officer's Movement (in which he was tangentially involved), he espoused a just dictatorship that would grant political liberties to the virtuous alone. Later he wrote that rule by Shari'a would require essentially no government at all. On the issue of Islamic governance, Qutb differed with modernist and reformist Muslims who espoused the democratic ideals of Islam by referencing the Quranic institution of Shura that supported elections and democracy. Qutb disputed this because the particular Quranic reference was revealed during the Mekkan period and therefore did not address the issue of governance. Qutb opposed the idea of Arab nationalism in the wake of the 1952 coup; shortly after Nasser took power, he and Qutb had a falling out and

Qutb was subsequently imprisoned and subjected to the regime's practices of arbitrary arrest, torture, and oppressive violence. He believed the corruption of good men in positions of power could only be avoided through the institution of Shari'a. Qutb is referenced less in support of the argument to implement Shari'a than in the justification of armed conflict as a means of doing so. His most recent writings claimed that Muslims should resist any system where men are in servitude to other men, be they religious or political figures, because it violates God's sovereignty; a truly Islamic polity has no rulers because Muslims require neither judges nor police to obey divine law. Qutb believed much of the world to exists in a state of ignorance (*jahiliyyah*) and that submitting to Islam was a political, social, psychological, and spiritual liberation. The way to bring about this freedom was for a revolutionary vanguard to combat ignorance through preaching and through abolishing the organizations and authorities of all un-Islamic systems by *jihad*. This movement would spread across the Islamic homeland and ultimately throughout world, with all attaining freedom through submission to Islam.[29]

In 1965, Qutb was arrested again, charged with conspiracy to overthrow the government, and hanged in August 1966. Authorities hoped the execution would extinguish the Islamist threat to the Nasser regime. However, the tactic did not have the desired effect.[30] The execution drove many Islamists underground and clandestine cells began to form among the restless, disillusioned, and disaffected youth, especially students. The Qutbist legacy was not only a binary worldview (or one that perceived the world as starkly divided between good and evil, the former being true Islam and the latter being the state of ignorance, *jahiliyya*, in which the rest of the world existed), but also the resurgence of *takfir*[a] (or declaration of apostasy) authority pronounced by the *umma* (Muslim community), no longer solely the prerogative of the *ulema* (religious scholars), *imam* (prayer leader), or *Shari'a*.[31] These teachings were embodied by a number of young Muslims who opposed both the government and the Muslim Brotherhood's perceived appeasement of the regime; both EIJ and EIG emerged from this Salafist community.

The Neglected Duty or The Absent Obligation, which has been retrospectively classified as the founding manifesto and the de facto operational manual of EIJ, was written by its founder, an electrical engineer named Mohammed Abd al-Salam Farraj. In this text, Farraj began by stating,

---

[a] The traditional sentence for apostasy under Shari'a was execution, amputation, or expulsion, thus requiring stringent evidence for such accusations and often requiring an Islamic court or a religious leader to pronounce a fatwa (religious edict) of takfir. This subordination or decentralization of authority to declare takfir now empowered devout Muslims to rationalize not only judgment of their fellow Muslims but also violent punishment.

"Jihad for God's cause . . . has been neglected by the Ulema of this age."[32] He expanded the interpretation of *jihad* to include the violent struggle that is a duty (*fard al-ayn*) incumbent on all Muslims. Farraj's argument was an extension of Qutb's with a more explicit advocacy for the consideration of jihad as the sixth pillar of Islam. He rejected the interpretation that inner spiritual struggle was the greater jihad and emphasized the role of armed combat. He believed that peaceful means could never bring about a truly Islamic society and so jihad was the only option. He built on Qutb's idea that modern Islamic societies represented *jahiliyyah* and used the ideas of Ibn Taymiyyah[b] to blame this on modern "apostate" Islamic rulers. Farraj's advocacy entailed the establishment of an Islamic vanguard, an elite cadre of pious Muslims with either academic or military credentials that made them pillars of society. The responsibility of this vanguard was to serve as a model for elites in other Muslim nations to emulate. He made the initial classification of the "near enemy" (the impious Egyptian government) and the "far enemy" (Israel), subordinating all Islamic goals to the fight against local apostates.[33] Farraj believed that a new Islamic state should be established in Egypt prior to initiating the campaign to reclaim lost Muslim lands because jihad under the banner of the then-current Arab regime would embolden the impious rulers (whom he blamed for colonial presence in the Middle East).[34]

EIJ saw its organization not only as the vanguard of Qutb's vision for an Islamic revolution but also as an entity with requisite political and religious authority to declare all those who did not meet their requirements for piety essentially non-Muslim, regardless of what the individuals professed to believe. EIJ operationalized Qutbism by exemplifying piety in its members, as well as the organization's oppositional stance against the Egyptian government. Nowhere was this more apparent than in EIJ's commitment to conducting attacks against Egyptian and Western interests, despite the increasing difficulty of such attacks and the EIJ's relative lack of success in the early to mid-1990s.[35]

During the mid-1980s, much of the EIJ underground relocated to Peshawar, Pakistan, to escape the Egyptian government and to participate in the Afghan resistance to the Soviet Union's occupation. Although some EIJ members recognized the still-imprisoned Abud al-Zumur (a former lieutenant colonel in the Egyptian Army implicated in the Sadat assassination) as emir, others began to consider Sayyed Imam al-Sharif

---

[b] Taqi ad-Din Ahmad ibn Taymiyyah was an Islamic scholar, theologian, and philosopher and was born in Harran (the region that is now southwest Turkey). He lived during the period of the Mongol invasion and subsequent rule of the Middle East. He was a member of the school founded by Ahmad ibn Hanbal, a Salafi scholar. His writings did not have considerable impact during his lifetime; however, they have been influential in Salafist ideologies from the nineteenth century through today.

128

(Dr. Fadl) as EIJ's leader. Sharif, a physician who was administering to the refugees from Afghanistan under the Red Crescent, emphasized the importance of the Qutbist ideology and the condemnation of those who deviated from it. Sharif had established a series of guesthouses and training camps in Peshawar to receive, stage, and prepare to move and integrate young Muslims from the Middle East into the Afghan resistance. His was a parallel effort to that of Osama bin Laden and his spiritual advisor (a role Dr. Ayman Al-Zawahiri was eventually to assume), Abdullah Azzam, who had established Maktab Al Khidamat (MAK) or Services Office to facilitate the resolution of administrative problems for foreign Muslim fighters. At some point during the mid-1980s, Sharif's academic, professional, and ideologically like-minded colleague, Zawahiri, became the recognized leader of EIJ. The operational and ideological relationship between bin Laden and Zawahiri grew closer during their years in Pakistan and Afghanistan. EIJ under Zawahiri was never financially stable and became increasingly reliant on stipends from bin Laden to retain a capability. [36]

The Iraqi annexation of Kuwait in 1990 and the subsequent deployment of U.S. and coalition forces to the region resulted in an update to the Salafist insurgent ideology. Bin Laden saw the stationing of U.S. troops in Saudi Arabia, home to Islam's holiest sites, as an unforgiveable offense. The Royal Family's dismissal of bin Laden's offer to provide security using his hardened Arab Afghan fighting force rather than receiving the American military delegation compounded his indignation. The Americans' increasing role in the region, signified by repeated bombing campaigns in Iraq, the United States' central role in the sanctions regime, the increasing visibility of U.S. military forces, and its bank-rolling of many despotic Arab regimes had now largely outstripped the residual goodwill it had earned by abstaining from colonization and supporting the Afghan resistance to the Soviets. In 1996, Al Qaeda announced its intention to expel foreign troops and interests from what they considered Muslim territory. Bin Laden issued a *fatwa*, entitled "Declaration of War Against Americans Occupying the Land of the Two Holy Places," a public declaration of war against the United States and any of its allies, and began to refocus the organization's resources toward large-scale psychological operations. The *fatwa* represented an overall shift in focus on behalf of the EIJ and the other cells comprising Al Qaeda from the near enemy, or Muslim "apostate" governments, to the far enemy, or the United States.[37] In 1998, Zawahiri issued a joint fatwa with bin Laden and Rifi Taha (EIG) under the title "World Islamic Front Against Jews and Crusaders" which read:

> The ruling to kill the Americans and their allies—
> civilians and military—is an individual duty for every

Muslim who can do it in any country in which it is possible to do it, in order to liberate the al-Aqsa Mosque[c] and the holy mosque from their grip, and in order for their armies to move out of all the lands of Islam, defeated and unable to threaten any Muslim. This is in accordance with the words of Almighty Allah, "and fight the pagans all together as they fight you all together," and "fight them until there is no more tumult or oppression, and there prevail justice and faith in Allah."[38]

In June 2001, although difficult to distinguish for years, Al Qaeda and EIJ merged, forming "Qaeda al-Jihad,"[39] and hereafter all activities and operations of former EIJ members were done under bin Laden's banner. At this point, Zawahiri was presumed to be the deputy to bin Laden and the leader of EIJ. While the charismatic leadership of bin Laden was by now well known among the mujahidin, Al Qaeda's strength and appeal did not lay solely in its sophisticated theological discourse but also in its ability to comprehend, co-opt, and exploit modern grievances. This narrative combination resonated with extremists and moderates alike, regardless of whether an individual approved of the means by which Al Qaeda sought to accomplish its goals. Al Qaeda's leadership was not composed of highly trained religious scholars, and its religious rhetoric was far from complex or nuanced, making it broadly accessible across the socioeconomic spectrum. The specific messages within the larger narrative rarely focused on citing authoritative texts (beyond selective interpretations of previous theorists reinforced by Quranic quotes without context) but rather relied on the application of general religious or ethical principles to modern political and social problems.[40]

Despite the death of bin Laden in May 2011, Al Qaeda affiliates are still active throughout the globe, with the most prominent "franchises" being Al Qaeda in the Islamic Maghreb (AQIM) in Northern Africa, Al Qaeda in the Arabian Peninsula (AQAP) in Yemen, and Al Shabaab in Somalia. Although the members probably agree with the teaching of Qutb, that literature is unlikely to be the proximal cause of their rationale for joining.

---

[c] The Al Aqsa Mosque, considered Islam's third holiest site, is located in East Jerusalem, the proposed capital of an eventual Palestinian state.

Affiliative Factors

Affiliative rationales for joining a radical group are far more common than ideological rationales. Maslow identified that humans require belonging and acceptance (whether from large or small social groups); they need to love and be loved (sexually and nonsexually) by others. The absence of these elements may increase one's susceptibility to radicalization.[41] These needs often exacerbate issues of social identity, particularly in immigrant enclaves, leading individuals to question their role in a new culture and/or seek out others with similar ethnic, social, and/or professional backgrounds.[42] It is more likely that only after an individual joins and is indoctrinated that he or she comprehends and accepts the ideology upon which the group was constructed. In contemporary work on the Black Guerrilla Family (BGF) very few acknowledge the Marxist–Leninist–Fanonist influence of George Jackson or his advocacy of Black Separatism. Instead, most members choose to affiliate upon their first incarceration as a means of protection.

Sageman's work on the Al Qaeda social movement identifies similar motives for joining. In the study, he compiled biographies of 400 Al Qaeda-affiliated radicals from trial transcripts, press accounts, academic publications, and corroborated Internet sources. Of that sample, 162 were from the Maghreb (the predominantly Muslim countries of North Africa), 132 were from Arabian countries, and 55 were from southeast Asia. He further distinguished those 38 high-value individuals who were components of the Al Qaeda central staff. The vast majority had secular, not religious, educational backgrounds. Seventy-three percent were married, and most had children (all of the central staff and southeast-Asian members were married). The central staff was composed mostly of Egyptian Islamic militants who had been released from prison and who traveled to Afghanistan for the jihad against the Soviets. The central staff and Maghreb Arabs were upwardly mobile young men from religious, caring, and middle-class families. Many spoke three to four languages and possessed computer skills. Most found themselves abroad, separated from traditional and cultural bonds, and sought social interaction with those of similar backgrounds and psychological states. They adopted the set of radicalized beliefs after they established social networks with radicalized individuals. The organization (or social movement) was a bottom-up, self-organizing activity with no centralized recruiting mechanism. Of those interested in joining, only very few were actually accepted. Sixty-eight percent joined because of pre-existing friendships with members, and 20 percent joined because of familial ties with members; in 98 percent of the cases, social bonds preceded ideological commitment. There was no evidence of coercion or brainwashing; individuals acquired the beliefs of those around them. In each case, the individual joined the jihad through human bridges (acquaintances, relatives, and imams) and not electronic or bureaucratic ones.[43]

Reviewing the reasons stated by captured members for joining a movement, one finds a paucity and ambiguity of data and further difficulties in the interpretation of the data available. Nonetheless, certain conclusions may be stated.

- Multiplicity of motives. Usually, more than one motive is present when a member joins. A combination of factors is usually cited; no one factor is preeminent.

- Personal and situational factors. Most of the motives cited for joining tend to be related to situational or personal problems and to reflect the individual's immediate needs.

- Belief in the cause or political reasons. Only a minority admits that political reasons or sympathy with the ideology or organization are related to joining.

- Propaganda and promises. Few join because of propaganda or promises alone. These are apparently more effective when combined with situational factors.

- Coercion. Coercion alone is a small but important factor in joining.

- Coercion with other positive incentives. When combined with other positive incentives related to personal or situational factors, coercion accounts for a significantly large number of recruits.

- Government persecution. This factor appears to be a small but significant factor leading individuals to join the movement.

REASONS FOR STAYING

Once involved in an underground organization, individuals develop reasons for staying that are often very different from their original reasons for joining. Individual differences often disappear in the face of the powerful unifying forces of group and organizational psychology. Although there are few empirical studies of insurgent motives for remaining with the movement, a review of two studies of conventional military personnel provide insights into the motives of men in combat situations and into the sustaining role of ideology.

In a study of American soldiers during World War II, it was found that the soldier's willingness to fight was not significantly affected by indoctrination or ideological justifications or by receiving awards for exceptional valor. The concern for what his fellow men within the unit thought of him was an important influence on his performance and group effectiveness. It was concluded that most nonprofessional soldiers fight reluctantly and are probably motivated by status-group considerations. Another study, based on the collapse of the German army in World War II, found that in those units that did not surrender, values such as honor and loyalty had created a sense of obligation among the soldiers. Loyalty to their comrades was more important than ideology in their willingness to continue fighting to the end. Ideology, however, did play an indirect role. The type of leadership had a positive effect on the combat effectiveness and commitment of the individuals within a unit. When the men in the units accepted the leadership of officers and noncommissioned officers who were devoted Nazis, the units' performance was much more effective than that of units without ideologically oriented leaders.[44] If the leadership is ideologically oriented, the units seem to be more cohesive and effective, even if the members are apolitical.

In most military units, individuals fight less because they agree with the political system than because they feel a loyalty to their fellow soldiers. They develop an esprit de corps and, in spite of adversity, try not to let their comrades down. Many insurgents who have defected still have favorable memories of the comradeship and togetherness of the guerrilla camps or the underground cells.

In his study of the Philippine insurgency, Scaff concluded that although people joined the Huks for various reasons, there was a tendency for a person, once a member of the movement, to gradually develop new motives for staying.[45] Members stayed on because they were made to believe that the movement would bring about a better life for them and for the masses. Insurgents often are influenced by their own propaganda and agitation themes. The impact of agitational slogans was shown in one study of 400 captured Huk guerrillas: 95 percent asserted that their main reason for fighting was to gain land for the peasants.[46]

Psychological methods and morale-sustaining techniques are also important for maintaining loyalty. Because defections often occurred after serious losses, the Viet Cong went to elaborate lengths to keep up morale. Those killed in battle were carried away, often by special volunteers for that purpose, and buried with great ceremony. If it was not possible to carry the dead away immediately after the battle, the insurgents returned for them at night. This experience built up support for the movement through a desire to avenge the deaths of comrades and was apparently a significant psychological factor in keeping up morale.[47]

Insurgents, particularly new members, also know that they are under surveillance and will be punished for disloyalty. Most underground movements require recruits to take an oath promising to remain with the movement on the penalty of death. Terror or enforcing squads are often used to retaliate against defectors. Threats of revenge are especially effective when it is difficult to defect to safe areas. Atrocity stories about how the government mistreats defectors are also used.

The Communists' frequent criticism and self-criticism sessions act as a form of catharsis and permit members to voice fears and problems. These sessions give members a chance to speak out and be heard. No matter how limited and directed it may be, this process apparently serves as an outlet for emotions that might otherwise lead to defection. In addition, an individual who is disillusioned with the movement will find it difficult to conceal this in the frequent self-criticism sessions. Another significant factor that prevents people from leaving a subversive movement is the human tendency to inertia: to do what is customary and expected of them in spite of any displeasure with the organization.

Several conclusions can be drawn as to why insurgents tend to stay with the movement.

- Changing motives. Motives for remaining within the movement are usually quite different from those for joining. Indoctrination and propaganda expose the individual to new ideas, of which he may have been unaware before joining. New friends and organizational responsibilities are also motives for staying.

- Group norms. Insurgents are influenced by other members of the movement. They are probably more motivated by what their friends and comrades think of them than by any ideological considerations and tend to stay out of loyalty to them.

- Ideology. Ideology plays an indirect role. Units whose leaders are ideologically oriented are more cohesive and effective than those whose leaders are not, even when the members of the group are apolitical.

- Morale-sustaining techniques. Various psychological techniques are used to maintain morale such as special ceremonies and group discussions that give members an opportunity to air their emotional problems and receive group support and reinforcement.

- Surveillance and threat of retaliation. Continual surveillance and threats of retaliation from terror enforcement units keep many members within the movement.

- Inertia. Simple inertia and habit may be stronger than any inclination to leave. It is easier to continue a habit than to change it.

REASONS FOR LEAVING

Of vital interest is why an individual decides to leave a radical group. Psychological motivations to leave an organization are ideological, social, and behavioral but as diverse and idiosyncratic as the motivations to join.[48]

Factors influencing the decision to leave can be categorized into two factors: push and pull. Push factors are negative circumstances or social forces that render the status quo unappealing such as criminal prosecution, social disapproval, or effective violence from oppositional groups. [49] Leaving often entails a shift in the normative factors that comprise the moral and ideological foundation of the respective groups from the insurgent to another.[50] Some of those factors include losing faith in the group's ideology, frustration with the lack of success, and renunciation of violence.[51] If an individual joined and remained in a group for specific values the group espoused and upheld, a shift in or compromise of those values may push him or her away. Eamon Collins's account of his disillusionment with the petty criminality of some members of the PIRA is one such example. Collins, who believed strongly in the virtues of Republicanism and his identity as a "volunteer," describes a sense of moral betrayal when he witnessed individuals stealing money from a cash register after bombing a hotel. Collins's unit had taken steps to avoid civilian casualties, and he viewed thievery as a compromise of the PIRA's integrity.[52]

Pull factors are opportunities or social forces that attract an individual to an alternative course of action such as employment or educational opportunities, newfound social responsibilities, or an alternative political path.[53] Affective factors—the social and organization attachments within the group—are vital to both organizational cohesion and individual identity. Affective factors contributing to disaffection include failing group interaction and failing leadership.[54] Related to (and often an extension of) affective factors are individual maturation and competing social relationships that shift the pragmatic calculus away from remaining involved to the less violent "other."

Understanding the rationale behind disengagement requires primary interaction with those individuals; unfortunately, the empirical evidence is scant.

Defection

Disaffection may result in a person's leaving or defecting from an insurgent movement to the government side. A potential defector seeks the easiest and safest avenue of escape. If circumstances are such that he can simply leave, he will likely do so; if, on the other hand, the possibility of going to government forces arises first, and is relatively easy and safe, he may defect. The process of deciding to leave is an unseen cognitive phenomenon that systematically increases one's openness to possible disengagement through the consideration of alternative world-views.[55] For some, this process is gradual while for others it is an abrupt separation stemming from a specific catalytic event.[56]

In Vietnam, in January 1963, President Ngo Dinh Diem began the *chieu-hoi* (open arms) program. Viet Cong defectors were offered amnesty and assistance after a short indoctrination and retraining course. Between February 18 and June 25, 1963, 6,829 Vietnamese defectors took advantage of the *chieu-hoi* program. The defections from the Central Lowlands were primarily from Quang Ngai and Binh Thuan provinces, both sparsely populated; there were none from the city of Da Nang. The majority of the defections from the Central Highlands were in the rural provinces of Darlac, Lam Dong, and Phu Bon. There were no defectors reported from the city of Da Lat and only thirteen from Saigon. The largest number of defectors, more than twice as many as from the other areas combined, came from the western region. The majority of these defections were in the An Giang Province. The wide range in the number of defectors from the various areas probably reflects local political, social, and psychological conditions, as well as Army of the Republic of Vietnam's military strength in particular areas.

One explanation for the low defection rate in the more populated urban areas, where it would appear to be easier to defect, could be that underground members in the cities do not have to endure the hardships that units in the field do. Another explanation is that defection to the government is unnecessary to leave a movement; individuals can simply disappear into the urban setting.

An analysis was made of the relationship between Republic of Vietnam government appeals and the number of Viet Cong defectors. On the basis of this sample, there appears to be a low relationship between appeals for defection and the number of people who defect. Of those who defected as a result of government appeals, most heard of defection appeals indirectly from civilians and other insurgents. It is likely that many individuals decided to defect first and then became sensitive to propaganda appeals. A large number found their reasons for defection so compelling that they defected without ever having heard any appeals.

In an analysis of 382 Viet Cong defectors, figures for defectors from guerrilla units were higher than those for political defectors. This is significant in that there are usually far more underground members and liaison agents than guerrillas. There are several explanations for this disproportion: political units have a less rigorous physical existence than military ones; political units' day-to-day activities require them to reiterate propaganda themes and carry on persuasive arguments in favor of the movement, so they tend to be insulated from thoughts of defection. Nineteen defectors were liaisons. This relatively high number suggests that, in spite of insurgent efforts to place only the most reliable people in such positions, it is a highly vulnerable job. The liaison agent has unusual opportunities for defection because he usually travels by himself and goes into government-controlled areas. Amnesty offers probably influence liaison agents to take advantage of their chance to escape.

In October 1964, the *cheu-hoi* program reached its low point, with only 253 Viet Cong defecting during that month. Defection continued at this low level until April 1965, when 532 defected. Then the figures began to climb: 1,015 defectors were reported for May and 1,089 for June. The increase in the defection rate coincided with and was largely attributed to the stepped-up Viet Cong conscription program. The young men pressed into service did not have the ideological conviction of earlier recruits and, in many cases, resisted recruitment. Among the new conscripts who defected, personal hardships and the contempt shown them by the veteran Viet Cong were among the reasons cited for defecting.

A 1965 study of 1,369 men and women who defected from the Viet Cong attributed their defection to the harshness of material life in the Viet Cong. Food shortages and limited medical supplies were most often mentioned. Almost none of the defectors mentioned ideological factors. In addition to material and personal factors, the military situation also affected the decision to defect. Members of guerrilla units, for example, were found to be most susceptible to defection appeals immediately after a battle—especially if their unit had suffered heavy losses.

An analysis of defection from the Huk movement in the Philippines has also been made. Several motives for leaving were given by the ninety-five former Huks interviewed, just as they gave several reasons for their earlier entrance into the movement. Sixty-one percent gave physical hardship as their chief reason. In particular, they complained about the cold, hunger, and lack of sleep. Government forces contributed to these hardships by frequent attacks. Many of the interviewees said that they were tired of years of being fugitives and just wanted to live in peace. Forty-five percent said they defected because of the failures and disappointments of the Huk organization. Specifically, they resented the strict discipline in the movement and found orders distasteful, or had lost the feeling of progress and foresaw failure of the insurgency. Twenty-three percent surrendered because of promises and opportunities offered by the government. The most effective promise was that of free land. Mentioned almost as often was the promise that the surrendering men would not be tortured. Other promises cited were those of a job, of payment for surrendered firearms, and of freedom for those against whom no criminal charges were being held. Almost half (45 percent) of the defectors had heard of the government-sponsored Economic Development Corps (EDCOR), and most of these indicated that the program was influential in their decision to give up.

Some said the EDCOR program gave them hope for a new life. Not more than 5 percent said they surrendered because of pressure by their families. Of course, most of the Huks were unmarried young men who did not have many family responsibilities. In sum, there was no single overall motive for the defections. The hardships of existence and the constant pressure of pursuit, disillusionment with the Huk organization, and government promises appeared, in that order, to be the main reasons for surrender. Thus, the government effectively pressed the Huks toward surrender by maintaining steady pressure against them and by providing various promises and opportunities—in particular, EDCOR. Interviews with sixty former Communist insurgents in Malaya indicated that some of the reasons that led them to join the Communist movement were related to their later defections. Many joined as a means of personal advancement and security. They saw the party as a strong organization that would give them a voice in the future. But as they perceived the party to be growing weaker, they felt that they had made a mistake and wanted to extricate themselves as expeditiously as possible.

Most defectors gave no serious thought to leaving the movement during their first year, being too strongly involved in party work or still having high expectations. The critical phase for most came about a year and a half after joining the party. At this point they reappraised and gave critical thought to their current position and possible future. Most had made great sacrifices for the party, and it was increasingly clear that greater sacrifices were to be demanded even while the chances of victory grew slimmer. Many began to feel that the future was hopeless and passed through a period of doubt in which various "crises" arose that often triggered defections. One category of crisis centered on the member's inability to meet the requirements of party membership.

Generally, an individual who developed personal difficulties within the party simultaneously developed critical arguments against the party's goals and methods and the Communist cause in general. Most defectors specified Communism as "bad" in terms that were most meaningful in a setting of personality politics. If the party and its leaders were seen to be "corrupt," the defector could justify his own personal position and his subsequent defection.

Another category of crises resulted from the party's attitude toward sex. Party members were supposed to lead chaste lives even though 10 percent of those in the jungle were women. Even thinking about sexual matters was classified as symptomatic of "counterrevolutionary" attitudes. Permission to marry was generally refused. The party made death the penalty for rape, which was loosely defined and judged by party leaders rather than by the woman. Accordingly, the women tended to attach themselves to the party leaders, and members resented this departure from the party policy of "equality."

Another type of crisis appeared when the party failed to satisfy the defector's personal hopes, either not meeting his needs at all or doing so at too high a price. After World War II, the living standard of the general population increased and social stability improved in Malaya. In contrast, the party member often saw his life as rugged and unrewarding. Among the immediate problems cited by many ex-Communists were that they had to work too much, that life in the jungle was too boring, or that they underwent too much physical suffering. Almost one-third felt that their existence had become too dangerous. Some were pushed into a decision to defect by the death of a friend. In nearly all cases, the decision to defect took place after several minor crises. The likelihood of a crisis leading to defection was especially strong if earlier crises were not resolved. The psychological preparation for defection was complete when a member began to formulate general criticisms of Communism. Once members became disaffected, they sought to disengage themselves from the party as rapidly as possible. Although these interviewees, because they were defectors, are not necessarily representative of the whole party membership (many of whom might conceivably fight to the bitter end rather than surrender), the Malayan data strongly indicate that there is continuity in the defectors' motivations. At one point in time they joined the movement, at another they deserted it; the roots for both actions lay in the same purposes and hopes. As conditions changed, the attractiveness of the alternative paths for achieving these hopes also changed. As the overall prospects of the rebellion changed, the strategic calculus changed and many perceived their desires for personal safety and social advancement were better served by defecting.

Thus, apparently contradictory actions (joining and defecting) had a motivational consistency. Certain generalizations can be made about acts of defection among insurgents in general and among members of the underground and of guerrilla forces.

- Types and rate of defection. Once the individual becomes disaffected, he may stay in the movement but not participate actively, he may leave the movement simply by withdrawing, or he may defect to the government side. The rate of defection varies widely, with a high rate in some areas and a low one in others. Local factors chiefly determine the rate of defection.

- Multiplicity of reasons. Defectors usually give many interrelated reasons for their defection, usually involving some personal and situational factors.

- Conflict and crisis. Internal conflicts and personal crises within the organization usually precede defection. Conflicts usually arise over frustration of individual goals, harsh discipline, or lack of advancement.

- Time of defection. Young recruits who are forcibly conscripted tend to defect early; those who join for ideological reasons tend to reconsider and have second thoughts some months (approximately a year) after joining the movement. It is at this time that they are most susceptible to defection.

- Appeals. Although many defectors are unaware of government appeals and rehabilitation programs, these programs appear to be an influencing factor among those who do hear of them.

- Underground defection. There are some unique characteristics related to underground defection. There is less defection to the government side among members of the underground engaged in political work than among members of guerrilla units; similarly, there is less defection in the populated urban areas than in the rural areas. There are several reasons for this: political activities probably insulate underground members from thoughts of defection; the underground is not exposed to the rugged, harsh existence of guerrilla life; and while defection may be the only option of guerrillas, underground members may be able to simply withdraw or be passive.

- Guerrilla defection. Among the guerrilla units, the rigors and hardship of life in a guerrilla unit, such as bad weather and lack of food and sleep, are often cited as reasons for defection. Usually, however, a personal crisis involving individuals in the guerrilla force is the ultimate triggering force. Defection is also frequent immediately after battles, especially if there have been heavy losses among the guerrillas.

Disengagement

Disengagement is the cessation of terrorist activities.[57] Programs that focus more on disengagement tend to focus on affiliative factors and more generalized social regulation, rather than try to change the ideological mindset of the participants. In essence, the programs attempt to get participants involved in activities and networks that are not violent, rather than seek a conversion. The role of social networks in recruitment and in the process of radicalization is widely acknowledged. The relevant literature in a variety of disciplines indicates the central importance of affiliative factors in individual motivations for entry into and exit from the radical organizations.[58] One Saudi program seeks the involvement of participants' family members in the process and, reportedly, in some cases provides assistance in finding a wife. Others seek to reduce potential economic barriers to disengagement through the provision of social services such as job training or placement, educational opportunities, or modest cash stipends.[59] In each instance, the program adopts an approach that may be readily accommodated and implemented in the setting in which it is to be carried out (government-run detention facilities in the case of deradicalization, short-term abduction conditions in the case of deprogramming, and longer-term municipal and federal project structures in the case of de-ganging).[60]

Deradicalization

Both radicalization and deradicalization (the acceptance and subsequent abandonment of a violent political ideology) have been equated with spiritual experiences, analogous to religious conversion with the differences being that the deradicalization process is often an individual decision that isolates one from an in-group and typically does not require an individual to accept or incorporate a new ideology.[61] However, some deradicalization programs aimed at Islamic fundamentalists attempt to convert the radical's belief in the central ideology of the group or movement to which they belong. State-sponsored deradicalization programs have been implemented in Egypt, Yemen, Saudi Arabia, Jordan, Algeria, Libya, Tajikistan, Malaysia, Indonesia, Singapore, and the United Kingdom.[62] The programs in many predominantly Muslim countries consist primarily of discourse between program participants and moderate imams, who attempt to persuade participants, through religious discussion and debate, to abandon terrorist ideologies in favor of a more moderate, nonviolent understanding of Islam. Many programs also often include a counseling or psychological component, as well as a social outreach component that seeks to engage their pre-radical network. However, despite these efforts, there has yet to be a valid and reliable set of political, cultural, or psychological indicators of successful deradicalization.[63]

A study that evaluated programs in fifteen countries lends some insight on approaches to deradicalization. The existence of hierarchical command and control structures in the deradicalization program may enable radicals to surrender their arms and even renounce violence while retaining the inherent dignity of their identity as a solider/fighter/warrior. When political concessions form part of a negotiated agreement between the state and the radical group, deradicalization should be a component of the said agreement. However, the process requires tangible goals, effective management, and patience.[64] This multinational study also recommends using a combination of counter-ideological education (or religious re-education) with vocational training. It recommends employing credible and empathetic interlocutors who can relate to prisoners' personal and psychological needs and not simply confront their beliefs with an unsympathetic set of alternative beliefs. It also recommends re-establishing a social network in mainstream society through pre-existing (but nonviolent) connections. This process is obviously confounded when the radical is collocated with other violent radicals. Material inducements are often helpful, but as a supporting effort only. The main effort requires a sophisticated holistic approach to simultaneously satiate one's hierarchy of needs by credibly incentivizing positive behaviors.[65]

Measuring the effectiveness of deradicalization programs is rather difficult, and differences in cultural contexts as well as program eligibility requirements have produced data that are difficult to assess and nearly impossible to compare. Although several studies are ongoing, reliable data on recidivism rates of terrorist behavior and/or longitudinal studies on the long-term effectiveness of deradicalization programs are scarce. This problem is compounded by the inconsistent inclusion criteria and/or inaccurate claims of success by many nations who have implemented Islamic deradicalization programs. An example is the Saudi program that released 700 of the 2,000 prisoners who participated in their deradicalization program and the subsequent claim that only nine have been rearrested. The Saudi criteria for what constituted a "radicalized" militant have never been made public and thus their recidivism rates are difficult to decipher.[66] A Jordanian program boasts similar recidivism rates, however, some graduates of the program were arrested or killed in Iraq supporting the various Islamic militant groups contributing to the anti-coalition insurgency.

Despite the challenges of and approaches toward deradicalization, some themes have emerged from the literature. The focus on social networks and recreating (or repairing) familiar ties broken through joining a violent group is consistent with recommendations from social scientists. Incentivizing "pull" factors are more successful than disincentivizing "push" factors. This trend appears to extend to criminal youth organizations (gangs) as well. A Canadian report identified similar "pull" factors as prevalent in the decision to leave a violent organization but indicated that barriers associated with the negative identity of being a former gang member inhibit the transition process.[67] While positive incentives require more nuanced targeting, social and economic development policies aimed at weakening support for violent extremist organization have shown some, albeit limited, success.[68]

Reintegration

A trend in insurgent warfare since 1962[69] has been the intervention of supranational organizations and international nongovernmental organizations (NGOs) to broker peace agreements and facilitate the transition to civil authority.[70] The international community's involvement in demilitarization (or disarmament), demobilization, and reintegration (DDR) has occurred in dozens of conflicts and is often a planning consideration for military staffs tasked with the variety of mission-sets under peace operations.[d] Central to DDR is demobilization/reintegration (transitioning combatants to productive civilian life) for four reasons: humanitarianism, compensatory justice, potential contribution, and jeopardizing peace.[71]

Humanitarian perspectives consider some insurgents, despite being perpetuators of violence, as victims. This is particularly appropriate when those individuals are children because the reintegration of child soldiers has profound psychological challenges. There are estimates that approximately 300,000 children in thirty-three countries (in Africa alone) have been forcibly conscripted into regular or irregular armies.[72] The United Nations (UN) considers those less than 16 years of age to be "children" and thus requires a different process of treating

---

[d] Joint Publication 3-07.3 Peace Operations defines peace operations (PO) as crisis response and limited contingency operations that normally include international efforts and military missions to contain conflict, redress the peace, and shape the environment to support reconciliation and rebuilding and to facilitate the transition to legitimate governance. PO include peacekeeping operations, military peace enforcement operations, peace building post-conflict actions, peacemaking processes, and conflict prevention.

(psychologically and/or medically), training, and reintegrating these juvenile combatants.

Reintegrating former combatants presents psychological challenges because most insurgents consider their service a duty to their in-group (be it a clan, ethnicity, or nation) and may expect compensatory justice.[c] The subjective humiliation of grievances being ignored can lead not only to psychological disturbance, suicide, or addiction, but also to socially maladaptive behaviors and/or continued or political disaffection.[73] Truth and Reconciliation Commissions (TRCs) are often established to ensure past wrongdoings (from oppression to war crimes) are made public and redress is afforded the aggrieved. The TRC organization and process is designed so that the resultant society can move forward without the underlying resentment of one group toward another. While some commissions are internal and independent, others operate in conjunction with the International Criminal Court (ICC), which prosecutes perpetrators of war crimes. Over twenty nations currently have active TRCs to address a variety of political grievances. One of the more prominent TRCs was that established in South Africa[74] after the abolition of apartheid in 1994 with a mandate to "bear witness to, record and in some cases grant amnesty to the perpetrators of crimes relating to human rights violations, as well as reparation and rehabilitation."[75] While the South African TRC was not initially viewed well by South African blacks[76] or whites,[77] it has since served as a model reintegration program (although of an oppressive majority, not necessarily a violent insurgency). The international perception of the South African TRC was largely favorable; however, that might be more a function of the charismatic leadership (and concomitant popularity) of Nelson Mandela more so than a triumph of compensatory justice.

The threat posed by ex-combatants is tangible given the nascent state of security forces (in the event of a regime change), particularly amid fractionalization of the movement itself.[78] It behooves the governing body, be it that of the status quo antebellum, the public component of the insurgency, or an international caretaker regime, to provide an opportunity for ex-insurgents to address their grievances in a nonviolent forum. Reintegration of former combatants can also provide a positive force in the reconstruction and rehabilitation of regions damaged by war.[79]

---

[c] Compensatory justice is strictly conceptualized (in a legal context) as requiring unitary and discrete event that produces a clearly defined injury, a clear causal connection between an indefinable defendant's conduct and identifiable plaintiff's injury. The concept requires the legal system to restore the plaintiff to the position that he would have occupied if the unlawful conduct had not occurred.

The challenges of reintegration are evident in the policies implemented as part of the Irish Good Friday Agreement (GFA) of 1998. The Early Release Scheme offered political prisoners the conditional opportunity to participate in a peaceful resolution to The Troubles so long as their respective group did not violate the cease-fire. Prisoners on both sides experience challenges finding employment in Northern Ireland and securing travel visas to find work elsewhere. The Northern Ireland Association for the Care and Resettlement of Offenders (NIACRO) was eventually established to provide education and vocational training to reintegrate all prisoners into Northern Irish society.[80] Prior to the GFA, a charitable trust (the Educational Trust) was specifically dedicated to education and resettlement support of political prisoners. The program was later expanded to their families; of the 308 persons to take advantage of that program, over one hundred were children of prisoners.[81] While still criticized, the Early Release Scheme was viewed as both successful and vital to the peace process in Northern Ireland.

Colombia's Reincorporation Program focused on the group (vice the individual) to demobilize the Fuerzas Armadas Revolucionarias de Colombia (Revolutionary Armed Forces of Colombia, or FARC). The first step in the process required paramilitary commanders to provide a list of names to the Office of the High Commissioner. That office verified the names with the Ministry of Defense. The individuals specified went to a location selected by the government to be debriefed and registered with Colombia's Technical Investigative Body. The government registrar supplied them with identification classifying them as demobilized individuals to avail them of government benefits. Each demobilized guerrilla ceremonially surrendered his or her weapon to government representatives; representatives of the Organization of American States witnessed and verified the process.[82]

A focus of the International Security Assistance Force (ISAF) has been reintegrating former Taliban insurgents into Afghan society. Funded by both coalition partners and the Government of the Islamic Republic of Afghanistan (GIRoA), reintegration programs have increased in urgency since 2009, however, with mixed results. An analysis of thirty-six reintegration cases in Afghanistan between 2001 and 2011 showed that 36 percent reintegrated because of a perception that their group was losing the war; in 33 percent of the cases, coercion was a critical factor; and in 71 percent of the cases, insurgents reintegrated because of intra-group grievances.[83] This research suggests that "push" factors more so than "pull" factors influenced the decision to reintegrate into Afghan society.

INSURGENT TRANSITIONS

Sometimes insurgencies succeed, resulting in a different kind of transition, from that of a mobile and flexible opposition either to the role of a legitimate political party or directly to the role of a governing coalition, with all its implied expectations and responsibilities. From the demobilization of the guerrillas to the establishment of the necessary machinations of civil bureaucracy, the transition from armed movement to governing institution is one not easily made. The trend in modern revolutions has seen more conflicts conclude at the negotiation table than in the battlefield. The decision by insurgent movements to opt for negotiations is viewed as a positive, politically oriented decision and not necessarily an admonition of military inferiority or the renunciation of violence.[84] As identified in the previous section, the process must make accommodations for the reintegration of the guerrillas into the emerging political environment. From the revolutionary perspective, plans must be made to ensure political and/or military gains (regardless of the means by which they were made) are not lost in the emerging government. To this end, the underground must shift their attention (and possibly their resources) from the guerillas to the public component and/or ensure the integration of the shadow government into whatever institutions will be preserved or created.[f] In some cases, reintegration is not desired and the programs put into place are aimed at punishment and not reconciliation. The 1975 North Vietnamese victory saw little interest in reintegrating former Army of the Republic of Vietnam (ARVN) soldiers and South Vietnamese civil servants. South Vietnamese were put in re-education camps by the hundreds of thousands—often for months to years (some for decades).[85] The North Vietnamese perspective held that the South Vietnamese regime was corrupt beyond rehabilitation and that all those associated with it should be purged from society. Furthermore, the revolutionary leadership needn't retain a tangible enemy.[86] The persistence of the negative identity of the corrupt yet defeated adversary of Communism not only validated the years of war but also served to intimate those who might consider resisting the new regime.

The ongoing Quetta Shura Taliban insurgency took place during two periods that were interspersed with a phase during which the Taliban was the de facto (though not universally recognized) government of Afghanistan. The first phase of the insurgency took place from 1994 to 1996 against the ruling mujahidin. The phase second was from 2001 to 2009 after U.S. and coalition forces toppled the regime and

---

[f] For more information on insurgent shadow governments and the public component, refer to Chapter 9 of *Undergrounds in Insurgent, Revolutionary, and Resistance Warfare.*

helped establish the Government of the Islamic Republic of Afghanistan (GIRoA). The Taliban began their rise to power in October 1994 when close to 200 guerrillas attempted to secure the commercial road from Kandahar to Pakistan. The group conducted a series of operations against other Afghan factions competing for power in the wake of the Soviet withdrawal. Eventually they captured Spin Boldak and moved to Kandahar where they unseated Mullah Najibullah. As a result of securing these strategic positions and weapons acquisitions (including some remnants of the Afghan army's armor and aviation assets), they were able to disarm local armed groups and secure the military garrison and administration in Kandahar. They gained control of the commercial road between Kandahar and Pakistan while maintaining security on that road for goods and travelers. This was their first demonstration to the Afghan people that they were committed to combating corruption and establishing order in Afghanistan, a foothold to legitimacy. The Taliban established their administration in Kandahar and began to organize their military campaign for the surrounding towns and provinces. They would send a delegation of *ulema*, a group of educated Islamic legal scholars, to engage local militia commanders, asking them to implement Shari'a law and promote peace by surrendering their weapons and ammunition. Acquiescence resulted in access to social welfare programs as well as to government funding. Rejection resulted in further engagement with tribal leaders and a Taliban representative. Only if the second delegation failed was violence employed. The Taliban effectively "conquered" and governed southern Afghanistan (Kandahar and Helmand provinces), but their inroads to Herat in the west and in Nuristan in the east were not as effective. They were not able to gain access to and influence in the Northern provinces whose militia (the Northern Alliance) became their principal rival. In 1997, the Taliban was officially recognized as legitimate rulers by only Pakistan and Saudi Arabia. The Taliban ruled brutally yet consistently, imposing a harsh interpretation of Islamic law that has drawn criticism from many international human rights organizations as well as other nations. They became a target of U.S. and coalition Operation Enduring Freedom (OEF) forces in the wake of the September 11, 2001, Al Qaeda attacks. OEF resulted in the rather rapid unseating of the regime by the end of 2001 and their displacement to the Northwest Frontier Provinces in Pakistan. From this sanctuary, the Taliban reconstituted its capability and eventually began increasing its influence in the Southern region of Afghanistan in much the same manner as it had a decade earlier. The Taliban leadership under Mullah Omar as well as that of Al Qaeda has been the focus of a controversial program run by the U.S. Central Intelligence Agency that has employed unmanned aerial systems (UAS) strikes in Pakistan to eliminate their interference in the nascent GIRoA.[87]

The Taliban has proven resilient and rather effective at transitioning between armed insurgent group, governing entity, and back. Throughout they have employed multipronged diplomatic, subversive, ideological, and terroristic methods in currying popular support, intimidating potential opponents, and forging alliances with state and substate actors. It is unclear whether these transitions were a planning consideration; however, the diversity of their tactics and use of Islamic emissary practices (the elevation of political and religious engagement seeking conversion over military conquest) may have prepared the organization better for its assumption of power. While Omar's brief tenure as the leader of Afghanistan was fraught with civil, social, and international failures, he has been able to remain in a position of authority and influence for nearly two decades, despite being one of the most hunted men in the world.

SUMMARY

Information from a wide range of insurgencies, across different settings and time periods, presents a picture of why insurgents and members of the underground join movements, why they stay, and under what circumstances they may leave via defection or other means.

People join insurgencies for a variety of reasons. Ideology is most important in the earliest stages of insurgencies, when danger is most imminent and trust most critical; ideology and propaganda remain important but other factors also enter when insurgencies are more successful and move to mass recruitment. Studies of former insurgents with the Huk rebellion in the Philippines and Communist insurgents in Vietnam illustrate how personal, situational, and intellectual reasons combine to lead to insurgent involvement. Personal circumstances such as trouble with the law, trouble with relatives, or friendship with someone else who joins are often the most salient factor. Belief in the cause seems helpful but not sufficient or even strictly necessary. This belief comes in two types: a more complex cognitive commitment to the Communist cause, and a simpler response to propaganda promises on tangible issues like land reform. Coercion also played an important role, often combining with other factors to be an effective short-term means of building a capability.

Very different motivations for recruitment are seen in the urban insurgencies of the Provisional IRA and EIG and EIJ. Coercion was quite atypical in each of these. The PIRA drew from a well-established identity group of Republican sympathizers. While run-ins with the police or firsthand exposure to violence was often the immediate cause for wanting to join, the PIRA attempted to select those who would be

148

sound, long-term members, not academics, revenge or thrill seekers, the well-connected, or the rich. EIG/EIJ's effective campus recruitment presents another noncoercive model. These organizations entered the lives of young students—often from rural backgrounds—entering universities and filled their needs for support, community, and connection to their religious roots.

Once recruits have joined an insurgency, motivations for staying often develop that are different from their original motivations to join. Some recruits develop the more complex ideological commitments that they initially lacked. But the most important factor for continued engagement of both insurgents and soldiers seems to be the loyalty, obligation, and connection to fellow soldiers and to leaders.

Leaving such movements can also involve a complex interplay of factors. Hardship, military losses, and loss of confidence in the cause can lay the groundwork for defection, after which individuals may opportunistically take advantage of the proximity of enemy lines, promises of amnesty, or promises of reintegration programs.

There is currently a great deal of interest in reintegration and "deradicalization" programs in Middle Eastern countries, particularly programs targeted at domestic religious extremists. Reintegration is a core concept—reintegration is the process by which individuals who had isolated themselves or were isolated in an insurgent group renew family and moderate friendship ties, or are introduced to new contacts, including moderate religious leaders. The incentivizing "pull" factors seem to be generally more successful than coercive "push" strategies (although participation in the programs themselves is generally not voluntary).

Reintegration of child soldiers, particularly in Africa, presents a different set of challenges. Reunification with family and moderating influences is again central to the efforts, but addressing the psychological and developmental damage is also required. Unfortunately, many reintegration programs are insufficiently resourced and therefore inadequate in addressing some of the underlying psychological, social, and/or economic underpinnings that prompted individuals to join a violent group.

Large-scale reintegration of insurgents also requires addressing the underlying grievances that motivate the movement, as well as addressing specific atrocities. Truth and Reconciliation Commissions, most famously implemented in post-apartheid South Africa, are an approach that brings past injustices to light without necessarily seeking criminal charges or retributions against individual perpetrators. Other programs that have seen some success are Ireland's Early Release programs for political prisoners, Colombia's Reincorporation program aimed at FARC insurgents, and ISAF's Taliban reintegration programs.

ENDNOTES

[1] Alvin H. Scaff, *The Philippine Answer to Communism* (Palo Alto, CA: Stanford University Press, 1955).

[2] Ibid.

[3] Ibid.

[4] Ibid.

[5] Ibid.

[6] Ibid.

[7] Chuck Crossett and Summer Newton, "The Provisional Irish Republican Army: 1969–2001," in *Casebook on Insurgency and Revolutionary Warfare, Volume II: 1962–2009*, ed. Chuck Crossett (Alexandria, VA: U.S. Army Publications Directorate, in press).

[8] Ibid.

[9] Ibid.

[10] Ibid.

[11] Jason Spitaletta, "Egyptian Islamic Jihad," in *Casebook on Insurgency and Revolutionary Warfare, Volume II: 1962–2009*, ed. Chuck Crossett (Alexandria, VA: U.S. Army Publications Directorate, in press).

[12] Ibid.

[13] Ibid.

[14] Ibid.

[15] Georges Sorel, *Reflections on Violence* (Cambridge: Cambridge University Press, 1999), 14.

[16] Sara Savage and Jose Liht, "Mapping Fundamentalisms: The Psychology of Religion as a Sub-discipline in the Understanding of Religiously Motivated Violence," *Archive for the Psychology of Religion* 30, no. 1 (2008): 75–91.

[17] Angela J. Yu and Peter Dayan, "Uncertainty, Neuromodulation, and Attention," *Neuroanatomy* 46, no. 4 (2005): 681–692.

[18] Michel J. Dugas, Mark H. Freeston, and Robert Ladouceur, "Intolerance of Uncertainty and Problem Orientation in Worry," *Cognitive Therapy and Research* 21, no. 6 (1997): 593–606.

[19] Michel Dugas, Mary Hedayati, Angie Karavidas, Kristin Buhr, Kylie Francis, and Natalie A. Phillips, "Intolerance of Uncertainty and Information Processing: Evidence of Biased Recall and Interpretations," *Cognitive Therapy and Research* 29, no. 1 (2005): 57–70.

[20] Michel J. Dugas, Mark H. Freeston, Robert Ladouceur, Josee Rhéaume, Martin Provencher, and Jean-Marie Boisvert, "Worry Themes in Primary GAD, Secondary GAD and Other Anxiety Disorders," *Journal of Anxiety Disorders* 12, no. 3 (1998): 253–261.

[21] Keith Bredemeier and Howard Berenbaum, "Intolerance of Uncertainty and Perceived Threat," *Behavior Research and Therapy* 46, no. 1 (2008): 28.

[22] Magda Osman, "Controlling Uncertainty: A Review of Human Behavior in Complex Dynamic Environments," *Psychological Bulletin* 136, no. 1 (2010): 65–86.

[23] Todd B. Kashdan and Jonathon Rottenberg, "Psychological Flexibility as a Fundamental Aspect of Health," *Clinical Psychology Review* 30, no. 7 (2010): 865–878.

[24] Ibid.

[25] Ibid.

26 Dugas et al., "Intolerance of Uncertainty."

27 Spitaletta, "Egyptian Islamic Jihad."

28 Johannes J. G. Jansen, *The Neglected Duty: The Creed of Sadat's Assassins and Islamic Resurgence in the Middle East* (New York: Macmillan, 1986).

29 Ibid.

30 Spitaletta, "Egyptian Islamic Jihad."

31 Devin R. Springer, L. Regens, and David N. Edger, *Islamic Radicalism and Global Jihad* (Washington, DC: Georgetown University Press, 2009).

32 Jansen, *Neglected Duty*.

33 Fawaz A. Gerges, *The Far Enemy: Why Jihad Went Global* (New York: Cambridge University Press, 2005).

34 Jansen, *Neglected Duty*.

35 Spitaletta, "Egyptian Islamic Jihad."

36 Ibid.

37 Jason Spitaletta and Shana Marshall, "Al Qaeda," in *Casebook on Insurgency and Revolutionary Warfare, Volume II: 1962–2009*, ed. Chuck Crossett (Alexandria, VA: U.S. Army Publications Directorate, in press).

38 Ibid.

39 Muntasir Zayyat, *The Road to Al-Qaeda: The Story of Bin Laden's Right-Hand Man* (London: Pluto Press, 2004).

40 Spitaletta and Marshall, "Al Qaeda."

41 Chuck Crossett and Jason Spitaletta, *Radicalization: Relevant Psychological and Sociological Concepts* (Ft. Meade, MD: Asymmetric Warfare Group, 2010).

42 Aidan Kirby, "The London Bombers as 'Self-Starters': A Case Study in Indigenous Radicalization and the Emergence of Autonomous Cliques," *Studies in Conflict and Terrorism* 30, no. 5 (2007): 415–428.

43 Marc Sageman, *Understanding Terror Networks* (Philadelphia: University of Pennsylvania Press, 2004).

44 Edward Shils and Morris Janowitz, "Cohesion and Disintegration in the Wehrmachts in World War II," in *Public Opinion and Propaganda*, ed. D. Katz, D. Cartwright, S. Elderwold, and A. McLee (New York: Dryden, 1954).

45 Scaff, *Philippine Answer*, 116, 119.

46 Ibid., 121–122.

47 Seymour Topping, "Portrait of Life with the Viet Cong: A Defector's Own Story," *The New York Times*, May 23, 1965, E-3.

48 Darcy M. E. Noricks, "Disengagement and Deradicalization: Process and Programs," in *Social Science for Counterterrorism Putting the Pieces Together*, ed. Paul K. Davis and Kim Cragin (Santa Monica, CA: RAND Corporation, 2010).

49 Tore Bjorgo and John Horgan, *Leaving Terrorism Behind: Disengagement from Political Violence* (New York: Routledge, 2009).

50 Natalie J. Allen and John P. Meyer, "The Measurement and Antecedents of Affective, Continuance and Normative Commitment to the Organization," *Journal of Occupational Psychology* 63, no. 1 (1990): 1–18.

51 Kira J. Harris, *Review: Disillusionment with Radical Social Groups* (proceedings of the 1st Australian Counter Terrorism Conference), accessed December 2, 2011, http://ro.ecu.edu.au/act/4.

52 Eamon Collins and Mick McGovern, *Killing Rage* (London: Granta, 1998).

53 Bjorgo and Horgan, *Leaving Terrorism Behind*.

54 Harris, *Review: Disillusionment with Radical Social Groups*.

55 Ibid.

56 Ibid.

57 John Horgan, "Individual Disengagement: A Psychological Analysis," in *Leaving Terrorism Behind: Disengagement from Political Violence*, ed. Tore Bjorgo and John Horgan (New York: Routledge, 2009).

58 Madeline Morris, Frances Eberhard, Jessica Rivera, and Michael Watsula, *Deradicalization: A Review of the Literature with Comparison to Findings in the Literatures on Deganging and Deprogramming* (Institute for Homeland Security Solutions, 2010).

59 Richard Barrett and Laila Bokhari, "Deradicalization and Rehabilitation Programmes Targeting Religious Terrorists and Extremists in the Muslim World: An Overview," in *Leaving Terrorism Behind: Disengagement from Political Violence*, ed. Tore Bjrgo and John Horgan (New York: Routledge, 2009).

60 Morris et al., *Deradicalization: A Review of the Literature.*

61 Renee Garfinkel, "Personal Transformations: Moving from Violence to Peace," Special Report 186 (Washington, DC: United States Institute of Peace, 2007).

62 Noricks, "Disengagement and Deradicalization."

63 John Horgan and Kurt Braddock, "Rehabilitating the Terrorists?: Challenges in Assessing the Effectiveness of De-radicalization Programs," *Terrorism and Political Violence* 22, no. 2 (2010): 267–291.

64 Peter R. Neumann, *Prisons and Terrorism: Radicalisation and De-radicalisation in 15 Countries* (London: The International Center for the Study of Radicalisation and Political Violence, 2010).

65 Ibid.

66 Noricks, "Disengagement and Deradicalization."

67 Ross Hastings, Laura Dunbar, and Melanie Bania, *Leaving Criminal Youth Gangs: Exit Strategies and Programs* (Ottowa: Institute for the Prevention of Crime, 2011).

68 Noricks, "Disengagement and Deradicalization."

69 Chuck Crossett, ed., *Casebook on Insurgency and Revolutionary Warfare, Volume II: 1962–2009* (Alexandria, VA: U.S. Army Publications Directorate, in press).

70 Bill Rolston, "Demobilization and Reintegration of Ex-Combatants: The Irish Case in International Perspective," *Social & Legal Studies* 16, no. 2 (2007): 259–280.

71 Ibid.

72 Aaron Young, "Preventing, Demobilizing, Rehabilitating, and Reintegrating Child Soldiers in African Conflicts," *The Journal of International Policy Solutions* 7 (2007): 19–24.

73 Rolston, "Demobilization and Reintegration," 259–280.

74 "The TRC Report," accessed December 3, 2011, http://www.justice.gov.za/trc/report/index.htm.

75 Ibid.

76 The Centre for the Study of Violence and Reconciliation & the Khulumani Support Group, "Survivors' Perceptions of the Truth and Reconciliation Commission and Suggestions for the Final Report," accessed December 5, 2011, http://www.csvr.org.za/wits/papers/papkhul.htm.

77 Jay A. Vora and Erika Vora, "The Effectiveness of South Africa's Truth and Reconciliation Commission: Perceptions of Xhosa, Afrikaner, and English South Africans," *Journal of Black Studies* 34, no. 3 (2004): 301–322.

78 Rolston, "Demobilization and Reintegration," 259–280.

79 Ibid.

80 Horgan and Braddock, "Rehabilitating the Terrorists?"

81 Ibid.

82 Ibid.

83 Seth G. Jones, *Reintegrating Afghan Insurgents* (Santa Monica, CA: RAND Corporation, 2011).

84 Veronique Dudouet, "Negotiating Conflict Settlements: Lessons Learnt and Challenges," Berghof Project: Resistance/Liberation Movements and Transitions to Politics: Building a Network of Experience (Roundtable Meeting Report, March 7–9, 2008, Schwanenwerder, Berlin, Germany).

85 Andrew Wiest and Jim Webb, *Vietnam's Forgotten Army: Heroism and Betrayal in the ARVN* (New York: NYU Press, 2007).

86 Ibid.

87 Sanaz Mirazei, "Taliban 1994–2009," in *Casebook on Insurgency and Revolutionary Warfare, Volume II: 1962–2009*, ed. Chuck Crossett (Alexandria, VA: U.S. Army Publications Directorate, in press).

CHAPTER 6.

GROUP DYNAMICS AND RADICALIZATION

CHAPTER CONTENTS

Nathan Bos

This chapter is about the group dynamics that affect the course of underground and resistance movements, with particular a focus on the process of radicalization. Previous chapters have discussed related individual psychological factors, but current research suggests that group dynamics are at least as important, and perhaps more important, than individual psychology in understanding and predicting radicalization.

Radicalization is the process by which an individual, group, or mass of people undergo a transformation from participating in the political process via legal means to using or supporting violence for political purposes. Radicalization includes specific forms, such as terrorism, which is violence against the innocent bystander, or insurgency, which is violence against the state. It does not include legal and/or nonviolent political protest; such protest is more properly called activism. The chapter is organized around a set of mechanisms of group radicalization adapted from a framework by Clark McCauley and Sophia Moskalenko.[1]

For our purposes, we consider violence to be a necessary component of the definition of radical activity. The Polish Solidarity movement would be an example of an organization that was illegal but not radical. The organization became illegal when the Polish government decided to outlaw the movement in the 1980s, driving the labor union to become an underground protest organization. Although they conducted illegal activities, the Solidarity union would not fit our definition of a radical movement because they did not routinely or purposefully engage in violent opposition to the government.

Many people associated with radical groups do not themselves commit violence, however. Only a subset of people belongs to the armed components of these movements; others are involved in the underground or public component. We consider the group to be radicalized if any part of it commits violence. Members who do not directly participate in violence must still be supportive of the actions at some level, and the group as a whole must go through a process of radicalization before it can regularly take on violent action.

Organizations often start out emphasizing activism and only later turn to violence. Often, the decision is made incrementally, as part of a process whereby actions, decisions, and behaviors move toward becoming more illegal or violent. At other times, the decision to cross the line between activism and radicalism is abrupt, decisive, and emotionally jarring. Understanding the circumstances under which groups turn toward more violent tactics is of great importance. By radicalization, we also mean the process by which already-violent groups become more violent, for example, by endorsing new tactics—such as violence against civilians or suicide attacks—that they had previously rejected.

We will give a number of historical examples of these group radicalization mechanisms and will use one particular example repeatedly. The pathway to violence between ethnic Sinhalese and Tamils in Sri Lanka is a well-documented escalation of group conflict that illustrates several of the common mechanisms of group radicalization. The Tamil people did not have a pre-existing culture or history of violence. Yet beginning in 1956, when the Sinhalese exclusionist Sri Lanka Freedom Party (SLFP) took control of the national government, and over the course of less than two decades, a large part of this population became radicalized. The group at the spearhead of this was a particularly violent and extreme insurgent group, the Liberation Tigers of Tamil Eelam (LTTE), commonly known as the Tamil Tigers. This single case of large-scale radicalization illustrates in-group/out-group conflicts as well as radicalization through isolation, radicalization under threat, and the role of martyrs.

SOCIAL PSYCHOLOGY OF GROUP CONFLICT

Most of the mechanisms described in this section are based on group dynamics. Some background in the social psychology of intergroup conflict will be useful, including an understanding of what social identity is, how and why in-groups form, and what is known about out-groups and intergroup polarization.

Social Identity Groups: Categorization and Salience

Humans have natural inclinations to form into cohesive groups that serve the purposes of information exchange, provide social support (caring for ill, community raising of children), allow collective action (hunting, building projects), and enable mutual protection. People seek affiliation with groups more strongly when faced with anxiety or conflict. Modern research is uncovering how social behavior has deep roots in our cognitive and biochemical makeup and affects a wide variety of human behaviors. Persistent groupings are called social identity groups.

All individuals have a sense of belonging to multiple identity groups. Since the 1950s, psychologists have used the simple "Twenty Statements Test" to gauge self-concept.[2] Participants make twenty statements in the form of "I am _____." Responses tend to fall into different types, one of which is social categorization, or social identities. Social identity responses might be "I am Christian," "I am American," or "I am a Teamster."[a],[3] Individuals may have many social identities along dimensions of ethnicity, religion, politics, economics, and ideology, among others.

Patterns of identities are often studied by presenting people with lists of possible identity groups and asking them to choose and rank identities according to their importance. (Sometimes the question is also left open-ended and responses are categorized later by the researchers.) This method showed, for example, that in Northern Ireland, the number of Protestants who identified as "Irish" declined precipitously in the decades preceding "The Troubles." In these surveys, Catholics often self-identified as "Irish" while Protestants more typically identified themselves as "British."[4] A shift such as this shows a larger group (residents of Northern Ireland) recategorizing themselves into two mutually exclusive, and in this case hostile, in-groups.

Knowing an individual's identity affiliations can be the key to understanding attitudes and opinions, as individuals tend to adopt opinions compatible with their salient identity groups.[5,6] Identity can help explain the day-to-day behavior of individuals when rituals, mores, practices, or more subtle behavior patterns are associated with identity groups.[7] Understanding the pattern of identities in a population is a key to understanding conflicts, predicting both where conflicts are most likely to occur and how groups are likely to align in a conflict situation.

Stability

Identities vary in their stability, or how much they shift over time. Some groups, such as military units or sports teams, may command a great deal of attention and commitment for a relatively short time. Others, such as religious, ethnic, and social class, may represent lifelong identity components. But even the most seemingly unchanging identity groupings, such as ethnicity, do shift and redefine themselves over time. Ethnic groups often propagate narratives to create the impression of having existed unchanged for thousands of years, but these rarely hold up to historical scrutiny. Ethnic groups can merge with each other when incentives are aligned. For example, the Gungawi group in northern Nigeria, finding itself surrounded by the higher-status Hausa group, began to intermarry (paying a premium bride price to marry a Hausa woman), began to convert to Islam at high rates, and generally experienced an intentional blurring of boundaries; over time, many members

---

[a] Interestingly, people in "collectivist" cultures tend to give more group-oriented answers to this test (I am a sister, I am a member of this mosque), whereas people in individualistic cultures tend to cite personal qualities in their responses (I am a doctor, I am a good basketball player). This is only one of a wide range of differences between cultures being identified in the ongoing study of culture and cognition.

simply became Hausa.[8] Arab tribes have been observed undergoing similar processes, often inventing a shared lineage to reinforce a newly shared identity. And the opposite process, fissioning—driven by drifts in geography, language, or culture —is equally common. Fissioning can also be intentionally brought about by leaders to maximize their own influence over a subgroup. In the 1990s, as The Troubles waned, young, educated, and middle-class people in Northern Ireland, particularly Protestants, increasingly self-identified as "Northern Irish," an identity chosen to avoid the divisiveness of previous labels.[9] Religious, economic, and all other social identity types have similar dynamics.

Salience

Identities also fluctuate widely in their salience. All individuals have multiple possible identities that can become more or less important (salient) depending on the situation. Ethnic and religious identities that are central to life in rural villages often take a back seat to ideological, political, or professional identities when young people leave their hometowns for better jobs in urban areas. Divisions within groups, such as rifts between religious factions, often become more or less salient in different settings; for example, emigrants who are practitioners of distinct religious factions (Sunnis and Shiites, Baptists and Episcopalians) often worship together and downplay differences in countries with a different majority religion.

Conflict

Conflict is often the factor that piques identity salience, as will be discussed below in reference to out-group conflict. Studies of voting and ethnicity in ten African countries show that the strength of peoples' ethnic identification seems to increase in importance during election years, because post-colonial national elections have often been seen as competitions between ethnic groups.[10] Some identities, such as professional identity, may have almost no salience until a conflict arises; but when a land use conflict arises between "farmers" and "herders" or a workplace dispute arises between "workers" and "management," these professional identity groups may coalesce quite quickly.

Conflict can also make identity differences less salient—rival groups may join together against a perceived common enemy. It is not uncommon for a leader facing domestic opposition to exaggerate an external threat or even engage in a foreign war in order to unite the country against such an enemy. Saddam Hussein, who throughout his rule feared opposition from his country's Shiite majority, may have sought out conflict with Iran—which is Shia, but ethnically Persian—to unite his country against the "Persian" enemy.

160

Finally, social identity can be used flexibly within conflicts to activate different narratives or bring differences into relief. An interesting finding from the conflicts in Northern Ireland is that each identity group described the conflict with reference to a different identity category. Protestants in the Northern Ireland conflict often identified themselves with that religious label, which evoked a narrative of victimization at the hands of the Catholic majority in Ireland and Northern Ireland. The Irish Republican Army (IRA), while predominantly Catholic and having many informal ties with the Catholic Church, nevertheless did not usually refer to their in-group as Catholics. The salient identities for this group were the Irish ethnic identity and the Republican political identity. Their opponents were always referred to as British Royalists; the relevant narrative evoked was that of the Irish Republic struggling against the politically dominant British. In this way, both groups' dominant narrative was of a minority group struggling against a numerically superior foe: the Protestants felt persecuted by the Catholic minority in Ireland, while the Irish Republicans claimed to be long-suffering victims of the British, who were militarily and numerically superior in the British Isles. This is sometimes referred to as a "double minority," and parallels can be found in many other settings.

In-Group Formation

How do identity groups form, and what are their typical modes of interaction? Much of the classic research on negative intergroup relations was begun after World War II in an attempt to understand the virulent intergroup hatreds that fed the conflict. Henri Tajfel remains one of the most influential researchers in this area. A Polish Jew, Tajfel left Poland for France because of discrimination in his home country. He fought for the French in World War II and was captured by the Germans, but he survived life as a prisoner of war. After the war he found that most of his family and friends had been killed in the Holocaust and spent a number of years working with Jewish orphans and refugees.

Tajfel later began a program of research to understand how intergroup conflicts arise and was surprised to find how easy it was to induce groups of strangers to form into semiexclusive groups. Tajfel had planned a series of laboratory interventions intended to see what factors would cause such groups to form. What he found was that essentially anything he did to encourage subgroup formation worked, even assigning meaningless and arbitrary labels. This type of experimental manipulation became known as the "Minimal Groups Paradigm,"[11] and the technique became the basis for hundreds of follow-up studies on the effects of group divisions. These cohesive groups became known experimentally as "in-groups."

In-Group Biases

Once they form, for whatever reasons, groups begin to show a typical set of in-group biases. When given the chance to distribute resources, in-group members very quickly begin to favor members over nonmembers.[12] Individuals also begin to trust in-group members more, attribute more positive attributes to in-group members than others, pay more attention to in-group members, and show more empathy to in-group members.[13] These findings about newly formed in-groups are robust but also somewhat benign. When it is pointed out to individuals that they are favoring in-group members unfairly, they usually adjust their allocations to fit their values about fairness. In-group members usually police themselves more closely when, instead of favors, they are asked to allocate obligations or penalties.

In-group biases serve to help bring groups together and can be good for both group functioning and individual self-esteem. Military units, businesses, families, and almost all other organizational types strive to use positive emotions, trust, and biased attention to create cohesive groups. At the level of societies, pride in one's ethnic group, religion, profession, etc., are all generally positive forces.

Out-Group Stereotypes and Discrimination

Over time and in the wrong circumstances, however, positive in-group dynamics can turn into more destructive in-group/out-group hostility. In large, diffuse identity groups, it is sometimes unclear where the boundaries of the group are, so in-groups often begin to define themselves in contrast to one or more out-groups. A political party that brings together a number of interest groups may try to increase internal unity by emphasizing sharp disagreements with selected other groups, e.g., not "liberal" or not "capitalist." Members on the boundaries feel pressure to clarify their identities by choosing one side or the other. (In Northern Ireland, these identity borders correspond closely to geographic borders. Donnan's study of Protestants living near the border with the Republic of Ireland documents how collective memory of conflicts and "atrocities" associated with specific border locations is particularly important for Protestants for maintaining both their geographic and social identity boundaries.[14])

Once out-groups are identified, a number of negative social dynamics can be set in motion. In-group members create and reinforce stereotypes of out-group members. Out-group members tend to be seen as more similar to each other than in-group members, whose individual characteristics are more distinctive.[b] Out-group members are associated with more negative traits than in-group members. The so-called "fundamental attribution error" also affects group perceptions. It is well established that individuals tend to explain the actions of themselves and people close to them as a result of situations and explain actions of others as being the result of permanent personal dispositions. Attributions toward out-group members can show a sinister twist on this bias[15]: when out-group members exhibit negative behaviors, people see them as inherent qualities of the out-group, but the same behaviors are blamed on circumstance for in-group members. The reverse is true for positive behaviors.

Negative stereotypes of out-groups may be spread by opportunistic in-group members for their own purposes. Leaders or would-be leaders of in-groups also have reasons to want to demonize out-groups. By forcing members to choose between being "in" and "out" of a group, leaders increase the cohesion of the group. Identification of a common enemy can cause groups to "pull together" and minimize internal conflicts, increasing the leader's power and at the same time increasing salience of the group for members (salience is discussed below). Peripheral or low-status members of an in-group, who do not feel secure in their membership, are often more negative about out-group members in an attempt to show loyalty and solidify their identity as one of "us" versus "them."

Fault lines can exacerbate in-group/out-group polarization. Fault lines occur when the dividing line between two or more identities overlay each other; the most common example is ethnicity and religion. In Sri Lanka, the Sinhalese language majority was mostly Buddhist while the Tamil language minority was Hindu with some Muslim and Christian elements. This double divide made reconciliation more difficult. In contrast, during the height of the U.S. civil rights conflict, the shared religious background of white and African-American groups was a source of common ground; the Reverend Dr. Martin Luther King, Jr. was a Baptist preacher whose values and rhetoric were easy for white Americans to understand and identify with.

Competition Between Groups

Competition between groups is one of the most important factors that tend to highlight differences between groups and lead to negative in-group/out-group dynamics. Competition also tends to mobilize and energize groups and force on-the-fence individuals to choose sides. In

---

[b] This may be less strong or even reversed in collectivist cultures, where different attitudes toward uniformity and individuality come into play.

a famous field experiment with ten- to twelve-year-old boys at a series of summer camps,[16] in-groups were created by randomly assigning groups and isolating them, but out-group hostility did not occur until a series of athletic competitions. After this point the boys began to call each other names, lead "raids" on each others' barracks, and avoid each other socially.

In real-world political contexts, competition is more likely to be over resources or governance. Control of a nation's central government is the most common source of escalating competition. Other common points of conflict include control of key export resources (such as oil), farmland, and government-appointed jobs. Corrupt governance and lack of free market opportunities can exacerbate this struggle because they increase the importance of government control and reduce the available means of acquiring wealth through the private sector.

MECHANISMS OF GROUP RADICALIZATION

Escalation of In-Group/Out-Group Conflicts

Marginalization and persecution of one or more identity groups is one of the strongest risk factors for violent resistance, especially if this prejudicial behavior is conducted and condoned by the central government.[17]

The escalation of hostility between the Tamils and Sinhalese in Sri Lanka[18] is a prototypical example of escalating in-group/out-group conflicts. The island has a Sinhalese majority across most of the island (approximately 75 percent of population) and Tamil minorities in the north and east (approximately 20 percent). These two identity groups are defined primarily by language, although they also differ by religion. Sinhalese speakers today are mostly Buddhist, while the Tamils, more connected to the Indian mainland, include Hindu, Christian, and Muslim subgroups. Under British colonial rule, the groups were more politically separate than before. The Tamil minority grew disproportionately in political power and were over-represented in the British civil service, creating a pent-up reservoir of resentment within the Sinhalese population.

The groups worked together to achieve independence from Britain in 1948, but soon afterward, the Sinhalese majority began systematically discriminating against non-Sinhalese. Extreme nationalist Sinhalese political parties gained power against both moderate Sinhalese and minority parties in the parliamentary government. In 1956, a coalition of these groups took control of parliament and began demanding that the Sinhalese-Buddhist majority have its "rightful" share of economic and employment benefits.[19] Sinhalese replaced English as the official language, removing one of the Tamil's key cultural advantages. Tamils responded with protests and riots. The nationalist Sinhalese Prime Minister Solomon Bandaranaike was assassinated in 1959 and was replaced by his even more polarizing wife, Sirimavo Bandaranaike, who instituted more blatantly pro-Sinhalese educational and employment policies. Mostly nonviolent protests and international pressure led to another attempt at reconciliation, which broke down by the early 1970s. By this time, the once-privileged Tamils had been almost entirely removed from the civil service and the army, anti-Tamil university quotas were established, and in 1972, anti-Tamil policies were solidified in a new constitution.

Tamils reacted by forming a united political opposition, the Tamil United Front, and a number of militant resistance groups also formed. The 1974 attack by police on the Fourth International Tamil Conference in Jaffna killed eleven Tamils and accelerated the move toward militant resistance. By the late 1970s, violence was escalating; an LTTE attack killed thirteen government soldiers in 1983, prompting anti-Tamil riots that killed an estimated 2,500 Tamils.

One of the most radicalizing events for Tamils was the 1981 burning of the Jaffna Library, which held 90,000 Tamil books and manuscripts and was the most important repository of Tamil culture. This event radicalized many Tamils, even though it did not involve mass fatalities, new government policies, or have much direct economic impact. But the burning of the library had great symbolic value, which is best understood from the perspective of social identity theory. This event convinced many Tamils that their identity itself was under attack and that nothing less than unified, militant resistance could preserve it. The ensuing intergroup conflict was destructive for both sides, leading to decades of loss of life, international sanctions, loss of prestige, and economic and political stagnation.

Radicalization Through Isolation

Group isolation can be a force that leads to radicalization. The groups in this case can range from small peer groups to entire subpopulations of a country. Another well-known social psychological phenomenon can help explain how isolation can lead to radicalization. The phenomenon called "risky shift"[20,21] describes a tendency for groups of like-minded people to, as a group, make riskier decisions than they would individually. Closely related is the phenomenon of ideological polarization in isolated groups. For example, groups of moderately profeminist women tend to be more strongly profeminist after group discussion with like-minded individuals.[22]

Polarization and risky shift have two similar causes. The first is called persuasive argumentation[23] and has to do with hearing a biased set of opinions within a group without any dissenting opinions. Before joining a group, individuals presumably were aware of how their opinions compared with those of society and were exposed to contrary ideas from a variety of perspectives. But when a group of people who are on the same side of an issue communicate more with each other than with the more diverse outside world, each individual's more extreme views are confirmed and reinforced, and the group mean may move toward a point that is more extreme than the original mean. (In American politics, this tendency to listen only to agreeable viewpoints is sometimes referred to as a political "echo chamber.") A second cause of risky shift is the social comparison process,[24] which is similar to the mechanisms of polarization through identity differentiation. Particularly in groups that define themselves ideologically, e.g., Marxists or environmentalists, people do not want to be seen as being on the periphery of the group and do not want to be seen as weak members. For both identity and power-seeking reasons, most would rather be positioned in the center on the vanguard (extreme edge) of such a group. The result over time of all members trying to move toward the vanguard-center is, of course, that the entire group tends to move away from the center, even to extreme positions. The combination of a move toward more extreme views and a tendency to take riskier actions as a group can be a powerful mechanism of radicalization.

Why and how do groups become isolated? There are four types of reasons: circumstantial, self imposed, leader imposed, and externally imposed.

Expatriate communities are an example of circumstantial isolation. Recent immigrants or foreign nationals often have strong associations with other immigrants, living in the same neighborhoods and interacting around religious or social institutions. For first-generation immigrants, language is often part of the barrier between them and the larger society. These enclaves can form isolated groups that may be more vulnerable to radicalization than less isolated groups in the home country. There has been a great deal of attention paid to possible radicalization of immigrant groups from Arab nations living in Europe. Marc Sageman makes the argument that these groups became more isolated after suspicion was cast on them following the 9/11 attacks in the United States and the Madrid subway attacks in Spain;[25] isolation then increased tendencies toward radicalization.

Groups can also self-impose isolation. Religious devotion is a common reason. Seeking greater devotion and piety, groups cut themselves off from corrupting influences of society by severing secular contacts and setting up their own social structures, schools, and economic networks. For groups moving toward a religious or ideological extreme, often the groups nearest them are seen as the greatest threat; they may go to great lengths to separate themselves from "lukewarm" or "secularized" groups. Self-imposed isolation is also a danger sign for individual radicalization.

Isolation may be encouraged or demanded by group leaders. The reasons given for this would be to purify the group ideologically or to avoid exposure to and infiltration by perceived enemies. Isolation also reinforces leadership's power by increasing cohesiveness, increasing social and economic dependence on the group, removing competing influences, and isolating the leader from many potential rivals. The American cult leader Jim Jones used isolation techniques in the extreme when he transported his group to a camp in Guyana that was surrounded by jungle; the isolation was part of a regime of manipulation that turned a group of middle-class Americans into a radicalized group that committed multiple murders and later committed mass suicide.[26] For larger groups, isolation is more likely to be enforced ideologically, through demonization of out-groups, attacks on in-group members accused of collaboration or fraternization, and constant warnings about infiltration by out-group members.

Isolation can also be imposed from the outside. Apartheid in South Africa was a very large-scale example of the isolation imposed by others; in this case, the isolation extended to neighborhoods, professions, and public and educational facilities. Refugee camps are a second isolated environment that is believed to breed resentment and radicalization. Palestinian refugee camps in Jordan and elsewhere are a key recruiting ground for the Palestinian Liberation Organization (PLO) and Ḥarakat al-Muqāwamah al-'Islāmiyyah (Islamic Resistance Movement, or HAMAS). Prison is another isolating environment that has often been shown to breed radicalization. In an interview with a Marine Corps researcher, Sheikh Ahmad Bezia Fteikhan al-Rishawi, who took over leadership of the Iraqi (Anbar) Awakening Party after the assassination of his brother, said that the United States' sending of low-level insurgents to Bucca prison was a mistake: "It was better to have referred those criminals to the Iraqi justice system than to Bucca, because they started having a school, and they indoctrinated other detainees."[27] (Chapter 5 discussed the importance of social reintegration as part of successful demobilization programs.)

Radicalization Under Threat

Uniting against a common enemy is one of the most powerful activators and unifying factors for identity groups, increasing the group's salience to members, decreasing internal dissent, and increasing a sense of unified purpose. External threats are often the catalyst for identity groups to radicalize toward violent action.

Government crackdowns against opposition are often the catalyst for radicalization. Most internal opposition groups attempt to work within the political system first, use nonviolent protest second, and turn to violence as a last resort. Even when group leaders, for their own reasons, might prefer violent struggle from the beginning, it is difficult to recruit enough support among the population if other options seem to be available. But governments often aid the radicalization process by engaging in disproportionately harsh or violent responses to opposition.

As already described, the escalation of aggressive tactics between the Sinhalese majority and Tamil minority in Sri Lanka after independence is a prototypical case of group conflict. However, until the mid-1970s, large-scale armed struggle was not inevitable. Tamils continued to pursue both violent and nonviolent opposition tactics. Political compromises were forged in 1957 (the Bandaranaike–Chelvanayakam Pact related to language rights) and 1965 (the Senanayake–Chelvanayakam agreement), and a coalition government ruled briefly from 1965 to 1968. But increasingly effective attacks by violent groups, including the Tamil Tigers, drew increasingly harsh responses from the government. The "Jaffa massacre" in Sri Lanka seems a clear example of disproportionate government response forcing opposition groups to either radicalize or capitulate. In January 1974, when police attacked the peaceful Fourth International Tamil Conference in Jaffna, killing eleven Tamils, the mood among the Tamils, especially the youth, turned from alienation and protest to overt defiance and militant action. Armed resistance became seen as the only effective avenue by military-age Tamils, leading the way to an increasingly active insurgency.

McCauley and Moskalenko[28] refer to government-fueled resistance as "jujitsu politics," because it is the government's own response that causes the violent opposition that may bring it down. It is not clear that violent repression is always an irrational response, however. The Sinhalese government in Sri Lanka seemed to have made a tactical decision that violent suppression of Tamils was preferable to political accommodation, perhaps gambling that they could win any subsequent armed struggle (which they did, but decades later). Autocratic governments that serve the privileged elite or pursue exclusionary, prejudicial policies may have little hope of attaining support in a fair and open contest for power. The most stable autocracies effectively use secret police, political arrests, and every other available means to crush opposition before it can pose a serious challenge.

Governments that seek the consent of the governed, even partially, must be more cautious about proportionate use of force. The Bloody Sunday attacks in Northern Ireland that catalyzed decades of subsequent conflict would be a clearer example of a jujitsu effect. British paratroopers killed thirteen unarmed protestors during a largely peaceful Catholic demonstration. This was an unintentional provocation on the part of the British. In retrospect, military leaders regretted assigning a unit of paratroopers known for thuggish qualities to crowd control, a mission for which they were completely unsuited. Prior to this, Catholic protest groups had welcomed the presence of British troops as protection against sectarian violence. However, this event turned the Irish Catholics against the British soldiers and energized the violent factions within the IRA, accelerating a decline into The Troubles.[29]

Insurgent groups usually understand the dynamics of jujitsu politics and may manipulate them to their own ends. Opposition leaders wishing to unite and radicalize their followers may undertake attacks for the sole purpose of drawing a disproportionately violent response. Placing a few violent actors within a peaceful crowd is a time-honored tactic: one protestor throws a rock at police from within or near a crowd, fires a shot, or simply creates an explosive noise; police or troops perceive an attack from within the crowd and respond with force; the crowd perceives that a peaceful demonstration has been attacked; and a catalyzing event is created.

Provoking a disproportionate response as a means of radicalizing a movement is not a new tactic. The Brazilian Marxist Carlos Marighella, in his 1969 *Minimanual of the Urban Guerilla*, advocated this tactic as a means of cultivating popular support. Instead of compelling the populace to join an insurgent cause, the insurgents would provoke the state to react violently and indiscriminately, thus driving the populace into the arms of the insurgents.[30]

A study by RAND sought to identify effective counterinsurgency tactics across a number of historical case studies and reduced the findings to a small list of "Good" and "Bad" factors. Among these are a number of recommendations to prevent jujitsu effects by well-intentioned governments. Governments that avoid collateral damage, disproportionate use of force, use of collective punishment, and forced resettlement tactics fare better in the long run than those that do.[31]

Radicalization Through Condensation or Splitting

Another somewhat ironic mechanism of radicalization (from the government's perspective) is radicalization through condensation. The basic mechanism is that a large percentage of a group is suddenly removed. From the outside, the movement appears to be ending, but sometimes what is left is a more committed and isolated core, which subsequently undertakes more radical actions.

Removal of part of the group could happen when there is a government crackdown on a largely peaceful group—a demonstration is attacked, or dissidents are arrested—leading to widespread defection and distancing by less committed members of the group. In this case the ones that remain may be more extreme, more committed, somewhat desperate to regain momentum, and motivated to seek revenge.

Another situation where groups may find themselves isolated and condensed is when a large proportion of a larger organization accepts a political compromise or decides to go mainstream. In 1969, the IRA appeared to be transitioning to a legitimate, and largely peaceful, political organization. They reorganized away from a military structure, and military training was de-emphasized as the group's leadership transformed the Republican institution into a "revolutionary citizen army" capable of economic resistance and direct political action, as well as military operations. More significantly, at the IRA Convention of December 1969, the group leadership voted overwhelmingly 39–12 to begin participating in rather than abstaining from the government in Northern Ireland and also to form a united national liberation front with radical leftist elements, including Marxists. To most observers, this appeared to be a standing down from armed confrontation. However, a splinter group formed, calling itself the Provisional IRA, as differentiated from the "official" IRA. This group was smaller but more committed to resistance. The Provisionals re-established the military brigade and battalion structures. And it was this condensed and radicalized group that grew into the violent wing behind The Troubles of the following decades.[32]

Radicalization in Competition for the Same Base of Support

Insurgent movements often face competition from parallel or splinter groups. One way to gain public support in the face of such competition is to be seen as the most dedicated and the most radical. The Tamil Tigers again provide a good example. As persecution and marginalization of Tamils in Sri Lanka accelerated in the late 1970s and 1980s, a large number of opposition and resistance groups began to form among the Tamils. There were as many as thirty-six militant groups along with other nonviolent resistance groups. Within a few years of their formation in 1976, however, the Tamil Tigers were the dominant militant force. They attracted support from other groups, including less militant groups, because of their effectiveness, dedication, and ruthlessness.

From very early on, the LTTE leadership has demanded that all their personnel should take an oath of loyalty expressing a readiness to die for their cause, even by their own hand. This commitment was embodied in the cyanide vials (kuppi) carried around their necks—a ready instrument for use if captured. This practice, adopted around 1983–1984, immediately garnered admiration among all classes of the Tamil population, giving an edge to the LTTE in the competition with other militant groups for recruits and supporters. Michael Roberts interviewed a Christian Tamil octogenarian in Adelaide about the Tigers, at a time when other groups with stronger Christian elements were still active, but this interviewee stated that the "devotion that the Tigers showed was unmatched."[33]

The Tamil Tigers did not restrict themselves to competing for supporters; they also engaged in direct attacks on competitors, wiping out competition. It is believed that during their rise to power, the Tigers killed more Tamils than Sinhalese.[34] Radicalization for the Tigers also included attacks on influential Tamils holding centrist positions; moderate politicians and policemen were frequent early targets.

The dynamics of radicalization through competition may also apply to the competition between the PLO and HAMAS to be the spokespersons for the Palestinian cause. While the PLO gained political advantage and a number of concessions in the Oslo Accords with Israel, they also lost credibility with radical supports and opened the door for HAMAS. In this case, it is important to note that the groups were not just competing for local support; at least as important was external support of wealthy Muslims using Palestinians as proxies in an ongoing war against Israel. These supporters might see little value in political accommodations of the type negotiated by the PLO in Oslo.

Radicalization Through Hate

Hostility and suspicion between an in-group and one or more out-groups can take a malevolent turn toward out-group hatred. In this dynamic, hatred, suspicion, conspiracy theories, and planned attacks on the out-groups become the focal point of group energy.

Hate fills similar functions as other in-group/out-group dynamics. It provides a common enemy that draws together the in-group,

171

de-emphasizing any within-group rifts, and gives members a sense of belonging. Emphasizing the negative qualities of the out-group and the positive qualities of the in-group creates a sense of pride. And the imminent threat posed by the out-group creates a sense of urgency and purpose.

Scapegoating refers to out-group hatred targeted at a group that is a less politically and socially powerful group than the haters. Scapegoating of Jews in Nazi Germany and hatred of African Americans and foreigners by American white supremacist groups are two examples. To sustain a high level of fear and hostility against less-powerful groups often requires some ideological contortions to magnify their perceived power. These could involve conspiracy theories that claim the scapegoated group secretly has more power than is evident. Or, the scapegoated group could be accused of conspiring with and opening the door for powerful international enemies. A third form accuses the out-group of degrading the purity and culture of the in-group with negative effects disproportionate to the actual social or political power of the out-group.

Is there really a difference between the dynamics of hatred and more common out-group stereotyping and hostility? The differences are partly a matter of emphasis and degree. But there are also some characteristic patterns of groups moving toward hatred. The first pattern has to do with group rhetoric and dehumanization of hated groups as a way of psychologically preparing for more extreme action. The second is a pattern of observable activities that indicate a group is preparing behaviorally and tactically for violence.

Rhetoric of Hate: Dehumanization and "Selective Moral Disengagement"

Primo Levi asked a Nazi camp commandant why they went to extreme lengths to degrade their victims, whom they were going to kill anyway. The commandant chillingly explained that it was not a matter of purposeless cruelty. The victims had to be degraded to subhuman objects so that those who operated the gas chambers would be less burdened by distress.[35]

Dehumanization refers to attempts, through rhetoric or action, to portray a person or group in a way that strips them of their essential human qualities, thus justifying some kind of subhuman treatment. One of the classic experiments in this area was published by Albert Bandura and colleagues in 1975[36] and showed that small changes could lead to large differences in aggressive behavior. College-age males were assigned to be "supervisors" or "workers." Three-member teams of supervisors were supposed to deliver electric shocks to unseen three-member work teams when their work was judged inadequate. The "supervisors" were told when they had to deliver shocks but could freely choose the level of shock, from one to ten. Bandura varied two conditions related to actions of the supervisors and three related to descriptions of the workers (what experimentalists would call a "two-by-three" design). Supervisors were sometimes told they were delivering shocks directly to an (unseen) individual worker, and they were sometimes told that the shock levels assigned by the three supervisors were averaged. When shocks were averaged and therefore less personal (a means of dehumanization) in the second condition, supervisors tended to assign higher levels of shock. And supervisors' level of punitive shocks was also significantly affected by three levels of labeling. Subjects "accidentally" overheard comments from the administrators describing the worker group in neutral terms, dehumanizing terms ("animalistic"), or humanizing terms ("perceptive," "understanding"). Dehumanizing descriptions led to significantly higher punitive shock levels and humanizing to lower shock levels, as compared with the neutral condition. When the two "dehumanizing" conditions both occurred together, i.e., depersonalized administration and dehumanizing descriptions, the shocks delivered were particularly large, showing a statistical interaction.

Hate groups also employ these same tactics toward the targets of their enmity. Haslam's studies of dehumanization[37,38] showed that the characteristics we consider most uniquely human are the ones that recur in dehumanizing speech. He identified two distinct types: comparison to animals and comparison to machines.

Targets of hate are frequently compared to animals (dogs, rats) or assigned animal-like qualities. Targets are accused of lack of culture, coarseness, amorality, irrationality, and immaturity and denied uniquely human characteristics of civility and moral sensibilities. When the opportunity exists, targets are sometimes made to perform animal-like acts. Among the mistreatments perpetrated by American soldiers at Abu Ghraib prison in Iraq was forcing prisoners to wear dog collars and crawl on the ground.

Targets of hatred may also be compared to machines. Hate groups create enemy stereotypes of coldness, lack of emotion, and rigidity. This is a quite different set of characteristics than the animal comparisons: cold, rational, computer-like beings on the one hand and irrational beings controlled by passions on the other. But both serve the same purpose: to deny intended victims some qualities we consider most characteristically human.

Depersonalization is also an important component of dehumanization. Hate rhetoric avoids assigning individual qualities to most enemies and makes use of broad stereotypes.

Later in his career, Bandura referred to the process of preparing to commit violence against hate victims as "selective moral disengagement" and describes a number of other typical communication patterns, including sanitizing language, distancing of personal responsibility for actions, minimizing or disregarding harm to be committed, and blaming victims for bringing punishment on themselves.[39]

Why do hate groups need to engage in dehumanization? This process has both individual and group functions. Most individuals do not easily take to violent actions, as discussed previously. The crueler the acts, the more preparation and justification is required. People have individual standards of morality and a need to maintain a positive self-image by upholding those standards. Some individuals will find violence easier to acclimate to than others, and this affects what positions individuals will be assigned to within a group, such as whether they will be judged most fit for violent operations or underground support. But to convince mass movements, including many different kinds of people, to commit cruel acts, or to look the other way when they are committed, is a more difficult feat. Constant repetition of dehumanizing speech acclimates new members and reinforces messages.

Hate rhetoric can be a product of group radicalization, but it can also be produced by one or a few strong leaders driving the process for their own ends. Edward Glaeser, a Harvard economist, attempted to create an economic model of supply and demand for hatred as a political tool. He argues that ambitious and unscrupulous leaders will use hatred as much as they are allowed to, because when such leaders are out of power it helps unify opposition, and when they are in power it justifies economic exploitation of other groups to the benefit of constituents. The limits on hateful rhetoric may be mainly a function of the strength of various groups, how much a population will tolerate, and how much humanizing contact there is between groups, but have little to do with grievances: "the fact that demagogues form hatred by telling tales of past crimes shouldn't fool us into thinking that the level of hatred is actually a function of past injustice."[40]

Escalation of Violent Actions

Besides moral justification, groups also escalate their violent actions over time as they individually and collectively desensitize themselves to violence. Many law enforcement organizations have employed a staged approach to assess this mechanism. The Seven-Stage Hate Model[41] comprises the following:

- Stage 1: The Haters Gather. Like-minded individuals find each other and recruit others who may or may not initially share their ideology.

174

- Stage 2: The Hate Group Defines Itself. Individuals see themselves as members of a defined group; symbols and rituals are often chosen at this point.

- Stage 3: The Hate Group Disparages the Target. Groups define themselves with a narrative of grievances and conspiracies and develop an ideology of hatred.

- Stage 4: The Hate Group Taunts the Target. The group's ideology is revealed to outsiders. Graffiti, use of symbols, and insults shouted from cars are typical.

- Stage 5: The Hate Group Attacks the Target Without Weapons. Schafer and Navarro tie these types of attacks to "thrill-seeking" behavior; attacks tend to be more violent than comparable crimes. Defending turf is a typical rationale for gang-type groups.

- Stage 6: The Hate Group Attacks the Target with Weapons. Weapons can range from tools and belts to more lethal weapons. Some members may prefer up-close violence (knives as opposed to guns or explosives) for the personal and thrill-seeking aspects.

- Stage 7: The Hate Group Destroys the Target. Groups move to more lethal weapons with larger effect.

The personal story of Lars, a former member of Norway's right-wing extremist National People's Party, illustrates the process of one person working through these stages. His story is recounted in John Horgan's book, *Walking Away from Terrorism*.[42] Lars, who would later be imprisoned for bombing a mosque, began his involvement with no particular malevolence toward Muslims or other groups targeted by the National People's Party. Instead, he began as a lonely and disconnected teenager in Oslo. He contacted the party after seeing a bumper sticker and was unaware of their political views; he was invited to gather with other young people at a social event (Stage 1). "It was very informal, not even really a meeting . . . we just met, barbecued, and drank beer. And we had a very good time." After becoming involved socially, he became exposed to the group ideology of hatred. The ideology at this point seemed to serve as a marker of identity (Stage 2) more than a matter of political urgency. Gradually, he began to take part in small acts of aggression against opportunistic targets, such as small shops owned by Pakistani and Indian immigrants. Taunting (Stage 4) took the form of surveillance and putting glue in the door locks at night. Smashing shop windows was another small escalation (Stage 5). Only after these lines were crossed was Lars encouraged, indirectly, to make an attack with some dynamite stolen from a construction site (Stage 6). In a tragi-comic turn, Lars was supposed to attack a synagogue, but he got on the wrong bus and decided to bomb a conveniently located mosque instead. He intended to commit only property damage in this attack, but several people were injured. Presumably Lars might have escalated to more lethal attacks had he not been arrested and undergone an ideological transformation in prison.

Radicalization Through Martyrdom

Martyrdom seems to have a power to inspire radicalism in a way that other grievances do not. The first death, or a widely publicized death related to a cause, can be a psychologically jarring experience with several effects on sympathizers. First, it elevates the seriousness and urgency of the cause. Sympathizers to a cause who had previously held back may be both shamed and emboldened; the threats of arrest, persecution, economic loss, etc., seem comparatively less important. A psychological mechanism of "survivor guilt" may help create strong feelings of obligation. Supporters also often say they are inspired to ensure that martyrs' deaths be made meaningful.

Even when a number of victims have died for a cause, a single martyr is often symbolically elevated to personalize the phenomena. Many protestors were killed in the 2009 Iranian election protests, but Neda Agha-Soltan's became the face of the protests, largely because good-quality video of the entire sequence of events was serendipitously taken and then made available on the Internet. Neda was an attractive young woman and a peripheral participant in the protest. Video showed her being shot in the back without provocation,[43] making it difficult for government supporters to claim that her killing was justified for security reasons.

Martyrs do not have to be killed by the government or opposition forces to serve the purpose. Bobby Sands's death galvanized IRA support even though his death was caused by a hunger strike. The 2010–2011 protests in Tunisia were touched off by the suicide of Mohamed Bouazizi, a street vendor who lit himself on fire in protest after police confiscated his fruit and vegetable cart. In cases of martyr suicide, the means of death is likely important; death by self-starvation or public self-immolation conveys a message of protest and emboldens others to take actions that might lead to less painful ends.

Leaders' responses to martyrdom may also be viewed as a critical test of character by their followers. Respect for Hassan Nasrallah and the strength of his leadership in Hizbollah increased greatly because of his continued resolve after the death of his son.

For the LTTE and other groups such as Hizbollah and HAMAS, suicide martyrdom became an integral part of the group's ideological and motivational structures. The transition from offensive military or terrorist operations to suicide attacks represents an important psychological escalation both for insurgent groups and their opponents.

The LTTE's first suicide operation was in 1987, when an explosive-laden vehicle was used to spearhead an assault on an army camp. This tactic was similar to the attack by Hizbollah against the U.S. Marine barracks in Lebanon four years earlier. LTTE began utilizing both land and naval missions very effectively. Soon thereafter, suicide attackers made up their own branch of the LTTE, the "Black Tigers." This group was believed to be the only one directly overseen by the Tigers' charismatic leader Velupillai Prabhakaran.[44] There were noteworthy rituals surrounding LTTE suicide attacks. Before their mission, Black Tigers were given the rare honor of sharing a last meal and having a photograph taken with the seldom-seen Prabhakaran.

It was a female Black Tiger that assassinated Rajiv Gandhi, India's former prime minister, in 1991. That same year, the LTTE began commemorating martyrs on November 27, a holiday called Maaveerar Naal, "Heros' Day." The most significant ritual of this day was a candle-lighting ceremony at a site called tuyilam illam, where martyrs were commemorated. The site featured stone gravestones and caskets. The gravestones themselves were unusual because Hindus burn rather than bury their dead; tuyilam illam was considered a temple rather than a cemetery. (The site has since been destroyed by the Sri Lankan government.)

SUMMARY

An understanding of the social psychology of groups is useful for understanding the actions of resistance movements. Of particular interest is the process of radicalization, whereby some groups make the transition from nonviolent to violent tactics.

Humans have a strong tendency to associate with identity groups, be they teams, organizations, political parties, or religions. Most people have a set of social identities that are important to their understanding of self; these identities can wax and wane in importance over time and depending on circumstances. While mostly a positive phenomenon, identity groups are also at the root of in-group/out-group conflicts. This chapter uses the extended example of the conflict between the ethnic Sinhalese and ethnic Tamil groups in Sri Lanka as an example of how in-group/out-group distinctions can breed animosity and lead to violence. The long, violent insurgency of the Tamil Tigers was perhaps a predictable outcome of ethnic tensions played out through government persecution after the end of British colonial rule.

Drawing on work by McCauley and Moskalenko, we identify a number of mechanisms by which groups may be drawn or forced to cross the line between nonviolent and violent resistance. Some of these involve reactions to government crackdowns, such as radicalization under threat. Condensation and splitting are particularly interesting phenomena, because they occur when movements may appear to at their weakest—splitting involves a schism within a group, and condensation happens when groups lose large numbers of members. In both cases, the cleavages leave the most radical members isolated together, and group dynamics or desperation can lead the remnants to turn to violence, in some cases very effectively. Martyrdom, from either external attack or self-imposed as in a hunger strike, can also be a catalyst to radicalization.

Other mechanisms of radicalization are less a response to external pressures and more a result of internal dynamics, such as competition between groups for members or competition between leaders for influence within the group. Understanding the stages that hate groups go through to dehumanize their targets and desensitize their members to violence is particularly important because these often occur without outside pressure or apparent cause.

ENDNOTES

[1] Clark McCauley and Sophia Moskalenko, "Mechanisms of Political Radicalization: Pathways Toward Terrorism," *Terrorism and Political Violence* 20, no. 3 (2008): 415–433.

[2] Manford H. Kuhn and Thomas S. McPartland, "An Empirical Investigation on Self-Attitudes," *American Sociological Review* 19 (1954): 68–77.

[3] Hazel R. Markus and Shinobu Kitayama, "Culture and the Self: Implications for Cognition, Emotion, and Motivation," *Psychological Review* 98, no. 2 (1991): 224–253.

[4] Peter Bull, "Shifting Patterns of Social Identity in Northern Ireland," *The Psychologist* 19, no. 1 (2006): 40. Reporting on a series of British Social Attitude surveys from 1989 to 1995.

[5] S. Alexander Haslam and John C. Turner, "Context-Dependent Variation in Social Stereotyping 2: The Relationship between Frame of Reference, Self-Categorization and Accentuation," *European Journal of Social Psychology* 22, no. 3 (1992): 251–277.

[6] John C. Turner, Michael Hogg, Penelope J. Oakes, Stephen D. Reicher, and Margaret S. Wetherell, *Rediscovering the Social Group: A Self-Categorization Theory* (Oxford: Basil Blackwell, 1987).

[7] Rawi Abdelal, Yoshiko M. Herrera, Alastair Iain Johnston, and Rose McDermott, "Identity as a Variable," *Perspectives on Politics* 4, no. 4 (2006): 695–711.

[8] Frank A. Salamone, "Becoming Hausa: Ethnic Identity Change and Its Implications for the Study of Ethnic Pluralism and Stratification," *Africa* 45, no. 4 (1975): 410.

[9] Peter Bull, "Shifting Patterns of Social Identity," *The Psychologist* 19, no. 4 (2006): 40–43.

[10] Benn Eifert, Edward Miguel, and Daniel N. Posner, "Political Competition and Ethnic Identification in Africa," *American Journal of Political Science* 54, no. 2 (April 2010): 494–510.

[11] Henri Tajfel, M. G. Billig, R. P. Bundy, and Claude Flament, "Social Categorization and Intergroup Behaviour," *European Journal of Social Psychology* 1, no. 2 (1971): 149–178.

[12] Marilynn B. Brewer and Roderick M. Kramer, "The Psychology of Intergroup Attitudes and Behavior," *Annual Review of Psychology* 36, no. 1 (1985): 219–243.

[13] Miles Hewstone, Mark Rubin, and Hazel Willis, "Intergroup Bias," *Annual Review of Psychology* 53, no. 1 (2002): 575–604.

[14] Hastings Donnan, "Material Identities: Fixing Ethnicity in the Irish Borderlands," *Identities* 12, no. 1 (2005): 69–105.

[15] Thomas F. Pettigrew, "The Ultimate Attribution Error: Extending Allport's Cognitive Analysis of Prejudice," *Personality and Social Psychology Bulletin* 5, no. 4 (1979): 461–476.

[16] Muzafer Sherif, O. J Harvey, B. Jack White, William R. Hood, and Carolyn W. Sherif, *Intergroup Conflict and Cooperation: The Robber's Cave Experiment* (Norman, OK: University of Oklahoma Book Exchange, 1961).

[17] Jack A. Goldstone, Robert H. Bates, David L. Epstein, Ted Robert Gurr, Michael B. Lustik, Monty G. Marshall, Jay Ulfelder, and Mark Woodward, "A Global Model for Forecasting Political Instability," *American Journal of Political Science* 54, no. 1 (January 2010): 190–208.

[18] Michael J. Deane and Maegen Nix, "Liberation Tigers of Tamil Eelam (LTTE)," in *Assessing Revolutionary and Insurgent Strategies* (Laurel, MD: The Johns Hopkins University Applied Physics Laboratory, 2009), 52–53.

[19] Howard Wriggins, "Ceylon's Time of Troubles, 1956–1958," *Far Eastern Survey* 28, no. 3 (1959): 33.

[20] James A. F. Stoner, "Risky and Cautious Shifts in Group Decisions: The Influence of Widely Held Values," *Journal of Experimental Social Psychology* 4, no. 4 (1968): 442–459.

[21] Daniel J. Isenberg, "Group Polarization: A Critical Review and Meta-Analysis," *Journal of Personality and Social Psychology* 50, no. 6 (1986): 1141–1151.

[22] David G. Myers and Helmut Lamm, "The Polarizing Effect of Group Discussion," *American Scientist* 63, no. 3 (1975): 297–303.

[23] Eugene Burnstein and Amiram Vinokur, "Persuasive Argumentation and Social Comparison as Determinants of Attitude Polarization," *Journal of Experimental Social Psychology* 13, no. 4 (1977): 315–322.

[24] Glenn S. Sanders and Robert S. Baron, "Is Social Comparison Irrelevant for Producing Choice Shifts?" *Journal of Experimental Social Psychology* 13, no. 4 (1977): 303–331.

[25] Marc Sageman, *Understanding Terror Networks* (Philadelphia: University of Pennsylvania Press, 2004).

[26] Neal Osherow, "Making Sense of the Nonsensical: An Analysis of Jonestown," in *Readings about the Social Animal*, 9th ed., ed. Elliot Aaronson and Joshua Aaronson (New York: Worth Publishers, 2004).

[27] Gary W. Montgomery and Timothy S. McWilliams, *Al-Anbar Awakening: Volume II, Iraq Perspectives. U.S. Marines and Counterinsurgency in Iraq, 2004–2009* (U.S. Marine Corps: 2009).

[28] Clark McCauley and Sophia Moskalenko, "Mechanisms of Political Radicalization: Pathways Toward Terrorism," *Terrorism and Political Violence* 20, no. 3 (2008): 415–433.

[29] Summer Newton, "The Provisional Irish Republican Army: 1969–1998," in *Assessing Revolutionary and Insurgent Strategies* (Laurel, MD: The Johns Hopkins University Applied Physics Laboratory, 2009), 64–66.

[30] Carlos Marighella, *Minimanual of the Urban Guerilla*, trans. John Butt and Rosemary Sheed (Havana, Cuba: Tricontinental, 1970).

[31] Christopher Paul, Colin P. Clarke, and Beth Grill, *Victory has a Thousand Fathers: Sources of Success in Counterinsurgency* (Santa Monica, CA: RAND Corporation, 2010).

[32] Newton, "The Provisional Irish Republican Army: 1969–1998."

33 Michael Roberts, "Pragmatic Action and Enchanted Worlds: A Black Tiger Rite of Commemoration," *Social Analysis* 50, no. 1 (2006): 73–102.

34 Rajan Hoole, Daya Bomasundaram, and K. Thiranagama Sritharan Rajani, *The Broken Palmyra: The Tamil Crisis in Sri Lanka, an Inside Account* (Claremont, CA: The Sri Lankan Studies Institute, 1990).

35 Albert Bandura, "Selective Moral Disengagement in the Exercise of Moral Agency," *Journal of Moral Education* 31, no. 2 (2002): 101–119.

36 Albert Bandura, Bill Underwood, and Michael E. Fromson, "Disinhibition of Aggression through Diffusion of Responsibility and Dehumanization of Victims," *Journal of Research in Personality* 9, no. 4 (1975): 253–269.

37 Nick Haslam et al., "Subhuman, Inhuman, and Superhuman: Contrasting Humans with Nonhumans in Three Cultures," *Social Cognition* 26, no. 2 (2008): 248–259.

38 Nick Haslam, "Dehumanization: An Integrative Review," *Personality and Social Psychology Review* 10, no. 3 (2006): 252–264.

39 Albert Bandura, "Selective Moral Disengagement in the Exercise of Moral Agency," *Journal of Moral Education* 31, no. 2 (2002): 101–119.

40 Edward L. Glaeser, "The Political Economy of Hatred," *Quarterly Journal of Economics* 120, no. 1 (2005): 45–86.

41 John R. Schafer and Joe Navarro, "The Seven-Stage Hate Model: The Psychopathology of Hate Groups," *FBI Law Enforcement Bulletin* 72, no. 3 (March 2003): 1–9.

42 John Horgan, *Walking Away from Terrorism: Accounts of Disengagement from Radical and Extremist Movements* (London and New York: Routledge, 2009).

43 Nazila Fathi, "In a Death Seen Around the World, a Symbol of Iranian Protests," *The New York Times*, June 22, 2009.

44 Chris Smith, "Sri Lanka: The Continued Armed Struggle of the LTTE," in *Transforming Rebel Movements After Civil War*, ed. Jeroen de Zeeuw (Boulder, CO: Lynne Rienner, 2007).

CHAPTER 7.

PSYCHOLOGICAL RISK FACTORS

CHAPTER CONTENTS

Jason Spitaletta

INTRODUCTION:
IS THE DISEASE MODEL APPLICABLE?

Critics of the field of psychology identify the disproportionate focus on the abnormal (or behavioral extremes) at the expense of the average. Accounts of terrorist acts in the media tend to do the same because commentators often describe abhorrent acts in clinical terms. One needn't be mentally ill to commit atrocities; in fact, the vast majority of terrorists and/or insurgents are psychologically normal (e.g., not clinically psychotic). Although there are certainly exceptions, it is reasonable to assume that terrorists and/or insurgents do not necessarily possess any cognitive, affective, or neurological deficits, or at least that those deficits are not precipitating causes of their behavior. There is no experimentally derived or empirically based psychological or demographic profile that would indicate a predisposition toward joining violent extremist organizations. What follows are a proposed set of risk factors that seem to apply to individual and group radicalization. A risk factor is a variable associated with an increased risk of radicalization whose presence is neither deterministic nor implicitly characterological. The behaviors or attributes described below merely point to a possible increased likelihood of a willingness to participate in or eventual participation in political violence.

Social-science research on the underlying causes of terrorism have focused on three main areas: (1) the political, economic, and social conditions that correlate with increased incidences of politically motivated violence, (2) group dynamic processes that facilitate radicalization and an increased risk of violence, and (3) psychological traits and characteristics of group members that predispose individuals to seeking membership in violent organizations.[1] Although explanations at the level of individual psychology are insufficient, the incorporation of multiple subdisciplines within psychology does help one comprehend the phenomena associated with radicalization. Many consider radicalized individuals rational in that their actions (although violent) are derived from a conscious, calculated decision to take this particular type of action because they believe it is the optimum strategy to accomplish a sociopolitical goal.[2]

The violent acts perpetrated by terrorists, against civilian targets including women, children, and countrymen, or against themselves in the case of suicide bombers, beg the question of whether terrorists are mentally ill. The balance of research indicates that this is not the case, in the sense of being legally insane or socially dysfunctional in normal settings. One of the best datasets on mental illness and terrorism comes from Marc Sageman's study of Al Qaeda members. In a dataset of 398 profiles, he identified only four cases of possible thought disorder, one case of mild cognitive impairment (what used to be considered mental retardation), and no indications of pathological narcissism or excessive egocentricity. There were very few indications of traumatic histories. Sageman considered many to be overprotected youth who became well-adjusted, psychologically healthy adults.[3] To date, no terrorist profile has been discovered,[4] and there is no scientific evidence of any genetic role in determining why certain people become involved in terrorism. In general, radical organizations probably have a sufficient range of personality and cognitive profiles within their ranks to be indistinguishable from the surrounding population.

Severely mentally ill people usually have difficulty fitting in within teams and larger organizations, and this is true whether the organization is a corporation or a component of an insurgency. Effective underground groups weed out unfit volunteers (see Chapter 2: Recruiting in *Undergrounds in Insurgent, Revolutionary, and Resistance Warfare*), because they pose a risk to both the effectiveness and security of the organizations.

Having said this, an understanding of mental health issues can shed some light on insurgent psychology. It will be useful to understand the distinction between "Axis I" and "Axis II" disorders.[a] This distinction comes from the *Diagnostic and Statistical Manual of Mental Disorders, Fourth Edition* (DSM-IV),[5] which is the "bible" of clinical psychiatric diagnosis. Axis I refers to the major clinical illnesses, such as schizophrenia or major depression, that are most familiar to laypersons. Axis II refers to developmental or personality disorders that can be subtler and less likely to be debilitating but still have profound effects on behavior.[6]

AXIS I DISORDERS AND "LONE WOLF" TERRORISTS

There are some isolated examples of insurgents or terrorists who do exhibit symptoms of Axis I disorder that indicate profound thought disorders and distortions of reality. These tend to be lone wolf terrorists, partly because of the difficulty they have integrating into larger groups, and partly because of the unique features of their pathology. Anders Behring Breivik used a diversionary bombing to kill seventy-seven people in a combined attack in Norway in 2011. He then proceeded to personally massacre sixty-nine students attending a youth leadership conference on an isolated island. His reasons for his actions appear to be senseless to most, but in his mind, they were a necessary step to combat foreign influences in Norway, according to his detailed

---

[a] The DSM-IV-TR uses a multiaxial or multidimensional approach to diagnoses because rarely do other factors in a person's life not affect their mental health. Axis I (Clinical Syndromes) are those psychological disorders that are the focus of a diagnosis. Axis I disorders are divided into fourteen categories, including Anxiety Disorders, Childhood Disorders, Cognitive Disorders, Dissociative Disorders, Eating Disorders, Factitious Disorders, Impulse-Control Disorders, Mood Disorders, Psychotic Disorders, Sexual and Gender-Identity Disorders, Sleep Disorders, Somatoform Disorders, and Substance-Related Disorders. Axis II (Developmental Disorders and Personality Disorders) are long-standing chronic conditions that may affect the clinical syndromes listed in Axis I. Developmental disorders include autism and mental retardation, disorders that are typically first evident in childhood. Personality disorders are clinical syndromes that have enduring symptoms and encompass the individual's way of interacting with the world. They are divided into three clusters: Cluster A (odd or eccentric) includes Paranoid, Schizoid, and Schizotypal. Cluster B (overly emotional, unstable, or self-dramatizing) includes Antisocial, Borderline, Histrionic, and Narcissistic. Cluster C (tense and anxiety ridden) includes Avoidant, Dependent, and Obsessive-Compulsive. Axis III disorders are medical/physical conditions that play a role in the development, continuance, or exacerbation of Axis I and II disorders or other physical conditions such as brain injury or HIV/AIDS that can result in symptoms of mental illness. Axis IV are social and environmental stressors that may affect the clinical syndromes listed in Axis I. Events in a person's life, such as death of a loved one, starting a new job, college, unemployment, and even marriage can affect the disorders listed in Axis I and II. These events are both listed and rated for this axis. Axis V represents the highest level of functioning where a clinical rates the person's level of functioning at both the present time and the highest level within the previous year. This helps the clinician understand how the above four axes are affecting the person and what type of changes could be expected.

manifestos and journals. An evaluation panel later declared that Breivik was insane at the time of his crime. The concept of "insanity" is a legal term, not a clinical diagnosis from the DSM-IV, but in Norway, as elsewhere, it usually implies psychosis or some psychological break from reality.

Typical characteristics of these lone wolf attackers are delusions of grandeur and some associated narcissistic tendencies. In this presentation of paranoid schizophrenia with the presence of thought disorder, individuals may believe the government or other powerful groups are "out to get them." The personal identification with the adversary (e.g., the government is out to get "me") is a delusion of grandeur that can be considered an unconscious attempt to elevate self-esteem. Presumably, if a powerful group has a specific grievance with a single individual, that individual must be important. The subsequent elevation in self-esteem and self-importance may perpetuate the disordered thinking.

Self-efficacy (the belief that one is capable of goal-directed behavior to affect the operational environment) and self-agency (the sense that one is the primary creator and principal driver of a particular thought or movement) are typically high in individuals who act alone to remediate political grievances. Self-efficacy and self-agency are often correlated with narcissistic tendencies. However, their presence does not necessarily indicate the presence of the personality disorder or the associated leadership style.[b] There is a greater probability of some degree of psychopathology in this category than in any other,[7] although the psychopathology is rarely the proximal factor in the transformation from law-abiding citizen to violent actor. Rather, it is a combination of some underlying psychopathology (major or minor) along with specific environmental conditions that propel the individual toward radical behavior. In certain cases, such as paranoid schizophrenia with underlying thought disorder, the violence is a result of disordered cognition and a break from reality and not necessarily truly politically motivated.

AXIS II: DEVELOPMENTAL AND PERSONALITY DISORDERS

While the most serious psychiatric disorders may be rare, there are associated personality and other disorders that may correspond to violent radicalism. Most of the literature attributing clinical mental disorder to radicalism speaks of the remorseless personality type, psychopathy, or sociopathy.[8] The concept of "insanity" is often associated with terroristic behavior. However, "insanity" is not a behavioral science

---

[b] Refer to Chapter 4 of this book for a discussion of leadership styles.

term but a legal one, usually implying psychosis, although no official legal definition or objective criteria exist.[9] Common theories hold that certain individuals possess or lack certain personality traits that make them more susceptible to radicalization and performance of terrorist acts than those individuals who do not. One theory is that this type of personality is largely the result of a dysfunctional childhood that fosters an impoverished sense of self and hostility toward authority. This resentment to authority may be an outgrowth of unconscious hostility toward abusive or controlling parents and is later reflected in the adult terrorist's rigid mindset.[10]

Antisocial personality disorder (APD) is considered an Axis II disorder and is the current term used to describe a pattern of remorseless disregard for the rights of others.[11] First referred to as "moral insanity" in 1835,[12] then psychopathy up until the mid-1950s, and then sociopathy until 1980,[13] APD is hypothesized to occur in approximately 3 percent of the male population, is characterized by chronic disregard of social norms and laws, lack of remorse, impulsivity, and other traits and seems a plausible explanation for some terroristic behavior.[14] A subset of violent extremists would meet criteria for a diagnosis of APD,[c] although many others would probably exhibit traits associated with APD without meeting full diagnostic criteria. Many individuals with APD share certain characteristics with violent radicals, such as a sense of social alienation, early maladjustment, impulsivity, and hostility.[15] There are also those who exhibit an antisocial or sociopathic leadership style and probably do not meet the clinical criteria for the Axis II disorder.[16] The characteristics of a sociopathic leadership style include lack of empathy, absence of moral constraints, and the consideration of violence as a tool to accomplish a goal.[17] Observables of a sociopathic leadership style include a history of criminal activity not motivated by politics and the projection of personal desire for violent action.[18] (Psychological profiling of leaders is discussed in more detail in *Chapter 4: Leadership.*)

Another assessment used in conjunction with a clinical interview and case history evaluation is Hare's PCL-R.[d] The PCL-R consists of two factors: Factor 1, a personality marked by "aggressive narcissism"; and Factor 2, a case history of a "socially deviant lifestyle." Factor 1 includes a history or current evidence of glibness/superficial charm,

---

[c] Based on a comparison between Napoleoni's 2005 open-source reporting of biographical data on Abu Musab Al-Zarqawi and Hare's psychopathy checklist-revised (PCL-R), it can be determined that Al-Zarqawi would most likely score between thirty and forty, thus reaching Hare's criteria for APD.

[d] The PCL-R is composed of the twenty aforementioned elements scored along a three-point Likert scale (0, 1, or 2). A score of thirty to forty indicates psychopathy; however, the PCL-R must be administered by a fully trained and credentialed clinician in an appropriate environment for the assessment to be considered valid.

grandiose sense of self-worth, pathological lying, cunning/manipulativeness, lack of remorse or guilt, shallow affect, callousness/lack of empathy, and failure to accept responsibility for one's own actions. Factor 2 includes a history or current evidence of need for stimulation/proneness to boredom, parasitic lifestyle, poor behavioral control, promiscuous sexual behavior, lack of realistic long-term goals, impulsivity, irresponsibility, juvenile delinquency, early behavior problems, and revocation of conditional release. Two additional traits not included in either of the factors are many short-term marital relationships and criminal versatility.[19] The behavioral profile of the psychopath suggests a disposition wholly unsuitable for the stressors of the underground. Psychopaths are often unstable, unreliable, unconcerned with group objectives, and have difficulty maintaining goal-directed behavior. For these reasons, they often fail to meet the selection criteria (if existent) for the group or are employed as a component of the auxiliary when and where required.

These personality and behavioral disorders are also associated with ordinary criminality. There is reason to believe that typical insurgents are healthier than typical criminals. A 1986 study[20] examined 106 persons charged with murder in Northern Ireland between 1974 and 1984 who were referred for psychiatric assessment. Of the political murders (those affiliated with a political organization), only 16 percent met the clinical criteria for mental illness (compared with 58 percent of the nonpolitical murderers), 6 percent (compared with 34 percent) met the criteria for diminished capacity, and 13 percent (compared with 64 percent) met the criteria for mitigation of sentence resulting from their mental state.[c,21] The study illustrates the prevailing characteristics of murderers in general: they are predominantly male, they are usually in their 20s to 30s, and approximately one-third to one-half have prior criminal records.[22] The finding most pertinent in studying insurgencies, however, was the relative stability of the political murderers as compared with nonpolitical murderers. The political murderers were also far less likely to be intoxicated (or under the influence of narcotics) at the time of the offense or to have prior criminal records.

---

[c] Criteria for a ruling of not guilty by reason of insanity (NGRI) vary between and within nations. However, some of those criteria include both loss of control and impaired reasoning and thus meet the criteria for both insanity (his or her free will is undermined by a mental illness and is thus exonerated) and diminished capacity (his or her free will is affected, but not completely undermined, but he or she nevertheless acts with less than usual moral blameworthiness). The M'Naghten test requires an actor to be ignorant of the nature and quality of an act and/or must not know that the act is legally or morally wrong. The Model Penal Code of the American Law Institute (ALI) test incorporates both the cognitive element of M'Naghten and the volitional element of the irresistible impulse test and requires the lack of substantial capacity to appreciate the criminality or wrongfulness of one's conduct or to control one's impulses.

The Lyons and Harbinson study is one of the few examples in the literature of a direct comparison between those who commit violence with a political goal (e.g., consistent with the objectives of an insurgent group) and those who simply commit a violent act.

SUICIDALITY AND SUICIDE BOMBERS

The phenomenon of suicide attacks also naturally leads to questions about mental health. Are suicide attackers similar to people who commit suicide for other reasons? Are there other personality characteristics or pathologies associated with this phenomenon?

Suicide attacks are among the more historically, socially, and psychologically compelling tactics used in support of a political objective. The willingness to consciously act in a manner that will logically lead to one's death is considered individually illogical and operationally unexpected. Accounts of such heroism from Thermopylae to Bastogne to Khe Sahn have provided validity to cohesion narratives within many cultures. That same romanticism has been employed by modern nonstate actors to vindicate the few who stood against the many or liberate the oppressed. While the admiration for such individuals (regardless of one's agreement with their political or moral ideals) is well deserved, the underlying motivation of the behavior remains somewhat of a mystery. Analogous to clinical research on suicidality, access to the actor is prohibited by the act itself and thus theories require extrapolation from the antecedent data. The resultant behavior has proven difficult to deter tactically and contend with politically. Suicide attacks have been a hallmark of a number of modern groups including Hizbollah in Lebanon, the Liberation Tigers of Tamil Eelam (LTTE) in Sri Lanka, the Egyptian Islamic Jihad (EIJ), and Al Qaeda. The Irish Republican Army's use of hunger strikes also resulted in a form of suicide,[f] although the underlying psychology and selection process may have been different.

No fewer than four motives for martyrdom (the suffering of death on account of adherence to a cause and especially to one's religious faith) have been postulated: the self-punitive or suicidal motive, the aggressive or homicidal motive, the altruistic or political motive, and

---

[f] The question as to the morality of a hunger strike is beyond the scope of this chapter; however, this was a divisive issue among Christian theologians during The Troubles. Catholic priests in Northern Ireland considered the hunger strike an act of resistance and thus if an individual carried the hunger strike to its logical conclusion, then that individual would be afforded all rights and privileges of a Catholic burial. However, Protestant leaders in Northern Ireland and Catholic leaders in England postulated that a hunger strike was a suicidal act and thus a sin, therefore denying Catholic hunger strikers those rights.

the erotic or sexual motive.[23] In Durkheim's typology,[g] the altruistic suicide is characterized by consumption with, and subsequently being overwhelmed by, a group's goals and ideology.[24] This phenomenon occurs in organizations (and societies) with high integration, where individual needs are subordinated to that of the whole.[25] Many religious cultures, political extremist groups, and military organizations hold those who sacrifice themselves for the common good in great reverence and thus this mechanism can be considered a perverse exploitation of the dead as an unimpeachable source of credibility.[26] Culture cannot be divorced from the analysis of individual and group patterns of behavior; the cultural significance of self-immolative behavior as a means of addressing grievances was evident in the 1917 and 1920 (IRA) and 1981 (Provisional IRA [PIRA] and Irish National Liberation Army [INLA]) hunger strikes.[27]

Studying suicide terrorism is difficult for a number of reasons, including the danger surrounding contact with relevant groups and also the simple fact that successful suicide attackers are no longer available to question. One of the best available studies by Ariel Merari[28] and his colleagues compared three groups of Palestinians imprisoned for involvement in terrorist activities. One group consisted of higher-level organizers within their respective organizations. A second group consisted of individuals who had been involved in non-suicide attacks, and the third group consisted of "martyrs," or suicide attackers, who were sent on missions but either experienced equipment failure or were caught before they could detonate their explosives. This analysis did find some differences between the groups. The would-be martyrs had lower levels of ego strength as well as a dependent and avoidant personality style, a profile that made them more amenable to group, leader, and public influence. Significantly more martyrs than the other control group members displayed symptoms of depression, and some of the would-be martyrs but none of the control group and organizers group members displayed subclinical (e.g., not meeting DSM criteria) suicidal tendencies. These findings may have more to do with how the Palestinian groups selected people for different assignments than the psychological makeup of people willing to die for a cause.

To be effective, the social construction of martyrdom must rule out the possibility of the martyr being afflicted by some kind of psychological disorder, as it would risk categorization as another of the four motivations. Rather, higher-status martyrs (those amongst the in-group

---

[g] Durkheim proposed four subtypes of suicide: egoistic (reflecting a sense of social isolation resulting in apathy and depression, altruistic (being overwhelmed by an integrated in-group's objectives), anomic (a moral confusion resulting from socioeconomic turmoil), and fatalistic (an over-controlled hyper-regulation).

with higher social, academic, or hierarchal standing) have more credibility with the masses and thus their sacrifice carries more weight. This seems to be the trend identified in studies of the selection process that the PIRA used to select those who would participate in the 1981 hunger strikes in the H-Block of Long Kesh (later called the Maze) prison. Hunger strikers had to volunteer and be approved by the PIRA military council, and most of those selected were highly regarded within the organization. Hunger strikers appear to have possessed a rather high level of dispositional resilience, thus enabling them to persevere through extraordinary physiological and psychological anguish while serving as a suitable example to engender ideological or financial support.

There is controversy over including hunger strikers in the same category as suicide bombers; however, both approaches require an individual to value a particular cause (or perhaps an otherworldly goal) over their own existence. While the literature suggests this commitment is unlikely to be the result of major psychopathology, there is still debate over whether those willing to engage in self-immolative behavior in the process of killing others and/or destroying property can be considered psychologically healthy.[29]

EMOTIONAL VULNERABILITY

The existence of some temporary emotional state that predisposes the individual to greater openness to the use of, or support for the use of, violence[30] creates a window of opportunity or vulnerability. This could be brought on by any number of internal (cognitive and/or personality predisposition) or external (death or injury of a loved one) stimuli.

The case of Timothy McVeigh illustrates the role of both vicarious victimization, in his identification with members of the Branch Davidians killed in Waco, Texas, in 1993, and emotional vulnerability, related to previous life experiences and failures. McVeigh was the perpetrator of the 1995 bombing of the Alfred P. Murrah Federal Building in Oklahoma City, which was, at that time, the deadliest terrorist act on American soil. McVeigh was the middle child between two sisters and born into a middle-class Irish, Roman Catholic family in rural western New York. He possessed above-average intelligence, was by temperament an introvert, although not antisocial, and was considered personable. Despite his social skills, he was reluctant to trust others and did not form close relationships easily. His parents separated when he was eleven years old, causing him to repeatedly withdraw into fantasies about himself as a hero as a means of coping with the familial tension during his latency years. The dissolution of his parents' relationship increased his interest in survivalism and firearms while also appearing to lower his motivation for academic achievement and severely dampening his interest in women. The crystallization of McVeigh's personality in late adolescence focused on his conscious identification as the self-described "ultimate warrior." This fantasy was inferred from numerous behaviors and personal predilections including his interest in (and proficiency with) weapons and survivalism, his enlistment and success in the U.S. Army (to include combat actions during Operation Desert Storm), and his political interest in defending Second Amendment rights. This narcissistic fantasy, although grandiose, was for a time supported by reality and compensated for other disappointments. The fantasy progressed undeterred until he returned from Kuwait and attended the Special Forces Assessment and Selection (SFAS) Program in April 1991. McVeigh's personal failure, however, and voluntary withdrawal from training was probably a humiliating and dysphoric (distressed/uneasy) experience. His subsequent behavior suggested this event was significant enough to contribute to his alienation from the U.S. government. He remained in his own mind the "ultimate warrior" in search of another war, socially adrift and exhibiting symptoms associated with mild depression and posttraumatic stress disorder (PTSD), although he was never diagnosed with or treated for either disorder. McVeigh found another war through the politics of the Patriot (or militia) movement, although he never officially joined any of the affiliated organizations. Instead, he formed a leaderless cell composed of himself, Terry Nichols, and Michael Fortier.

Consequent events in the United States, particularly the burning of the Branch Davidian compound in Waco, Texas, on April 19, 1993, provided a rationale for his anger directed toward the government. He began operationalizing that anger approximately a year after Waco, when he began actively planning the Oklahoma City bombing. McVeigh used a variety of methods to maintain his pathologically narcissistic belief that his act of bombing was his destiny. First, he held an intense interest in *The Turner Diaries,*[h] a fictional account of a race war in which a federal building is bombed, and an affinity with the novel's protagonist. Second, he personally identified with American patriots such as Samuel Adams, Thomas Jefferson, and Patrick Henry, as evidenced by his collection and dissemination of their writings. Finally, he selected

---

[h] Considered the bible of the racist right by the Southern Poverty Law Center, *The Turner Diaries* is a 1978 novel written by William Luther Pierce (the former leader of the white nationalist organization National Alliance). The novel depicts a violent revolution in the United States that leads to the overthrow of the federal government, nuclear war, and, ultimately, to a race war leading to the genocide of all Jews and non-whites. In the novel, the protagonist, operating independently from any organization, destroys a federal building; the event serves as a catalyst for the race war.

April 19, 1995, as the date of the bombing to not only avenge the deaths at Waco in 1993, but also to commit what he saw as a revolutionary act on the birthday of the American Revolution, April 19, 1775, in Lexington and Concord, Massachusetts. The meticulously crafted and tested ammonium-nitrate-and-diesel-fuel mixture bomb placed in the rear of a rental truck killed 168, injured nearly 700, and damaged more than 300 buildings within a 16-block radius.

Understanding the emotional state of an individual at various points in the radicalization process is mostly useful after the fact; unfortunately, individual emotional states change too frequently and in response to too many different stimuli for them to be easily definable or have sufficient predictive validity.[31] Absent direct access as well as psychometric and demographic data, it is not only difficult to determine the psychological state of an individual but it is also nearly impossible to develop appropriately idiosyncratic countermeasures.

YOUTH

Young males in their mid-teens to late 20s are the most likely to be recruited by or interested in joining a radical organization.[32] Empirical data suggest the greater the concentration of activists between the ages of eighteen and twenty-five, the greater the risk for radicalization.[33] This is a nonlinear statistical predictor and not a psychological predictor; although the adolescent personality is still developing and thus more susceptible to persuasion, one cannot necessarily say (for example) that, controlling for other variables, a 19-year-old male is more vulnerable to recruitment than a 24-year-old male. Census operations are among the most important information-gathering tasks in counter-radicalization operations. From Galula to Killcullen, practitioner-theorists have advocated that accurate census taking will facilitate better administration, security operations, and ultimately pacification. Propensity for risk taking is influenced by both life experiences and neurocognitive functions. Some factors contributing to increased risk-taking propensity include younger age, lack of parental status, lower reproductive goals (e.g., lack of interest in sexual relationships with women), being the last-born, and lower subjective life expectancy. The tendency for risk taking is not necessarily a stable personality trait but rather varies by life-history traits and specific operational or occupational domains.[34]

In 2010, the New York State Intelligence Center (NYSIC) published a report documenting thirty-two major terrorism cases in the United States from September 11, 2001 through 2010. The research sample in the report comprised predominately males between the ages of eighteen and thirty-three who explicitly support or follow radical Islamic ideology. Over half of the individuals were legal U.S. citizens at the time of their arrest, with 70 percent natural-born citizens. These individuals operated in small groups that often formed within a community where common attributes and beliefs were shared publicly.[35]

PERSONAL CONNECTION TO A GRIEVANCE (POLITICAL OR OTHERWISE)

Identification with victims (actual or vicarious)[36] will predispose an individual to radicalization. The proximity and/or strength of connection of one individual to another who was victimized or perceived to be victimized by a political out-group will increase the likelihood for radicalization. Any political or military action will have unintended consequences, including individuals who suffer physically, psychologically, financially, or otherwise by a change in policy, a military operation, legislation, etc.

Shamil Baseyev's Riyadus-Salikhin Reconnaissance and Sabotage Battalion of Chechen Martyrs is often associated with employing female suicide operatives recruited from pools of women whose husbands were killed by Russian forces.[37] These "Black Widows" are often described as being particularly vulnerable as a result of this personal tragedy. Superficially this appears to be an ideal exemplar of the personal grievance mechanism. Reality, however, is more complex. The degree to which these individuals willingly participated is debatable; Baseyev's recruiters coerced some, others felt the social pressure of marginalization from an insulated culture of strict Islamic mores, and others may have been acting out of a sense of grief-induced hopelessness (along a depressive continuum).

VICARIOUS EXPERIENCE OF GRIEVANCE

The personal grievance need not be direct but can be experienced by proxy. This dynamic is particularly prevalent in modern recruitment strategies of radical Islamic groups, which emphasize victimization of Muslims at the hand of Westerners. Vicarious victimization can be experienced through self-study, media exposure, or accounts learned from members of the same social network. Al Qaeda's narrative, from Osama bin Laden's fatwas to Dr. Ayman Al-Zawahiri's speeches and their propagation throughout the information environment, continues to call for devout Muslims to rise up and serve the greater *umma* through jihad. This struggle (says Al Qaeda's distinctly Qutbist interpretation) should be aimed toward the apostate regimes throughout the Muslim world because they have been corrupted by the "seductive" West. The narrative implies that Western culture has victimized Muslim rulers who have in turn victimized Muslims. The appropriate response to this dual victimization is to support Al Qaeda's efforts to rectify this situation. The individuals with no direct contact to Al Qaeda leadership or even operatives (sometimes referred to as self-radicalized or super-empowered) become adherents to the narratives of violent extremists and radicalized to the point of contemplating terrorist acts. There are several variants of these grievances, including grievances against governments, a hatred for the perceived erosion of fundamental values, disaffection from society, anger over unequal economic opportunities, the desire to belong to something larger than self, and a desire to make a name and prove oneself.

The phenomenon of vicarious victimization is becoming increasingly prevalent in Muslims residing in the United Kingdom. The Report of the Official Account of the Bombings in London on July 7, 2005, cited that international conflict involving Muslims was widely interpreted as widespread war against Islam, that leaders of the Muslim world were perceived as corrupt and non-Islamic, and that some U.K. policies were viewed as evidence of a persecuted Islam. Despite the variation in socio-economic status among the relatively small sample, the report identified that the U.K. Muslim population had internalized the issues affecting Muslims globally and often used the first person plural to refer to Muslims in Palestine, Iraq, and/or Afghanistan. This justification included the idea that violence was a demonstration of commitment to Islam and that suicide terrorism (including the praise and admiration for martyrs) was a noble undertaking. Jihad is interpreted as a way of life or a permanent and individual obligation on all Muslims.[38]

Vicarious grievances are most effective when supported by ideological frameworks articulating perceived problems, a vision of the future, and a prescription for action. Cases of individual radicalization to political violence (when the individual acts alone rather than as part of a group) are relatively rare.[39] In such cases, the individual is likely to have some association with a larger intellectual community or social movement.[40]

HUMILIATION

Humiliation and the consequent internal pressure for revenge is a psychological factor that has been suggested to predispose one to violent behavior.[41] The experiences of grief, either personally or vicariously, are often accompanied by strong feelings of humiliation, which present a dangerous risk factor for radicalization. The greater the degree to which the radical group is subjected to physical repression or torture, or perceives itself to be humiliated by its enemies, the greater the risk that they will take revenge through political violence.[42] Intentionally or unintentionally robbing an individual of his or her inherent dignity provides not only a logical rationalization for radicalization but also a sociocultural motivation to defend the inherent dignity of all those within the in-group. Revenge is an emotion that is probably deeply rooted in the adaptive instinct to punish transgressors who violate the contracts of social species; hence, it is a motivator that often serves not only the goals of a vengeful individual but also the goals of the group.[43] Revenge is not always considered an antisocial behavior and can be considered normal and potentially useful in certain contexts. The humiliation and traumatization of political opponents can create an environment that stimulates regression to more violent behavior, thereby aggravating a conflict or escalating the level of violence. Political, ideological, and/or religious narratives may mediate between the collective identity and personal misery from humiliation, but they may also reinforce a victimization identity that contributes to increased potential for violent behavior.[44] The concept that feelings of humiliation or exploitation give rise to a passion for revenge is prevalent in forensic psychiatry and criminology, and it has been suggested that it may contribute to nonpolitical murders.[45]

Richard Nisbett has described what he calls "cultures of honor,"[46] which are cultures where there is a perceived imperative to preserve honor by avenging even minor slights. Particularly for males, failing to retaliate for an insult, attack, or property encroachment can be seen as a serious threat to the individual's honor and reputation. Examples of these cultures exist all across the world, including parts of the American South and the Middle East. (Nisbett speculates that these cultures developed from the unique vulnerabilities of herdsmen, who possess property that is difficult to defend because of its topography and who need to maintain a reputation of toughness and retaliation.) Not surprisingly, such cultures are vulnerable to high homicide rates, cycles of retaliation, and longstanding feuds. In such cultures, humiliation can be a cumulative pervasive internalization (often spanning generations) that becomes so ingrained that it becomes the definitive trait of an individual.[47] This theory, in the context of political violence, may manifest as the logical reaction to a personal or political grievance or a misapplication of the law of social substitutability. This anthropological concept holds that the killing of any member of the in-group is considered a group offense and can be avenged by the killing of any member of the offender's out-group.[48] This law may be perverted or manipulated by individuals to rationalize terrorist behavior. For example, members of the Abu Nidal Organization justified the skyjacking and execution of hostages on an El Al airline on the grounds that Israel required compulsory military service. The individuals were therefore members of the military and thus justifiable targets.[49] This rationalization has been further extrapolated by Al Qaeda and its affiliated networks to include American citizens who, as taxpayers, support the U.S. government's oppressive and exploitative policies.

MECHANISMS

It is radical behavior, not simply beliefs and/or feelings that is of greatest practical concern when analyzing psychological risk factors of a prospective insurgent. However, the underlying susceptibility, vulnerability, and concomitant feelings of an individual throughout the process do help develop a potential set of observable behaviors and/or conditions. It is important to note that the adoption of radical beliefs alone does not necessarily mean an individual will become violent; the transition from activist to violent radical is not inevitable. In addition, an individual need not personally suffer a transgression to seek out radicalization opportunities. Any individual with a degree of empathy, sufficient emotional vulnerability, and the opportunity to access informative materials (of which there is no short supply on the Internet) could potentially succumb to this mechanism. However, there may be underlying factors that predispose one to vulnerability, but those factors have not been studied with sufficient rigor to have generated a set of empirically determined criteria for susceptibility to radicalization. The risk factors and mechanism presented in this and the subsequent chapter discuss how and the reasons why an individual chooses to participate in violent radical behavior. Some become radicalized because of a personal or political grievance whereas others do so because of social or environmental pressure. These mechanisms, while not deterministically predictive, do have value in interpreting the behavior of individuals when direct assessments are impractical. An example of the application is included in Table 7-1, where the mechanisms discussed in this and the next chapter are applied to the Zawahiri case study included in Chapter 4.

Table 7-1. Radicalization mechanisms and their relevance to the Zawahiri case study.

Mechanism[50]	Relevance to Zawahiri
Radicalization due to personal grievance: Harm to self or loved ones can move individuals to hostility and violence toward perpetrators.	High: Zawahiri's close relationship with his maternal uncle (Sayyid Qutb's attorney)
Radicalization under threat: Threat or harm to a group or cause the individual cares about can move the individual to hostility and violence toward perpetrators.	Medium: Generalized predisposition to perceived threat indicative of intolerance of uncertainty (manifested both individually as well as generally on behalf of Egyptian Islamists toward the secular regimes of Gamal Abdel Nasser, Anwar Sadat, and Hosni Mubarak)
Small involvements in political conflict can create new forces that can move an individual toward radicalization.	High: Early days (1966) as a clandestine cell leader within Muslim Brotherhood faction provided sense of political identity
Radicalization through social networks: Love for someone already radicalized can move an individual toward radicalization.	Low: Zawahiri's radicalization resulted more from his admiration for the idealized, martyred Qutb than the social network argument put forth by Sageman and McCauley and Moskalenko
Radicalization by disposition: The attractions of risk taking and status can move individuals, especially young males, to radical political action.	Low: Zawahiri did not exhibit novelty- or sensation-seeking behaviors in childhood, adolescence, or adulthood; in fact, he tended to be more "bookish" than athletic or adventurous
Radicalization through isolation: Loss of social connection can open an individual to new ideas and new identity that may include political radicalization.	Medium: This is more of a perception emanating from an introversive tendency as both maternal and paternal sides of his family were well connected socially, academically, and politically
Discussion among like-minded individuals tends to move the whole group further in the direction initially favored.	High: Particularly so during the internment period subsequent to the Sadat assassination
Radicalization in competition for the same base of support: Groups are radicalized in competition with other groups.	High: Particularly so when comparing the Muslim Brotherhood, Egyptian Islamic Group (EIG), and EIJ (although more so EIG-EIJ) from the late 1970s through the 1990s

Mechanism[50]	Relevance to Zawahiri
Radicalization through condensation or splitting: The power of group dynamics is multiplied to the extent that group members are cut off from other groups.	High: Particularly during the period immediately after internment when Zawahiri and many EIJ leaders "escaped" to Peshawar and linked up with Abdullah Azzam and Osama bin Laden
Radicalization through jujitsu politics: Terrorists often count on government reactions to advance their causes.	Medium: This seems to have been a consideration in the Embassy operations; however, Al Qaeda's organizational reasoning was biased by the 1993 incident in Mogadishu and thus it miscalculated the U.S. response to 9/11
Radicalization through hate: In protracted conflicts, the enemy is increasingly seen as less than human.	High: Particularly so when analyzing the content of Zawahiri's speeches/press releases; his abrasive negativism has become more pronounced
Radicalization through martyrdom: A successfully constructed martyr can radicalize sympathizers for the martyr's causes.	Medium: The salience of Qutb's martyrdom remains with Zawahiri; however, he has demonstrated lack of empathy in considering the human effect (e.g., surviving family members, innocents, etc.) of martyrdom operations instead focusing on the enemy—bin Laden on the other hand, seems to have displayed more empathetic reactions toward the families of both suicide operatives and those killed in combat against the Soviets

SUMMARY

Acts of terrorism, while they are often bloody and irrational to outside observers, are nevertheless not usually perpetrated by individuals with severe, diagnosable psychological disorders. As elaborated in other chapters, most insurgents are indistinguishable from other members of their societies but have gone through a social and ideological process of radicalization and desensitization to violence. A study comparing individuals guilty of politically motivated murder (IRA members or opponents) to other murderers found the politically motivated set to have much lower rates of psychological problems.[51]

Radical organizations likely have a sufficient range of personality and cognitive profiles within their ranks to be indistinguishable from the surrounding population on either of these axes. Having said this, there are a number of negative psychological risk factors that are contributors to violent radicalization and should be understood as such.

Most functioning insurgencies effectively screen out potential members who are dangerously mentally ill, having diagnosable Axis I disorders such as major depression or schizophrenia. These people, untreated, have great difficulty working with others and present both social and operational risks. There is a phenomenon of lone wolf terrorists, however, who may be quite sick but nevertheless be capable of feats of individual mayhem, with Anders Behring Breivik being a notable recent example.

Axis II developmental or personality disorders are subtler and may be more common among violent insurgents. Antisocial personality disorder, or APD, is characterized by a lack of empathy, absence of moral constraints, and the consideration of violence as a legitimate tool to accomplish a goal; individuals with this pathology may find a place in insurgent movements and may even be effective leaders in the right circumstances.

Suicide bombers in general have a different motivational profile than other suicide victims and may be manipulated to act even though they have no wish to die. However, a study of failed suicide bombers shows that they do in fact have more characteristics of depression and suicidality than other members of the same organizations.[52] It is unclear whether they were self-selected for suicide assignments because of this, whether they were selected by others, or whether there are other common causes.

Emotional vulnerability, brought about by internal or external circumstances, is a risk factor for both recruitment and radicalization into violent groups. Youth is a risk factor that interacts with others and is often exploited.

Personal connections to grievances such as government-sponsored atrocities is an important circumstantial risk factor. Counterinsurgent actions can sometimes be turned into effective recruitment tools for the opposition. The vicarious experience of grievances through social connections or media can also be quite powerful. Ethnic and religious extremists groups use video and other accounts of atrocities committed against their members very effectively, with Al Qaeda being a prime example. The experience of humiliation in various forms is also a potent risk factor; humiliation can be experienced either as an individual or as a member of an aggrieved in-group. These vulnerabilities are hard to detect a priori in individuals, but an understanding of them as risk factors is important for understanding insurgent recruitment, indoctrination, and other social dynamics of these movements.

ENDNOTES

[1] Chuck Crossett and Jason Spitaletta, *Radicalization: Relevant Psychological and Sociological Concepts* (Ft. Meade, MD: Asymmetric Warfare Group, 2010).

[2] Ibid.

[3] Marc Sageman, *Understanding Al Qaeda and the New Terror Networks* (speech given at the Foreign Policy Research Center, October 4, 2010), accessed May 9, 2011, http://www.fpri. org/multimedia/20101004.sageman.terrorism.html.

[4] Jerrold M. Post, *The Mind of the Terrorist: The Psychology of Terrorism from the IRA to Al Qaeda* (New York: Palgrave McMillan, 2007).

[5] American Psychiatric Association, *Diagnostic and Statistical Manual of Mental Disorders*, 4th Ed. (Washington, DC: American Psychiatric Association, 2000).

[6] Ibid.

[7] Clark R. McCauley and Sophia Moskalenko, "Mechanisms of Political Radicalization: Pathways Toward Terrorism," *Terrorism and Political Violence* 20, no. 3 (2008): 415–433.

[8] Jeffrey Victoroff, "The Mind of the Terrorist: A Review and Critique of Psychological Approaches," *Journal of Conflict Resolution* 49, no. 1 (2005): 3–42.

[9] H. H. A. Cooper, "Psychopath as Terrorist," *Legal Medical Quarterly* 2 (1978): 253–262.

[10] Jordan Maile, Tali. K. Walters, J. Martin Ramirez, and Daniel Antonius, "Aggression in Terrorism," in *Interdisciplinary Analyses of Terrorism and Political Aggression*, ed. Daniel Antonius et al. (Cambridge: Cambridge Scholars Publishing, 2010).

[11] Victoroff, "The Mind of the Terrorist," 3–42.

[12] Nancy McWilliams, *Psychoanalytic Diagnosis*, 2nd Ed. (New York: The Guilford Press, 2011).

[13] Robert D. Hare, *Without Conscience: The Disturbing World of the Psychopaths Among Us* (New York: Guilford Publications Inc., 1993).

[14] Willem H. J. Martens, "The Terrorist with Antisocial Personality Disorder," *Journal of Forensic Psychology Practice* 4, no. 1 (2004): 45–56.

[15] Ibid.

[16] Jerrold M. Post, Kevin G. Ruby, and Eric D. Shaw, "The Radical Group in Context 1: An Integrated Framework for the Analysis of Group Risk for Terrorism," *Studies in Conflict and Terrorism* 25, no. 2 (2002): 73–100.

[17] Ibid.

[18] Jerrold M. Post, Kevin G. Ruby, and Eric D. Shaw, "The Radical Group in Context 2: Identification of Critical Elements in the Analysis of Risk for Terrorism by Radical Group Type," *Studies in Conflict and Terrorism* 25, no. 2 (2002): 101–126.

[19] Hare, *Without Conscience*.

[20] H. A. Lyons and H. J. Harbinson, "A Comparison of Political and Non-Political Murderers in Northern Ireland, 1974–84," *Medicine, Science, and Law* 26, no. 3 (1986): 193–197.

[21] Ibid.

[22] Ibid.

[23] J. Reid Meloy, "Indirect Personality Assessment of the Violent True Believer," *Journal of Personality Assessment* 82, no. 2 (2004): 138–146.

[24] Emile Durkheim, *Suicide: Study in Sociology* (New York: The Free Press, 1997).

[25] Ibid.

[26] Rona M. Fields, *Martyrdom: The Psychology, Theology, and Politics of Self-Sacrifice* (New York: Praeger, 2004).

[27] James Dingley and Marcello Mollica, "The Human Body as a Terrorist Weapon: Hunger Strikes and Suicide Bombers," *Studies in Conflict and Terrorism* 30, no. 6 (2007): 459–492.

28 Ariel Merari, Ilan Diamant, Arie Bibi, Yoav Broshi, and Giora Zakin, "Personality Characterisitcs of 'Self Martyrs'/'Suicide Bombers' and Organizers of Suicide Attacks," *Terrorism and Political Violence* 22, no. 1 (2009): 87–101.

29 Ariel Merari, "Psychological Aspects of Suicide Terrorism," in *Psychology of Terrorism*, eds. Bruce Bongar, Lisa M. Brown, Larry E. Beutler, James N. Breckenridge, and Philip G. Zimbardo (New York: Oxford University Press, 2007).

30 John Horgan, *Walking Away from Terrorism: Accounts of Disengagement from Radical and Extremist Movements* (New York: Routledge, 2009).

31 Crossett and Spitaletta, *Radicalization.*

32 Andrew Silke, "Holy Warriors Exploring the Psychological Processes of Jihadi Radicalization," *European Journal of Criminology* 5, no. 1 (2008): 99–125.

33 Post, Ruby, and Shaw, "Radical Group in Context 1," 73–100.

34 X. T. Wang, Daniel J. Kruger, and Andreas Wilke, "Life History Variables and Risk-Taking Propensity," *Evolutionary Human Behavior* 30, no. 2 (2009): 77–84.

35 New York State Intelligence Center, *The Vigilance Project: An Analysis of 32 Terrorism Cases Against the Homeland* (2010).

36 Horgan, *Walking Away from Terrorism.*

37 Troy S. Thomas, Stephen D. Kiser, and William D. Casebeer, *Warlords Rising: Confronting Violent Non-state Actors* (New York: Lexington Books, 2005), 157.

38 Crossett and Spitaletta, *Radicalization.*

39 Lou Michel and Dan Herbeck, *American Terrorist* (New York: HarperCollins, 2001).

40 McCauley and Moskalenko, "Mechanisms of Political Radicalization."

41 Crossett and Spitaletta, *Radicalization.*

42 Post, Ruby, and Shaw, "Radical Group in Context 1," 73–100.

43 Jeff Victoroff and Arie Kruglanski, *Psychology of Terrorism: Classic and Contemporary Insights* (Washington, DC: Psychology Press, 2009).

44 Sverre Varvin, "Humiliation and the Victim Identity in Conditions of Political and Violent Conflict," *Scandinavian Psychoanalytic Review* 28, no. 1 (2005): 40–49.

45 Victoroff and Kruglanski, *Psychology of Terrorism.*

46 Richard E. Nisbett and Dov Cohen, *Culture of Honor: The Psychology of Violence in the South* (Boulder, CO: Westview Press, 1996).

47 Ariel Merari, *Driven to Death: Psychological and Social Aspects of Suicide Terrorism* (Boston: Oxford University Press, 2010).

48 Joyce Marcus, "The Archaeological Evidence for Social Evolution," *Annual Review of Anthropology* 37 (2008): 252–266.

49 Post, *Mind of the Terrorist.*

50 Clark R. McCauley and Sophia Moskalenko, *Friction: How Radicalization Happens to Them and Us* (New York: Oxford University Press, 2011).

51 Lyons and Harbinson, "Comparison of Political and Non-Political Murderers," 193–197.

52 Merari, *Driven to Death.*

PART III.

UNDERGROUND PSYCHOLOGICAL OPERATIONS

CHAPTER 8.

INSURGENT USE OF MEDIA: TRADITIONAL, BROADCAST, AND INTERNET

CHAPTER CONTENTS

Nathan Bos

The Zapatista rebellion of 1994 was, from the perspective of the Mexican government, no different from many small-scale uprisings of indigenous groups, and there was no reason to expect the same counter-tactics would not work as they had in the past. Frustrated by lack of response to their demands for land reform, and threatened by new privatization of communal lands, the Ejército Zapatista de Liberación Nacional (Zapatista National Liberation Army, or EZLN) planned a coordinated action to commence on January 1, 1994. That date was chosen because it was also the date that the North American Free Trade Agreement (NAFTA) would take effect, which the EZLN considered a threat to local farmers. On January 1, as planned, 1,000–2,000 militants swept down from the hills and took over five villages and one city, San Cristóbal de las Casas, a city of approximately 100,000 in the remote southernmost province of Chiapas, Mexico. The Mexican government responded with a two-pronged approach that, at first, appeared to be going exactly as planned. The military arrived and within two weeks drove the EZLN out of the occupied towns and villages and back into the hills. The federal government, meanwhile, used the state-controlled media to contain the rebellion and quell possible support from other regions by limiting news coverage. Media control allowed the government to downplay the size of the rebellion and give their own narrative of events, which framed the rebellion as an externally connected threat to Mexican sovereignty, "conjuring the threat of a pan-Mayan movement embracing both Southern Mexico and much of Central America."[1]

However, events around the Zapatista rebellion quickly deviated from the expected script. Far from being contained in Chiapas, the previously unknown Zapatistas soon became the subject of discussion on five continents and benefitted from an unprecedented nongovernmental organization (NGO) mobilization and advocacy campaign. Both sides accepted a cease-fire, and the combat continued in another venue. As the Mexican Secretary of Foreign Relations Jose Angel Gurria later said: "Chiapas, please take note, is a place where there has not been a shot fired in the last 15 months. . . . The shots lasted 10 days, and ever since the war has been a war of ink, of written word, a war on the Internet."[2] For the EZLN, neither the military tactics nor the basic land disputes had changed from prior conflicts. What had changed was their ability to bypass the state-controlled media using the Internet, along with the ability of NGOs and individuals to help groups organize quickly across distance, also using the Internet. As will be described later in this chapter, the social networking capabilities of the Internet had made possible what Cleaver dubbed "the Zapatista effect."[3]

IMPORTANCE OF MEDIA TO INSURGENCIES

Media exposure is critical to insurgencies of all types. Insurgencies know that they cannot win isolated force-on-force contests against larger, more established forces. Contact with and support from non-combatants is a part of every insurgency, and this has been true since long before the advent of mass media and the Internet. What also has not changed are the set of audiences important to insurgents (below). There has, however, been a sea change brought about by technology as to the means, effectiveness, and reach of communication.

Every insurgency group needs to communicate with the following groups to a greater or lesser extent:

- Internal supporters. Insurgencies need to communicate internally; when the insurgency is large or spread out, or when direct communication is denied, media channels become a means for coordinating action, spreading news that may be suppressed by official channels, clarifying and reinforcing ideology, providing encouragement, and sending antidefection messages. This is sometimes referred to as "auto propaganda."[4] Internal supporters are sometimes referred to as the auxiliary.

- External supporters. Insurgencies rely on networks of sympathizers, ranging from local residents, whose main contribution may be forgetting what they have seen when questioned by police, to international sympathizers, who may provide money, relay messages, or advocate to foreign governments. This group is on a continuum from support to apathy and may also become hostile. Timely and frequent communication from insurgents is often critical to maintaining the disposition of supporters. Different messages and different media are required to reach external supporters than are required to reach internal supporters.

- Nonsympathetic supporters. Terrorist groups, in particular, also design communications for external audiences that may never support their cause yet play a role in its success. These audiences are the target of influence to include intimidation and terror; but affecting them in any way may show strength, increase publicity (good or bad), and keep the neutral media interested.

- The enemy. External communications may also be designed to demoralize, intimidate, or mislead enemy combatants. Communicating directly with the families of enemy combatants, where possible, may be a particularly effective means of sapping an enemy's will to fight.

TRADITIONAL MEDIA

Handbills[a] and newspapers have been important tools for under-grounds for centuries, and they remain so, even as new media augment them. The two have different purposes: handbills are used for short, often time-dependent and highly localized events; newspaper are aimed at somewhat broader constituencies and deal with broader issues.

The Frente Farabundo Martí para la Liberación Nacional (Far-abundo Marti National Liberation Front or FMLN) of El Salvador were known for having a very-well-organized media arm. They used both handbills and newspapers and a well-articulated communication strategy laid out in a document called *Concerning Propaganda*, which was analyzed in a RAND study called *Underground Voices*.[5] For leaflets, they produced posters and pamphlets on a geographically dispersed network of mim-eograph machines (pre-photocopying duplication machines). This allowed faster dissemination and a more resilient production capacity. Local producers were also expected tailor information to local needs.

The ability to distribute handbills can itself be a signal of strength for an insurgency group. Ḥarakat al-Muqāwamah al-'Islāmiyyah's (Islamic Resistance Movement, or HAMAS) distribution of handbills starting in 1988 was one of the first public signs of its challenge to the Palestinian Liberation Organization (PLO)'s leadership in the area. Handbills are an important means of communication in the occupied territories and are often used to organize strikes and demonstrations.[6]

Hand delivery of handbills and newsletters seems to be a key to their effectiveness. In the Philippines, Malaya, and Korea, handbills were passed surreptitiously from person to person by hand or by chain letter. Giving a handbill thus implied a proof of confidence, an honor, and a privilege. It was reported that people saw, read, and remembered more of the handbills handed to them personally than those received by mass distribution. In an experiment conducted in six Korean vil-lages, pro-government leaflets were dropped by air and Communist propaganda was circulated personally. It was found that the villagers had a more accurate memory of the wording of the Communist leaflets than of the government messages.[7]

The FMLN also organized an unusual letter-writing campaign aimed at the military. Supporters were asked to write letters to soldiers they knew encouraging them to desert. Supporters were even asked to write to commanders they did not know "with names and grades in respectful tones calling on them to reflect about the situation."[8]

---

[a] In the lexicon of Military Information Support Operations (PSYOP), small printed bills delivered by air are called leaflets, and those distributed on the ground are called handbills.

This tactic assumed a semipermissive environment where such contact would not lead to retribution.

Most undergrounds produce newsletters, and most also produce multiple newsletters for different audiences or representing different factions. FMLN published a newspaper called *Venceremos*. The content indicates that it was not intended for their broader audience of sympathizers but rather intended to guide the political actions of the FMLN underground auxiliary and armed component, which their communications plan referred to as their "middle sectors" and "special sectors." Reaching beyond their borders, FMLN distributed different print materials worldwide, with titles including "Why is FMLN fighting?" and "Women and war in El Salvador." They also employed an American public relations firm to arrange press conferences and issue press releases.[9]

For the Bolsheviks, newsletters also served a training purpose. Members of the organization were able to practice gathering and distributing information; they also learned to estimate the effects of political events on various sections of the population and to devise suitable methods to influence these events through the revolutionary party. In Lenin's words: "Arranging for and organizing the speedy and proper delivery of literature, leaflets, proclamations, etc., training a network of agents for this purpose, means performing the greater part of the work of preparing for future demonstrations or an uprising."[10]

If control measures are strict, various disguises for handbills and newsletters can be used. The Viet Cong sometimes used the government's own materials to disguise their propaganda. In a number of villages, security forces found Viet Cong propaganda booklets with the same covers as government booklets.[11] During World War II, the Belgian underground also used a variety of disguises for their propaganda.[12]

One method of disguising the transmission of propaganda materials into a target country is through the diplomatic channels of embassies friendly to the underground cause. In October 1962, the Chilean government intercepted a crate that had been shipped by the Cuban government to its embassy in Santiago. The 1,800-pound crate was labeled "samples of Cuban products and cultural and commercial material" but actually contained subversive propaganda addressed to various Chileans who had visited Cuba within the previous few months.[13] The Internet, electronic storage media, faxes, and cell phones have made the basic task of distributing content—even content intended for printing and distribution—much simpler.

Even when using traditional media, production quality matters. Insurgents must not only produce and distribute their propaganda, but they must compete for attention and interest with private newspapers as well as government counter-messages. As stated in the FMLN's "Concerning Propaganda" document:

> We must keep in mind that the highest quality propaganda is necessary to carry forward the expansion of political work into new and more densely populated zones, to establish contact with the urban and suburban masses having a more sophisticated political and cultural level. It is a mistake to content oneself with mediocre propaganda simply because resources are limited. We must make its quality superior as well as appropriate to its place. We must make it attractive, accompanying it when possible with well made and expressive graphics, with clear and precise texts, with sharp and visible titles, with clean and attractive print.[14]

GLOBAL BROADCAST MEDIA

Technological changes in broadcast media have been changing the game for insurgency groups for at least a century, with particularly dramatic changes around the development of satellite television in the 1980s. By broadcast media, we mean radio and television in the pre-Internet era. (In the Internet era, "streaming" media has brought about another revolution, which will be discussed later.) The defining characteristics of broadcast media for insurgents are as follows:

- Broadcast media provide more immediate access to larger audiences than traditional media. Audiences can be illiterate or unwilling to invest time in reading written material.

- Broadcast media are more entertaining to many and thus provide new ways of capturing the attention of audiences. However, providing broadcast media comes at a cost of requiring insurgencies to provide information at different rates and in different formats and also requiring insurgencies to increase their production values in order to compete with other media outlets.

- Broadcast media require expensive and vulnerable infrastructure in the form of production studios and broadcast towers; maintaining and protecting these may require a significant allocation of resources.

- Broadcast media are more difficult to tailor to individual tastes or localized concerns than other forms of communication.

The rise of international broadcast media did not just change the way insurgent groups publicized themselves; it changed their targets, their methods, and their goals as well. The Irgun, a small terrorist off-shoot of the Haganah paramilitary organization that was part of the Zionist movement to create a Jewish homeland, was one of the first to recognize and exploit the opportunities international broadcast media provided.[15] Irgun had a cause that could only succeed through international exposure. The key audiences were the British government, which occupied Palestine and could designate a Jewish homeland, and the United States, with a strong pro-Jewish lobby and the international clout to pressure Great Britain and the United Nations to support such an action. Led by Menachem Begin (later the prime minister of Israel who signed the Camp David Accords), Irgun undertook a series of bombings against high-profile symbolic targets in Palestine. Immigration offices were targeted because they limited the number of Jews who could emigrate. One of the largest attacks was a bombing that killed ninety-one people at the King David Hotel in Jerusalem, which housed the British secretariat. These targets were chosen partly to undermine the confidence local Palestinians had that Britain could control the country. But the more important audiences were overseas. The most successful attack they undertook was one of the smallest but most effective because of the gruesome, symbolic images that accompanied it. The event was the public hanging of two British sergeants in retaliation for the execution of three Irgun terrorists.

"Photographs of the grim death scene—depicting the two corpses just inches above the ground, the sergeants' hooded faces and bloodied shirts—were emblazoned across the front pages of British newspapers. . . . For both the British and the press, the murders seemed to demonstrate the futility of the situations in Palestine and the pointlessness of remaining there any longer than was absolutely necessary."[16]

The new state of Israel quickly became the target of a "terrorist spectacular" attack when eleven members of Israel's Olympic team were kidnapped, held for two days, and killed by the Palestinian Organization "Black September" at the Munich Olympics in 1972. The international attention garnered by the operation far exceeded the scope or lethality of the action itself. As a media event, it was well chosen first because of the proximity of broadcast media set up to cover the Olympics and second because of the attention already focused on the event. According to one of the organizers, the target was also chosen for its symbolic value:

> We recognized that sport is the modern religion of the Western world. We knew that the people in England and America would switch their television sets from any program about the plight of the Palestinians if there was a sporting event on another channel. So we

decided to use their Olympics, the most sacred ceremony of this religion, to make the world pay attention
to us. We offered up human sacrifices to your gods of
sport and television. And they answered our prayers.
From Munich onwards, nobody could ignore the Palestinians or their cause.[17]

The Black September attack did not, in fact, realize its tactical goal,
which was to arrange for a prisoner swap with Israel. It was also widely
criticized, and it discredited the Palestinian cause in the eyes of some
potential sympathizers. But it nevertheless set an example for capturing and holding international attention. In the words of Abu Iyad, the
organization's intelligence chief at the time, the operation "didn't bring
about the liberation of any of their comrades imprisoned in Israel as
they had hoped, but they did attain the operation's other two objectives: world opinion was forced to take note of the Palestinian drama,
and the Palestinian people imposed their presence on an international
gathering that had sought to exclude them."[18]

The Irish Republican Army (IRA) did not embrace new media as
readily as other groups. But the Long Kesh hunger strikes of 1981, which
resulted in the deaths of Bobby Sands and nine other IRA protestors,
does point out some features of media-savvy protests. Bobby Sands and
nine other imprisoned IRA members went on a hunger strike to protest
their loss of prisoner-of-war-type status within Long Kesh prison. Hunger strikes are by design drawn-out affairs compared with other forms
of suicide, allowing time for media coverage to disseminate and public
support to spread. Bobby Sands's hunger strike lasted long enough for
him to be elected to the British parliament during the strike. Hunger
strikes also create a sense of building drama as strikers near death.
IRA hunger strikers took water and salt, which drew out their deaths.
These strikers also intentionally staggered the starts of their strikes,
rather than beginning at once, to create a constant stream of news
both when each began their strike and again as each died. Although it
did receive international coverage, Bobby Sands's hunger strike did not
reach the same level of media coverage achieved by some contempora-
neous events partly because the British could and did control media
access to the prison. It was impossible for international television to
show daily updates on his emaciated condition on television or for print
media to obtain daily photos. (Today it is much easier to find emaciated images from Mahatma Gandhi's hunger strikes in India decades
earlier than images of Bobby Sands or the other nine who died in Great
Britain in 1981.) The results of the hunger strikes were a partial success for each side. The IRA received some concessions on prison conditions and, more importantly, galvanized support around the martyred

Bobby Sands. Margaret Thatcher's government demonstrated resolve by allowing ten strikers to die first and in the end made only partial concessions.

According to Hoffman,[19] the hijacking of TWA flight 847 in June 1985 and subsequent hostage taking represented a low point in the media's unwitting complicity with attention-seeking terror groups. Hizbollah gunmen hijacked the flight on June 14 with the purpose of exchanging hostages for a group of associates imprisoned in Israel. The hijacked flight originated in Egypt and was carrying 153 passengers en route from Athens to Rome. The plane was hijacked by two German-speaking Lebanese men and diverted to Beirut. Over the next several days, the hijackers took the plane back and forth between Algiers and Beirut, releasing non-American passengers along the way and killing one American Navy diver. On Monday, June 17, forty American hostages were taken from the plane and held hostage in the divided, militia-controlled city of Beirut until they were released on June 30. Israel subsequently released 766 Shia prisoners, most likely as part of a multinational deal to secure release of the hostages.

The crisis was the subject of constant attention by the three major U.S. television networks. "During the 17 day crisis, while Americans were held hostage in Beirut, nearly 500 news segments—an average of 28.8 per day—were broadcast by the 3 major U.S. television networks . . . on average, two thirds of their daily early evening 'flagship' news shows (fourteen out of twenty-one minutes) focused on the hostage story, and their regularly scheduled programs were interrupted at least eighty times over those 17 days with special reports or news bulletins."[20] Networks filled in for the lack of new developments and access to the hostages themselves with human interest stories on the hostage's families, creating stronger emotional involvement on the part of viewers. The presence of multiple networks highly invested in the story created competition between them to be the first to find small updates or unpublicized tidbits of information. Even if one network wanted to downplay the crisis, competition for viewership would make it difficult to do so. The drawn-out nature of the crisis again lent itself to more media exposure than a single attack or simpler hostage-taking event might have.

Many critics have since questioned whether the media went too far in this crisis. Bias toward compelling visuals and human interest stories, or to use Garrick Utley's phrase, "immediacy, exclusivity, and drama," may have obscured the larger picture, and constant coverage may have inflated the importance of the events.[21] Both would-be terrorists and news organizations have taken lessons from these events, as will be discussed.

214

An effect of the constant attention was to create great pressure on Ronald Reagan's administration to go back on its policy against negotiating with terrorists. The immediacy and personal aspects of television tend to outcompete long-range goals and policies for public support and sympathy. The media's effect may have been somewhat intentional: CBS White House correspondent Leslie Stahl said that "We are an instrument for the hostages . . . we force the Administration to put their lives above policy."[22]

Insurgent groups aspiring to broadcast media coverage could distill from recent events these lessons for managing the media:

- Go where the coverage is. Broadcast media coverage is much easier to gain in urban areas, and the largest cities are the most media saturated. Insurgent groups realized that, despite the dangers, media coverage sometimes demanded leaving the countryside for the city. This has changed somewhat in the current era of self-coverage.

- Guarantee access to the media. Savvy insurgent groups may go out of their way to develop relationships and guarantee access. The Chechen insurgents' savvy in this regard has been documented by British journalist Anatol Lieven. Former insurgent leader Shamil Basayev was particularly proactive; he openly courted the media, frequently gave interviews from his command post or living quarters, and had correspondents as guests in his home; Lieven at one time was a guest in the home of Basayev's aunt. Journalists need these relationship as well to gain access and guarantee their own safety; Basayev personally signed safe-passage documents for some correspondents, which were effective: "any suspicion (Chechen fighters) showed was instantly dispelled by a pass bearing Mr Basayev's personal red stamp with its emblem of the lone wolf."[23] Media who are under the protection of an insurgent group may of course be guided to a particular view of events; they may also be vulnerable to some forms of intimidation.

- Images matter. A lesser event with compelling video or imagery may have a greater effect than a larger operation conducted away from cameras. Airplanes and skyscrapers are frequent targets partly because of their visual appeal and ability to stir emotions. Gabriel Weimann, who has written a series of books on the subject, writes that "Modern terrorism can be understood in terms of the production requirements of theatrical engagements. Terrorists pay attention to script preparation, cast selection, sets, props, role-playing, and minute-by-minute stage management. Just like compelling stage plays or ballet performances, the media orientation in terrorism requires full attention to detail to be effective."[24] During the 2006 clash between Israel and Lebanon, a Lebanese man named Salem Daher was discovered to be carrying the body of a dead boy from one location to another, allegedly so that different camera crews would photograph it in different settings.[25]

- Sustain events over time. Broadcast media thrive on events that are drawn out over time. A hunger strike is better theater than a suicide; hostage taking can generate more publicity than an attack. The obvious drawback of these types of events is the vulnerability they create for the perpetrators. A prolonged, media-accessible event goes against insurgent doctrine of quick, unpredictable strikes. Nevertheless, in situations where insurgents can control territory or are willing to undertake high-risk operations, the benefits are clear.

- Follow news cycles and provide a constant trickle of news. Media-savvy insurgent groups can also sustain attention by providing a steady stream of developments or new information rather than a single rush of information. This may be done by releasing hostages slowly or by providing frequent contact with insurgents or spokespersons. Insurgent groups may also be savvy enough to take into account differences in time zones for different audiences.

Media Dilemmas and Responses

Covering insurgency activities, particularly terrorism, presents a number of dilemmas for the broadcast media.[26]

- When should new developments prompt interruption of normal programming, e.g., "special bulletins"? This is less of an issue for all-news media such as CNN, but the issue of prioritizing coverage remains.

- What are risks and benefits of live coverage? How should reporters deal with the possibility of showing graphic violence? When should the rumors and inaccuracies that accompany breaking stories be corrected?

- When and how is the media responsible for preventing the spread of panic?

- What images and information should not be broadcast? How should the personal privacy of victims and bystanders be protected?

- How can correspondents avoid being publicists for terrorists or other parties? How much consideration (protection, special access) can correspondents accept from insurgents, the government, or the police without becoming beholden to them?

- How accommodating should media be to government sources? Media often gives relatively unfiltered access to government and police to broadcast warnings and sometimes publicize "most wanted" information. What is the media's responsibility to confirm these sources during time-critical events?

Some broadcast media leaders have responded to the criticism leveled at them. Some provide training and guidelines for their correspondents. BBC published an instruction document titled *Terrorism and National Safety*, which in its preamble states that the "basic role of BBC in reporting about a terrorist situation is – to tell the truth, fast, carefully, responsibly and avoiding speculation."[27]

The Committee of Ministers of the Council of Europe adopted a declaration called "Declaration on Freedom of expression and Information in the Media in the Context of the Fight Against Terrorism"[28] that includes the following suggestions for the media regarding coverage of terrorism:

> Bear in mind their particular responsibilities in the context of terrorism in order not to contribute to the aims of terrorists; they should, in particular, take care not to add to the feeling of fear that terrorist acts can create, and not to offer a platform to terrorists by giving them disproportionate attention.

> Be aware of the risk that the media and journalists can unintentionally serve as a vehicle for the expression of racist or xenophobic feelings or hatred.

> Respect the dignity, the safety and the anonymity of victims of terrorist acts and of their families, as well as their right to respect for private life, as guaranteed by Article 8 of the European Convention on Human Rights.

> Set up training courses, in collaboration with their professional organisations, for journalists and other media professionals who report on terrorism, on their safety and the historical, cultural, religious and geopolitical context of the scenes they cover, and to invite journalists to follow these courses.

Despite some good intentions, the growth of global media and twenty-four-hours-per-day coverage have only increased the pressures on media sources to compete for timely information, compelling visual images, and inside access.

Insurgent-Owned Broadcast Media

To avoid the difficulties of managing externally controlled broadcast media, underground groups have on occasion set up and run their own broadcast media sources. The FMLN in El Salvador, Sendero Luminoso in Peru, and the Contras in Nicaragua all used radio to broadcast propaganda aimed both at sympathizers and the neutral population. The FMLN actually operated two stations that were run by rival factions: Venceremos, founded by the People's Revolutionary Army (Ejército Revolucionario del Pueblo, or ERP), and Radio Farabundo Marti, operated by the Popular Liberation Forces. Venceremos was considered one of FMLN's most important communications tools. For a time the station switched from shortwave broadcast to FM and operated from atop the Guazapa volcano outside of the capital, San Salvador. This location presented an excellent broadcast location but was also very vulnerable to attack, requiring a large commitment of manpower to protect the facility. FMLN's information campaigns were well integrated with other operations; they magnified the effect of their broadcasts and made them more locally relevant by having supporters organize local "listening groups."[29]

One of the more successful underground-run broadcasts is Hizbollah's Al-Mansar television station. Al-Mansar is primarily dedicated to promoting Hizbollah's point of view but is also notable for its variety of programming. Besides news, it broadcasts documentaries such as "My Blood and the Rifle," about Hizbollah's guerilla fighters, and "In Spite of the Wounds," dedicated to individuals who have been injured while fighting against Israel. There have been features such as "Reversal of Picture," which presents and debunks Israeli TV and sometimes newspaper stories, and a talk show called "The Spider's House," dedicated in part to uncovering the weaknesses of the "Zionist entity." Al-Mansar even has insurgent-themed music videos, which represent about 25 percent of its programming and "feature violent images and incendiary language."[30]

Setting up actual broadcast facilities presents considerable logistical challenges. If in-country broadcast towers are to be used, insurgents need to control territory, and towers are usually placed on high ground in highly exposed places. A broadcast tower also presents a fixed, high-value asset of their own that can be attacked. When the FMLN operated a transmitter atop the Guazapa volcano, near San Salvador, they needed a staff of approximately one hundred to operate and guard the facility.[31]

In some cases, broadcasts can be made from outside of the target country, as in the case of Algerian Frente de Liberación Nacional (National Liberation Front, or FLN) broadcasts over Radio Cairo

218

and Radio Damascus. The Communist Greek underground operated Radio Station Free Greece, which pretended to broadcast from within Greece, although it was probably located in Albania or Rumania. Latin-American subversive groups such as the Dominican Liberation Movement, the Peruvian Anti-Imperialist Struggle Movement, and the Guatemala Information Committee transmit hostile propaganda against their respective governments through the facilities of Radio Havana.[32] Another famous example was the use of the BBC to support communications from foreign governments in exile to their populations and associated underground movements throughout Nazi-occupied Europe during World War II. In its early period, the Viet Cong operated a weak transmitter in the south, apparently from a junk floating offshore. Subsequently, they used a much stronger transmitter for Liberation Radio to broadcast news features and commentaries in five languages: Vietnamese, English, Cambodian, French, and Chinese.[33] Radio Hanoi lent official support to the insurgents.

Insurgent-controlled broadcast media still have to compete for attention with other media. As in other media, production quality matters; even sympathetic audiences will be dissuaded by poor video or sound quality and may not be interested in talk formats that present only repetitive, ideological rants. Insurgent media is subject to the same demand for "immediacy, exclusivity, and drama" as other media; privileged access to their own operations is a competitive advantage but may not be sufficient. Broadcast media of all kinds also have to make choices between being locally relevant or appealing to a mass audience. It is logistically much easier to produce neighborhood-relevant pamphlets than locally tailored broadcast programming. Because of these limitations, insurgent-owned broadcast facilities are still a relatively rare phenomenon. The emergence of the Internet, however, has drastically lowered the barriers to providing the same kind of content without the need for expensive broadcast infrastructure.

INSURGENT USE OF THE INTERNET

The emergence of the Internet as a globally accessible communications network has changed, and will continue to change, the equations for insurgent communications. By Internet, we include a range of applications that use the same infrastructure, including World Wide Web sites, established communications methods of e-mail and chat, and emerging social media applications.

Properties of the Internet

These properties, in particular, make the Internet a propitious place for insurgents to operate:

Inexpensive

Newsletters, historical information, and press releases can be made accessible worldwide for a few dollars a month. For not much more groups can avoid the cost of transmission towers or distribution networks and stream audio or video clips on broadcast Internet channels with content similar to that presented on broadcast radio or television.

Decentralized

The Internet was developed by the U.S. military to be a communications network that was resistant, through redundancy and flexibility, to disruption at any one node. There is no single central directory of Internet sites, for example; the directory that tells a web browser how to reach a particular URL is replicated across thousands of domain name servers worldwide, and disrupted communications are easily rerouted across servers or across borders if need be. This decentralized property is a perfect match for modern, stateless insurgencies such as Al Qaeda.

Anonymous

The Internet was developed without strong identity management protocols. E-mail addresses are easily "spoofed"—there is no foolproof way of confirming a sender's identity without cooperation from the Internet service provider (ISP). Content can generally be accessed without identity confirmation, except that demanded by the website itself through password protection or encryption. Methods exist to trace these requests, but there are also counter-measures such as anonymization servers.

Uses of the Internet

The most important current uses of the Internet for undergrounds and insurgencies fall into these categories:

Publicity and Communications

Websites have augmented or supplanted handbills and newspapers for many undergrounds. Public components can maintain websites in the open if they separate themselves from underground components. Those that advocate terrorism more directly must engage in an ongoing shell game of shifting servers, changing web URLs, and redirecting e-mails as they are shut down directly or hosts are pressured to remove them.

The Tamil insurgency in Sri Lanka made particularly effective open use of the Internet for publicity through sites including Tamilnet. com[34] and eelam.com.[35] Support from the diaspora was critical for the Tamil cause, and the Tamilnet website provides a much faster and easier means of keeping information flowing to distant supporters than printed newsletters did previously. In Sri Lanka, the government suppressed stories about police abuse (newspapers often printed "blacked out" stories in protest) and denied foreigners, including journalists, access to Tamil areas in the north and east of the country. The militant insurgents (the Tamil Tigers) also prevented unfettered coverage and did not permit journalists to enter territories they controlled unless they were sympathetic to the cause. Underground websites including Tamilnet (which does not appear to have been closely tied to the Liberation Tigers of Tamil Eelam) were often the most reliable source of information on government/Tamil clashes during the latter years of that insurgency. More than one observer has noted that despite the violent content, Tamilnet maintained the detached tone of a Western news source. Stories followed journalistic conventions and generally did not mix facts with commentary or political rhetoric. Even in accounts of graphic torture, stories were carefully sourced and referred to "alleged" police misdeeds. This suggests that Tamilnet's target audience went beyond its supporters and that the site sought to reach a skeptical foreign audience including the press. By providing information in a credible format, the site maximized its chances of being used as a source by foreign media.

The ability to recruit talent and work from distant locations is also key advantage of Internet-based journalism. The Tamilnet.com site was begun in Sri Lanka but benefitted from the assistance of Tamils living overseas, including a computer programmer from Norway, a systems analyst in London, and "dotcom entrepreneurs" from the United States. Distributing the technology does not obviate the danger for local reporters; a Tamilnet.com reporter was killed by a grenade tossed through his study window in 2000. But the decentralized nature of the Internet does provide more mobility to content providers. Endangered local writers and editors can also more easily move from place to place while continuing to produce content. Translating, editing, and some writing can also be done from overseas. As an example of this, Mark Whitaker describes a time when the editor of Tamilnet.com filed a story from Canada.[36] The story was about an instance of police brutality against Tamil detainees. The story was leaked by a member the Sri

Lankan government's own Human Rights Commission and included accounts of prisoners being tortured with boiling water, being forced to eat cow dung, and a string of similar abuses. The editor crafting the story was sitting in his nephew's bedroom in a suburb of Toronto, surrounded by Star Wars paraphernalia, trying to tune out six noisy nieces and nephews. (The same editor would be brutally beaten after his return to Sri Lanka during the fall 2001 election campaign.)

The Zapatista uprising is another example of a group using the Internet for publicity and effectively creating a counter-narrative to bypass government control of communications. The Mexican government sought to compare the Zapatista movement to the ongoing destructive insurgencies in Central America that Mexican citizens wanted badly to avoid. The Zapatistas were able, with the help of foreign NGOs, to communicate an effective counter-narrative. At a press conference soon after their January 1 uprising, Subcomandante Marcos declared that the EZLN (1) renounced Marxism and (2) did not aspire to overthrow the central government. These were understood to be important differentiators between Zapatistas and groups such as Nicaragua's Sandanistas and El Salvador's FMLN.

Clifford Bob[37] cites a number of factors as important for the EZLN's success in broadcasting their message internationally. First, the symbolic choice of the date of NAFTA's enactment was well chosen; many groups were already ready to mobilize around this issue. Second, the EZLN were proactive in seeking domestic and international attention. They benefitted from a charismatic and verbose spokesperson who went by the name Subcomandante Marcos. The group did not have a great deal of technical capability or savvy but received help in these areas from NGOs and academics who were. They gained important exposure by having their communiqués translated and distributed on an e-mail listserv, ANTHAP, hosted by Oakland University (Detroit) Applied Anthropology Research Computer Network. Information was also disseminated via a number of websites, including Chiapas95, which was established by an economics professor at the University of Texas at Austin.

The Zapatistas were also willing to adapt their message quickly to new audiences. Besides disavowing revolutionary intentions, they made a critical decision to refocus their message away from economic class struggle and onto the rights of indigenous peoples in Mexico, which proved to be much more in sync with international priorities. The EZLN's progressive feminist stance and strong female leadership, against the backdrop of a highly patriarchal indigenous culture, also proved to be an almost accidental boon; they drew support and attention from feminist groups that they might not have thought to appeal to directly.

The Zapatista case may also illustrate the limits of Internet-only communications. Bob argues that, despite the amplifying effects of the Internet, face-to-face connections were also critical to EZLN's success. EZLN leaders held in-person press conferences and traveled both within Mexico and internationally to promote their cause. They also encouraged supporters to come to them. More than 140 NGOs sent representatives to the area during the period immediately after the 1994 uprising.[38] EZLN actively supported and encouraged foreign sympathizers to visit Chiapas and made it possible for them to meet insurgents in person, an innovation dubbed "Revolutionary Tourism," similar to the very successful "ecological tourism" industries established elsewhere in Central America.

Targeting the Enemy

A less common goal of Internet communications seems to be in terrorizing or manipulating the opponent. The Internet offers too much choice and control to be manipulated in the same way that commercial broadcast media can sometimes be. There are some examples of this usage, however. In 1999, Hizbollah revealed on its website that a coffin returned to Israel contained body parts of a number of soldiers, not just one.[39] Following this gruesome revelation, some parents of Israeli soldiers reluctantly began to monitor Hizbollah.org because of the possibility it would have information on attacks not available elsewhere.

Recruitment and Radicalization

According to Rita Katz, director of the Search for International Terrorist Entities Institute Intelligence Group, "We know from past cases—from captured al Qaeda fighters who say they joined up through the Internet—that this is one of the principal ways they recruit fighters and suicide bombers." [40]

Videos of successful terrorist attacks or guerilla missions are popular features of sites focused on recruitment. Videos often show extreme graphic violence, sometimes accompanied by audio or text commentary. Producing these videos is important enough that a number of violent groups, including Hizbollah, the Chechen resistance, and Al Qaeda, routinely include a videographer as an essential part of an operational team. These videos serve several functions. First, they attract attention and excite passions of sympathizers, particularly young males who may be recruited to perform these types of actions. Second, they create mental imagery, allowing recruits to imagine themselves as successful operatives, and repetitive video reinforces the message that attacks are likely to be successful. Reading about a successful attack is not as compelling as watching one unfold in real time from the vantage point of someone who was involved and lived to deliver the video footage. Third, they begin the process of desensitizing recruits to violence and the parallel process of dehumanizing opponents. (See Chapter 6 on Group Dynamics and Radicalization.)

Hizbollah used the Internet and another emerging media for similar effects by creating and distributing a video game called "Special Forces."[41] The game is in the "first person shooter" genre. Players can take "target practice" by shooting at an avatar of then-Israeli prime minister and Israeli Defense Force chief of staff. Players can then play in recreated missions against Israeli tanks, helicopters, and fortified positions. Instructions for play are available in Arabic, English, French, and Farsi; the game claims to have sold more than 10,000 copies.

Many recruitment sites focus on adolescent, mostly male, recruits. One HAMAS website called Al-Fateh ("The Conqueror") goes even further, explicitly targeting children. The site features cartoon-like graphics and (as the website describes) "pages discussing Jihad, scientific pages, the best stories, not to be found elsewhere, and unequalled tales of heroism." Despite its audience, Al-Fateh does not shy away from violent imagery; the site at one time featured a photo of the decapitated head of a young female suicide bomber, with praise for her act.[42]

At some point in the recruitment and radicalization process, recruits usually make personal contact with mentors or organizational liaisons to further training or coordinated activity. Technology facilitates this process by allowing users to move from viewing a website to sending an e-mail, posting on a discussion board, or joining a real-time chat room. A 2010 analysis of the Irish Republican Board (http://admin2.7.forumer.com/) found invitations to take part in marches, invitations to support prisoners by writing letters, and links to online petitions, as well as more direct invitations to join Republican organizations. [43] The board also includes a chat feature, offering communication with like-minded members that is both more immediate and less public.

The transition from passively viewing to interacting involves increased risk for both parties. Revelation of personal details or concrete biographical facts increases the chances of being identified and geolocated. The Internet's anonymity provides partial protection but also means that both recruiters and recruits can be "spoofed" by opponents or law enforcement playing a role.

Providing contact information on illicit sites requires frequently changing that contact information and redirecting e-mail. In 2004, a U.S. researcher with the PRISM center[44] attempted to trace an e-mail contact for Ansarnet.ws, a group supporting global jihad. The address, sout@mail4all.us, was posted by an Algerian using the nickname of

"Sulayman ibn Al-Qa`qaa." The e-mail address redirected through www.mail.com, and the mail4all.us domain was registered to a New Jersey address through GoDaddy.com. The poster stated in his instructions that the e-mail would be changed from time to time, and he cautioned Saudis not to use a phone line or e-mail account that could be traced when contacting the group.

Training

A variety of insurgent groups are experimenting with use of the Internet as a channel for training in operational techniques. Self-training manuals such as *The Terrorist's Handbook*, *The Anarchist's Cookbook*, and *The Mujahadeen Poisons Handbook* are available online. *The New York Times* interviewed a Palestinian called Abu Omar who had been employed as a trainer in Iraq, teaching foreign fighters how to make bombs and stage roadside attacks; at the time, he was working with two cameramen, who were videotaping his bomb-making classes, to produce Internet instructional videos.[45] Marc Sageman's book *Leaderless Jihad*[46] reviews a number of attempts at completely independent self-training. One example of a simple technique that has been taken up by disconnected amateurs is "tree spiking." This technique of driving spikes into trees to destroy loggers' chainsaws was propagated by the Earth Liberation Front as one of a number of independent "monkeywrenching" techniques that have been used by individuals with no explicit connection to the front organization.

It is unclear to what extent it is realistic to expect that insurgents can self-train to a level of competency to make a serious contribution to an insurgent effort. Bomb-making techniques, for example, require considerable practice and expertise. In other complex domains, successful e-learning usually requires some personal interaction, contact, and feedback from experts. Self-training is probably more effective for disseminating new techniques or counter-measures to already-trained operatives than for training absolute novices.

Fundraising

E-commerce is an important aspect of private Internet usage, although it is not as often a part of insurgent fundraising. The first generation of insurgent websites included explicit appeals for online donations; however, subsequent legislation preventing fundraising for terrorist organizations forced this activity underground. Aboveground websites sometimes make money by selling souvenirs and may imply that the money will support the insurgent cause. The 32 County Sovereignty Movement, a group associated with the Real IRA (RIRA) at one time joined Amazon.com's "Associates" program and received a cut from book sales when they redirected visitors to buy those books at Amazon, but the company removed the PIRA from the Associates program when it was informed of the group's insurgent ties.[47]

Financial transactions are easier to track than other kinds of information that flow over the Internet, making fundraising more difficult for insurgents than recruitment or publicity. Finding new ways of soliciting funds over the Internet is another fast-changing cat-and-mouse game between insurgents and their opponents.

Command and Control

The Internet can also facilitate command and control. In the leaderless jihad model previously mentioned, groups may attempt completely decentralized command and control, for example, by openly suggesting targets and tactics and hoping that self-managed groups will enact them. The website irelandsown.net once published specific information related to Prince William's boarding school, including a suggested location for an attack.[48] This brazen type of suggestion, of course, also tends to alert law enforcement and potential targets.

One of the boldest uses of Internet for command and control is Hizbollah's dedicated fiber optic network. This network was emplaced parallel to Lebanon's legitimate cable television and Internet lines. It was a sign of Hizbollah's political influence that when the government threatened to dismantle the network, Hizbollah was able to pressure them to back down and leave the network in place.[49]

Cell phone-based text messaging has been successfully used to coordinate anti-government protests. The World Trade Organization (WTO) protests in Seattle pioneered the use of social media for this purpose. Strategic movement of crowds as a protest tactic has been used for many years but previously tended to rely on pre-planned sequences and formations similar to how military movements were once limited; cell phones, text messaging, and Internet-based communications allowed much greater flexibility in the movement of demonstrators and diversionary forces in response to police movements. (See Chapter 10: Nonviolent Resistance for a more detailed discussion of the so-called "Battle in Seattle.")

Internet and cell phone technology are thought by some to have played an important mobilizing role in the Iranian protests after the disputed 2009 elections, although this claim is controversial. The use of Twitter during these protests received a great deal of attention.[50] Twitter is a very flexible text messaging service that can be used either to broadcast to a large audience or to send personalized messages among friends, and the messages can be broadcast using either the Internet or SMS (cell phone based). It is clear that Twitter, along with other services such as the YouTube video-sharing service, were closely monitored by people outside the country who wanted to follow events. Some authors have questioned whether Twitter played an important role in mobilizing and organizing the protests themselves.[51]

On June 16, during these protests, the U.S. Department of State contacted Twitter to ask them to delay a scheduled server upgrade that might have disrupted Twitter traffic. Later, the Iranian government intentionally disrupted Twitter traffic by shutting down or throttling (slowing) both Internet and cell phone services in sections of Tehran.

There is a clearer case to be made that the Internet played a role in mobilization of popular protests in Egypt, which forced the resignation of Hosni Mubarak in 2011. One group at the center of the movement was the 6th of April Youth Movement, whose main outreach mechanism was a publicly accessible page on Facebook. The group had organized earlier demonstrations on April 6, 2008, with minimal impact, but had continued to gather followers, mostly among younger and better-educated Egyptians. This group was also able to communicate with and learn nonviolent protest tactics from the Serbian student group Otpor.[52] The January 25 protests, however, were not preplanned; they were an opportunistic response to events unfolding in Tunisia. A protest was hastily scheduled for the nearest available holiday, which ironically was "Police Day." Such rapid, coordinated response requires open and accessible communication channels, which the Internet provided. Mubarak's government shut down almost all Internet access to the country on January 26 in an attempt to control the movement. This was done relatively easily by pressuring four major telecommunications providers in Egypt. The move was ineffective at dispersing the already-assembled crowds, however.

It is important to note that the movement was not entirely organized and conducted online. Only a fraction of Egyptians have Internet access, and the movement did not engage in the more complex crowd maneuvers of the WTO protestors. But in a country where media is controlled and public gatherings attract police attention, the initial group assembly, instruction in tactics, and opportunistic mobilization would have been very difficult via any other available media.

More traditional media benefit from Internet distribution. A twenty-six-page pamphlet with instructions for protestors played an important role in the Egyptian movement and was distributed in either print or pdf format from person to person (Figure 8-1). Instructions on the front of the pamphlet urged that it not be posted on publicly accessible Internet sites, however, saying: "Please distribute through e-mail, printing and photocopies ONLY! Twitter and Facebook are being monitored. Be careful not to let this fall into the hands of the police or state security." The page reproduced from this handbook shows one of the more important tactics used by Egyptian protestors, which was to assemble groups outside of Tahrir Square. This allowed them to swell numbers by recruiting local residents, and, more importantly, it allowed them to aggregate people in the square more quickly and in a manner that was harder to prevent or disperse. (See also *Chapter 10: Nonviolent Resistance.*)

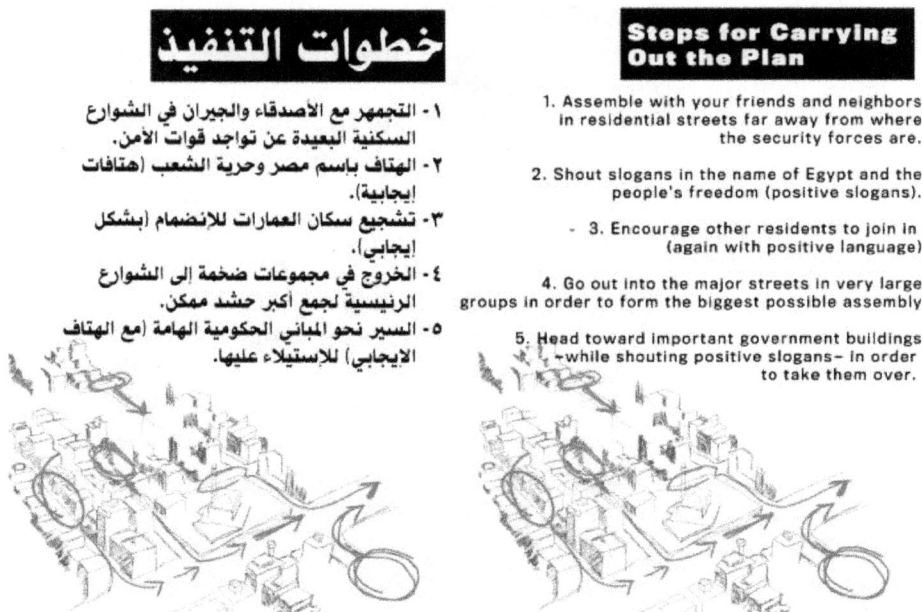

خطوات التنفيذ

١ - التجمهر مع الأصدقاء والجيران في الشوارع السكنية البعيدة عن تواجد قوات الأمن.

٢ - الهتاف باسم مصر وحرية الشعب (هتافات إيجابية).

٣ - تشجيع سكان العمارات للانضمام (بشكل إيجابي).

٤ - الخروج في مجموعات ضخمة إلى الشوارع الرئيسية لجمع أكبر حشد ممكن.

٥ - السير نحو المباني الحكومية الهامة (مع الهتاف الايجابي) للاستيلاء عليها.

Steps for Carrying Out the Plan

1. Assemble with your friends and neighbors in residential streets far away from where the security forces are.

2. Shout slogans in the name of Egypt and the people's freedom (positive slogans).

- 3. Encourage other residents to join in (again with positive language)

4. Go out into the major streets in very large groups in order to form the biggest possible assembly

5. Head toward important government buildings –while shouting positive slogans– in order to take them over.

Figure 8-1. Instructions for crowd assembly from Egyptian protestors' pamphlet. The page reproduced shows how the Egyptian protestors assembled groups outside of Tahrir Square, which allowed protestors to recruit local residents and aggregate people in the square more quickly and in a manner that was harder to prevent or disperse.

SUMMARY

Insurgencies do not survive on military strength alone; the ability to effectively convey a message using available media is critical to their success. Communication is typically an underground function. Undergrounds must have means of communicating with a number of audiences, including both internal and external supporters who provide recruits, funding, and support; they must also at times be able to reach nonsympathetic supporters and the enemy, who are the objects of intimidation or demoralizing messages.

Successful undergrounds have been opportunistic in their use of whatever media is at hand, up to and including the modern era. Handbills and newsletters have been used to good effect; effective delivery channels and production qualities are keys to success.

Manipulation and management of broadcast media became critical functions in the post-World War II era. Global broadcasts using near-real-time satellite transmissions changed the game for insurgent groups starting in the 1970s. Middle Eastern groups became particularly adept at staging "terrorist spectaculars" such as the PLO's Munich Olympic kidnappings in 1972 and the prolonged hijacking/hostage crisis perpetrated by Hizbollah in 1985. Groups discovered still-relevant principles for attracting and holding media attention, such as providing media access and drawing out the event over time. Even when these were tactical failures they were judged public relations successes and encouraged use of terrorist tactics against high-profile targets. Many accused broadcast networks of being too easily manipulated by such tactics, and over time the media have responded with guidelines and practices to avoid becoming inadvertent spokespersons for terrorists. Competition between news sources for immediacy and airtime works against these self-imposed limits, however.

The Internet era has ushered in a great deal of new capabilities that undergrounds are still learning to exploit. Groups such as Mexico's Zapatistas and the erstwhile Tamil resistance in Sri Lanka found that the Internet greatly increases the ability to both communicate with and garner support from foreign sympathizers. Extremist groups use both public and illicit websites to spread messages of hate, promote violent tactics, and attempt to provide training for would-be terrorists. The Internet is also becoming useful for mobilization and coordination of protest movements, such as the 6th of April Youth Movement that ultimately ousted Hosni Mubarak of Egypt. Governments have not been idle in the face of these changes in tactics; regimes in Iran and China have been particularly proactive in Internet counter-strategies, in an ongoing technological arms race between undergrounds and those that they seek to overthrow.

ENDNOTES

[1] Harry M. Cleaver, "The Zapatista Effect: The Internet and the Rise of an Alternative Political Fabric," *Journal of International Affairs* 51, no. 2 (1998): 621.

[2] Rodolfo Montes, "Chiapas Is a War of Ink and Internet," *Reforma*, April 26, 1995, translator unknown.

[3] Cleaver, "Zapatista Effect," 621.

[4] Bruce Hoffman, *Inside Terrorism* (New York: Columbia University Press, 1998).

5   Christina Meyer, *Underground Voices: Insurgent Propaganda in El Salvador, Nicaragua and Peru* (Santa Monica, CA: RAND, 1991).

6   Jonathan Schanzer, "The Iranian Gambit in Gaza," *Commentary* 127, no. 2 (2009): 29–32.

7   Fred Barton, *Salient Operational Aspects of Paramilitary Warfare in Three Asian Areas (ORO-T-228)* (Chevy Chase, MD: The Johns Hopkins University, Operations Research Office, 1963).

8   Meyer, *Underground Voices*.

9   Ibid.

10 Vladimir Ilyich Lenin, *What Is to Be Done? Collected Works*, vol. 5 (Moscow: Foreign Languages Publishing House, 1961), 20.

11 Seth S. King, *The New York Times*, March 17, 1965.

12 George K. Tanham, "The Belgian Underground Movement 1940–1944" (Ph.D. diss., Stanford University, 1951): 221–226.

13 Edward Martin, *Communist Subversion in the Western Hemisphere* (Washington, DC: U.S. Department of State, 1963).

14 Meyer, *Underground Voices*.

15 Hoffman, *Inside Terrorism*.

16 Ibid.

17 Christopher Dobson and Ronald Payne, *The Carlos Complex: A Study in Terror* (New York: Putnam, 1977), 15.

18 Hoffman, *Inside Terrorism*, 69. Original quote from Abu Iyad's *My Home, My Land: A Narrative of the Palestinian Struggle*.

19 Hoffman, *Inside Terrorism*.

20 Hoffman, *Inside Terrorism*, 69.

21 Garrick Utley, "The Shrinking of Foreign News: From Broadcast to Narrowcast," *Foreign Affairs* 76, no. 2 (1997): 2–10.

22 Hoffman, *Inside Terrorism*, 175.

23 Anatol Lieven, *Chechnya: Tombstone of Russian Power* (New Haven, CT: Yale University Press, 1998).

24 Gabriel Weimann, *Terror on the Internet: The New Arena, the New Challenges* (Washington, DC: United States Institute of Peace Press, 2006).

25 Marvin Kalb, "The Israeli—Hizbollah War of 2006: The Media as a Weapon in Asymmetrical Conflict," *Harvard International Journal of Press/Politics* 12, no. 3 (2007): 43.

26 List adapted from Gabriel Weimann and C. Winn, *Theater of Terror: Mass Media and International Terrorism* (White Plains, NY: Longman Publishing Group, 1994).

27 Anita Perešin, "Mass Media and Terrorism," *Medij. Istraž* 13, no. 1 (2007): 323, http://hrcak.srce.hr/file/28073.

28 Committee of Ministers of the Council of Europe, "Declaration on Freedom of Expression and Information in the Media in the Context of the Fight Against Terrorism," http://www.coe.int/t/dghl/standardsetting/media/doc/cm_EN.asp.

29 Meyer, *Underground Voices*.

30 Avi Jorisch, *Beacon of Hatred: Inside Hizballahs Al-Manar Television* (Washington, DC: Washington Institute for Near East Policy, 2004).

31 Meyer, *Underground Voices*.

32 Martin, *Communist Subversion*, 350.

33 Douglas Pike, *The Communication Process of the Communist Apparatus in South Vietnam* (Cambridge, MA: Center for International Studies, Massachusetts Institute of Technology, 1964).

34 Mark P. Whitaker, "Tamilnet.Com: Some Reflections on Popular Anthropology, Nationalism, and the Internet," *Anthropological Quarterly* 77, no. 3 (Summer, 2004): 469–498.

[35] Pradeep Jeganathan, "Eelam.Com: Place, Nationa, and Imagi-Nation in Cyberspace," *Public Culture* 10, no. 3 (1998): 515.

[36] Whitaker, "Tamilnet.Com"

[37] Clifford Bob, *The Marketing of Rebellion: Insurgents, Media, and International Activism* (New York: Cambridge University Press, 2005).

[38] Ibid.

[39] Maura Conway, "Reality Bytes: Cyberterrorism and Terrorist 'Use' of the Internet," *First Monday* 7, no. 11 (2002).

[40] Weimann, *Terror on the Internet*, 120

[41] Hoffman, *Inside Terrorism*, 209.

[42] Gabriel Weimann, "WWW.AL-QAEDA: The Reliance of Al-Qaeda on the Internet," in *Responses to Cyber Terrorism*, ed. Centre of Excellence Defence Against Terrorism (Amsterdam, The Netherlands: IOS Press, 2008), 61.

[43] Lorrain Bowman-Grieve, "Irish Republicanism and the Internet: Support for New Wave Dissidents," *Perspectives on Terrorism* 4, no. 2 (2010).

[44] Reuven Paz, "Who Wants to Email Al-Qaeda?" *PRISM Series of Global Jihad* 2, no. 2 (July 2004), http://www.e-prism.org/images/PRISM_no_2_vol_2_-_Who_Wants_to_Email_Al-Qaeda.pdf.

[45] Michael Moss and Souad Mekhennet, "An Internet Jihad Aims at U.S. Viewers," *The New York Times*, October 15, 2007.

[46] Marc Sageman, *Leaderless Jihad: Terror Networks in the Twenty-First Century* (Philadelphia, PA: University of Pennsylvania Press, 2008).

[47] Weimann, *Terror on the Internet*.

[48] Myra Philip and Dennis Rice, "Omagh Killers Target William," *Daily Express*, October 14, 2000, http://www.mail-archive.com/kominform@lists.eunet.fi/msg03972.html.

[49] Weimann, *Terror on the Internet*.

[50] Lev Grossman, "Iran Protests: Twitter, the Medium of the Movement," *Time Magazine*, June 17, 2009, http://www.time.com/time/world/article/0,8599,1905125,00.html.

[51] Maximillian Forte, "America's Iranian Twitter Revolution," *Open Anthropology* (blog), June 17, 2009, http://openanthropology.wordpress.com/2009/06/17/americas-iranian-twitter-revolution/.

[52] Dusan Stojanovic and Jovana Gec, "Serbian Ousters of Milosevic make Mark in Egypt," *Associated Press*, February 22, 2011.

CHAPTER 9.

PSYCHOLOGY OF INFLUENCE

CHAPTER CONTENTS

Nathan Bos and Jason Spitaletta

This chapter is about influence and persuasion. Understanding how human opinions and subsequent behaviors are shaped is a topic of fundamental importance to anyone concerned with "winning hearts and minds," including underground organizations struggling against superior forces, governments seeking to gain or bolster legitimacy, and military forces engaged in counterinsurgency or asymmetric warfare.

This chapter adds to and complements other chapters in this book that discuss specific aspects of influence, including Chapter 8 on insurgent use of media, Chapter 7 on psychological risk factors, and Chapter 6 on group dynamics and radicalization. This chapter will focus more narrowly on trying to connect the academic literature on influence to the problems relevant to undergrounds.

This chapter will review research by cognitive psychologists and communication theorists who have studied how individuals process persuasive messages; by social psychologists and sociologists who have studied various forms of social influence; by political scientists who have studied the effectiveness of influence campaigns, the effects of coercion, and the effects of economic variables; and by marketing and negotiation researchers who have studied the effects of different influence strategies. The chapter concludes with a study of narratives as justifications for insurgent activity, and an analysis of how different types of insurgent groups employ different narratives.

PSYCHOLOGICAL OPERATIONS

Few insurgencies have been won or lost by large, decisive military battles. More commonly, insurgencies are won by a combination of military and political means. Much of the political leverage involved in such settlements is derived from effective psychological operations that have structured the environment necessary for a political solution. To insurgents, influencing opinion and attitudes is not an end in itself, but only a means to enhance their ideology and/or organizational work among broad elements of society. Sometimes the goal is recruitment, sometimes support or complicity, and sometimes the objective of psychological operations is to create social disorganization and conditions of uncertainty. The resultant unrest and confusion are used as a cover to carry out underground operations.

Underground psychological operations are conducted in a variety of forms: mass media and face-to-face persuasion; leaflets and theatrical performances; programs for local civic improvement; and threats, coercion, and terror. Although the substantive content of psychological operations during any phase of a campaign is likely to be determined at the highest echelon of the organization, successful implementation depends in large part on the ingenuity of the operators at the local level. In attempting to influence mass action, and to develop mass support, psychological operations are directed primarily at specific audiences or groups within a population. Occupational, religious, ethnic, and other social groups are often singled out as "target audiences," and tactics are tailored to be effective within a particular group. The purpose of underground propaganda may be to win support among the neutral and uncommitted; to raise morale and reinforce existing attitudes and beliefs among underground members and their supporters; to undermine confidence in the existing government; and to lower the morale of government forces and personnel.

There is an enormous amount of academic research on the process of influence, but applying this research is sometimes difficult because most of the research is done in Western countries, often on college campuses, and far from the contentious politics and threats of violence that characterize underground and insurgent struggles.[1] Nevertheless, from the aggregation of prior work, we will attempt to identify fundamental principles relevant to both undergrounds and those who oppose them.

ASPECTS OF INFLUENCE

The ability to exert influence over the populace is among the most critical of all the requirements of an underground. Rogers's seminal book, *Diffusion of Innovation*,[2] begins with the story of what seems to be a simple persuasion challenge: A social worker named Nelida, aided by the government and medical authorities, spends several years attempting to convince rural villagers in Peru to begin boiling their water before drinking it. Given the toll that typhoid and other water-borne diseases takes on the population, and the relatively low cost of this innovation, this would seem like an easy sell. However, after spending two years and making hundreds of visits to family homes, she achieved only modest success—eleven families have begun regularly boiling their drinking water. Furthermore, very little of her success was due to her scientific and medical explanations of germ theory and water-borne contaminants. Instead, families were persuaded to change their behaviors for a variety of other reasons. For most families, boiling water violated local norms: "local tradition links hot foods with illness. Boiling water makes water less 'cold', and hence, appropriate only for the sick. But if a person is not ill, the individual is prohibited by village norms from drinking boiled water."

A woman referred to as "Mrs. B" adopted water boiling because she was an outsider to the village and thus less beholden to the "hot" and "cold" food norms. Mrs. B also felt somewhat shunned, and the social worker, Nelida, had become an important social contact for her. "Anxious to secure social prestige from the higher-status Nelida, Mrs. B adopted water boiling, not because she understood the correct health reasons, but because she wanted to obtain Nelida's approval." For most villagers, however, Nelida's status as an "outsider" meant that her message would probably be ignored and advice unheeded.

Another resident who did adopt water boiling, known as Mrs. A, suffered from a sinus infection and was regarded by her peers and herself as "sickly." Mrs. A accepted the practice because the idea was compatible with her belief system ("hot" foods are for sick people) and the associated behavior was compatible with her self-concept. Neither Mrs. A nor Mrs. B was an influential person in the village, however, so their adoption did not much affect their neighbors. For the small number of healthy, local people who did adopt water boiling, they did so only after somehow translating or rationalizing the message received so that it would be in line with their own belief system.

This example illustrates several central principles of influence:

- New ideas are most readily accepted when they are connected to, and compatible with, existing beliefs.
- Self-concept is critical: new ideas, and especially new behaviors, are evaluated as to whether they are compatible with an individual's self-concept and will likely be rejected if they are not.
- Social networks influence both beliefs and actions. Messages are evaluated on the basis of the status and identity of the source as much as for their information content.
- Social identity associations (in-group versus out-group distinctions) also influence both beliefs and actions.
- Changes in beliefs occur slowly, in stages, especially when changes in behavior are required.

INFLUENCE AS A SOCIAL PROCESS

Influence is fundamentally a social process that occurs on a number of levels. Influence can happen face-to-face, within small groups of people, through larger social networks, or through one-to-many broadcasts. In all of these social settings, charisma, nonverbal communications, and similarity of opinions have a strong effect on whether minds are changed. When influence happens through social networks of dozens or hundreds of individuals, the structure of the networks and opinions of key people in the network are critical. Influence also happens on larger, less personal scales. All humans see themselves as members of larger groups, including social identity categories such as ethnicity, religion, and ideology. These identity groups can come to be associated with each other, and with certain opinions or attitudes, and these associations affect individuals in subtle yet powerful ways. A predominant function of undergrounds is to determine the best methods to exert influence over their selected target audiences. In the nascent stages of a revolution, this is done almost exclusively through clandestine means, using pre-existing social networks.

Interpersonal Influence

What makes a certain individual more influential than others in a face-to-face setting? The ability to influence others is sometimes called charisma, which has many components. Charismatic leadership, discussed in Chapter 4, is often a boon to a nascent group; it can also serve as a vulnerability if the leader is effectively targeted. In certain cases, most notably Che Guevara, charisma can extend beyond the lifespan of the individual. Nevertheless, charisma in any of its manifestations is a valuable asset to any revolutionary.

Physical attractiveness gives a measurable advantage to people attempting to influence others, at least when they are physically present or visible on a broadcast medium. There are several underlying cognitive mechanisms that give attractive people their advantage. Attractive people can simply command more attention, increasing the likelihood that a message will be heard and understood. Attractive people may evoke positive moods, indirectly increasing idea acceptance through association. Attractive people may also be more well liked and evoke a desire in the listener to be liked in return.

The ability to communicate energy and positive emotions is a strong component of influence; much of this communication is nonverbal. Humans naturally synchronize both verbal and nonverbal behaviors. The degree to which they synchronize behavior is a strong indicator of whether they are establishing rapport or not. Humans quickly and nonconsciously imitate facial expressions of those with whom they interact.[3] Studies of brain activation measured with functional magnetic resonance imaging (fMRI) suggest that perceiving facial expressions activates the neural patterns for those facial expressions in the perceiver. That is to say, when a person perceives an expression of a recognizable emotion (happiness, sadness, fear, disgust), their own neural areas for those emotions are activated, and they tend to feel the corresponding changes in emotions. This often happens below conscious awareness, such that one may not even realize why one's emotions are being changed. The tendency to mirror the emotional states of people nearby is called emotional contagion.[4]

Besides facial expression, a number of other nonverbal signals show similar mimicry effects, including gestures, head movements, and indirect indicators of arousal. Emotions are also conveyed through use of language. People who are working well together tend to adopt each others' vocabulary and phrases and develop shared ways of referring to objects of mutual interest. Emotional contagion affects groups as well as pairs of people,[5] can show effects over long periods of time, and has even been shown to occur through text-only communication, in the absence of facial cues.[6]

Some people are naturally more expressive than others, and these individuals have been shown to have more influence over others' emotional states. Experiments show that expressive people exert this unequal influence even when subjects are sitting silently together and not interacting.[7] This quality is a component of charismatic leaders[8] and is one of the qualities that distinguish skilled actors. In general, synchronization of verbal and nonverbal patterns between conversational partners leads them to like each other more. Synchronization at these different levels is also strongly related to perceptions of competence, confidence, social skills, and negotiation success, and it has also been related to persuasiveness. Negotiators have a critical advantage if they learn to use mimicry and synchronization effectively.

Nonverbal communications can also be used to assert control or dominance. In conversation the listener normally tends to maintain eye contact while the person speaking does not. Violating this norm by continuing to hold eye contact while speaking is a signal of dominance that is referred to as "gaze control." Higher levels of eye contact in general are also associated with credibility, competence, and social skill.[9] Voice volume, voice pitch (low), and control of conversational turn-taking are other common nonverbal expressions of dominance.

Many other components of charisma, beyond physical and nonverbal characteristics, have been identified. A few of these are:

- Vision and intelligence: studies of leadership have shown that command of a situation and a vision for action are key differentiators of effective leadership.

- Similarity: individuals pay more attention to and are much more influenced by people who hold similar opinions and have similar social characteristics such as the same ethnicity.

- Liking: personality and communication style affect how much a person is liked, thereby strongly affecting that person's influence.[10]

- In some cultures, being male, being older, and being wealthier have a strong affect on influence, although others can often achieve indirect influence.

Social Pressures and Social Networks

In the study of water boiling in rural Peru, cited by Rogers and discussed at the beginning of this chapter, one variable was more important than all others in predicting whether a certain person had adopted the practice: their specific village of residence. The social worker Nelida worked in a village that had, for the most part, rejected the practice; in other villages adoption was nearly universal. For individuals, the predominant view of their social network tended to supersede their individual characteristics. This can be understood as an effect of conformity and a result of the structure of their social networks.

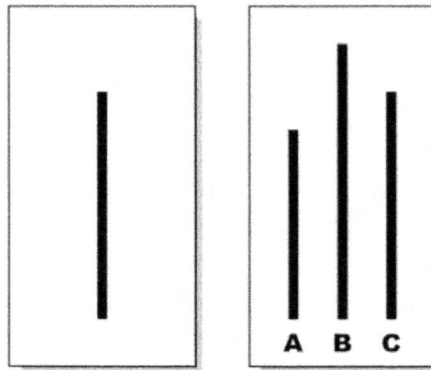

Figure 9-1. Which line segment matches the one on the left? In Asch's famous experiment on conformity, subjects were asked to state which of three line segments (A, B, or C) is the same length as the line on the left. When subjects performed this task alone, they almost always answered correctly. But when Asch placed people on a panel of supposed peers, who were actually accomplices of the experimenter, and peers unanimously gave an incorrect answer, respondents would go along with the crowd and give answers that were obviously incorrect about 37 percent of the time. Image from Wikimedia Commons: http://en.wikipedia.org/wiki/File:Asch_experiment.png.

Asch's famous experiment on conformity,[11] described in most introductory psychology textbooks, is nevertheless still compelling. Asch recruited subjects to sit on a panel and make a very simple judgment: to determine which of three line segments was the same length as a fourth. The task was made easy enough that when the subjects performed the task alone, subjects were correct more than 99 percent of the time (see Figure 9-1). But when Asch placed people on a panel of supposed peers, who were actually accomplices of the experimenter, and peers unanimously gave an incorrect answer, respondents would go along with the crowd and give answers that were obviously incorrect approximately 37 percent of the time. It is important to note that there was no overt social pressure to conform in these studies, or penalty for disagreement; the pressure to conform was unstated yet strongly felt. The power of conformity is an important part of the indoctrination used by all military and paramilitary forces. Conformity is much stronger when opposing opinions are unanimous; when Asch's subjects had just one other person who agreed with them, they were much less likely to conform (about 75 percent less). This is why social or physical isolation is also a hallmark of indoctrination. Underground movements have been described as "normative-coercive" organizations, because they employ both persuasive group pressures and overt coercion. They are normative in that institutional norms and mores secure behavioral conformity to certain rules and in that group membership satisfies certain individual needs and desires. Coercive power is applied at the same time through the threat or application of physical sanctions or through the deprivation of certain satisfactions.

Younger people are typically more vulnerable to conformity pressure than older people. There are at least two reasons for this. First, they are simply less experienced in processing messages and resisting unwanted influence. Second, they are developmentally at a stage where peer opinions and questions of self-identity are of paramount importance. The Viet Cong had a standard technique of separating young people from their elders to utilize this effect. According to a Viet Cong recruiting document, the first step was to organize large gatherings of young people for a celebration, a political rally, or a cultural event, segregating them from the older people of the village. During the gathering, recruiters made heated speeches denouncing the government. Earlier recruits were planted in the audience to applaud the speeches and volunteer to follow the Viet Cong to the mountain bases. Other young people, emotionally aroused by the speeches and seeing fellow teenagers "volunteering," also volunteered to go, sometimes en mass. Other young peoples' movements, such as "Hitler youth" programs and modern suicide bombing recruitment programs, use the same basic techniques of isolation and peer pressure.

Studies of social networks and influence continue to reveal how powerful networks are in shaping all kinds of opinions and behavior. A long-term study of health behaviors conducted in Framingham, Massachusetts, has found, for example, that close associates have a strong effect on peoples' individual decisions to quit smoking. Individual smokers were 67 percent more likely to quit smoking in a given year if their spouse quit smoking; 25 percent more if a sibling quit; 36 percent more if a friend quit; and 34 percent more if a coworker in a small company quit. These effects also propagated through the social networks; statistical effects of one person's behavior could be observed in presumed strangers several degrees of separation distant from them. The same authors found similarly strong effects on obesity and happiness, neither of which were previously assumed to be strongly determined by social influence.[12, 13, 14]

There is often a large discrepancy between rural and urban settings in the amount of pressure to conform. In rural villages, the social networks tend to be densely interconnected and the pressure of conformity very high. In small, rural villages, Communist recruiters often abandoned their strategy of trying to win over sympathizers a few at a time, assuming that they could not do this secretly or effectively, and instead adopted a strategy of converting the entire village using more coercive techniques.[15]

In larger, more urban and more diverse settings, social networks and conformity still matter, but there is more diversity along most dimensions, more competing influences, more complex patterns of affiliations, and less monolithic conformity to any idea. In these settings, *social identity* becomes an increasingly important factor in individual opinions.

Social Identity and Influence

Social identity refers to categorizations of people that affect, often on a deep level, an individual's self-concept, beliefs, and behavior. The dynamics of social identity group formation and activation and their role in conflict are discussed in more detail in Chapter 6, which addresses group dynamics and radicalization. The social identity categories that are typically most relevant for political conflicts are race, ethnicity, tribe, religion, social class, gender, ideology, political loyalty, and regional affiliation. Social identity has a number of implications for the study of influence, some obvious and some less so.

Not surprisingly, social identity usually has a direct effect on influence. In general, a speaker who is perceived as belonging to the same category as the listener will be more persuasive than one who is perceived as being different. It is commonly understood that, for instance, an expatriate will be more persuasive than a foreigner in psychological warfare; female spokespersons are generally better for reaching a female audience; co-religionists are helpful as advisors on crafting messages on ethical or religious topics, etc.

Understanding social identity's effects is complicated, however, by a few factors. First, most individuals have numerous identity categories that are relevant to them. A typical resident of the United States might have a gender identity, a trade union or professional identity, a sense of pride in being an American, a political party preference, a religious affiliation, and an ethnic heritage inherited from immigrant ancestors. Second, these identities are not equally important to every member of a group on every issue. An ethnic group may have strong norms for dress but none for diet, while another group may have the opposite profile. For many Westerners, their religious identity has an effect on a variety of ethical issues but little effect on political positions. In many other settings, religion and politics are intertwined; examples can be found in (some) evangelical Christian groups in the United States, Catholic and Protestant groups in Ireland; Hindu and Buddhist organizations involved in the Sri Lankan conflict; and conservative Islamic groups across the Middle East. In many parts of the world there are strong alignments between ethnic groups and particular occupations, which often comes as a surprise to Westerners.[16] The connections between identity groups of all types are highly variable, culture-specific features. An important part of the process of understanding any cultural landscape is identifying that identity groups have unified or officially sanctioned positions in which spheres of influence. These positions may be highly localized or idiosyncratic, and making assumptions based on one's own culture is a frequent source of error. For example, religious leaders are key political actors in some towns, whereas other towns that are not far away keep religion and politics separate.

On the basis of their relevant social identity and associations with that identity, individuals may have a strong attraction or aversion to ideas or attitudes that is invisible to outsiders. One anecdote involved an American in England who bought tickets to a soccer (football) match and invited some British friends. He knew them to be fans of the team but met with strange, awkward reluctance to attend in person. Further conversation revealed a strong association between game attendance and a hooligan culture (an identity group) that the friends did not want to be associated with. An experimental example of this effect, which is called *normative pressure*, comes from a study at the University of Texas A&M, whose sports teams and students are referred to as "Aggies."[17] Students were asked for their opinions on two issues, one about illegal drug use and one about sexual equality in the workplace. When students were told that "Aggies" on average had a fairly extreme opinion on one of these issues, they had a strong tendency to agree with this self-relevant group, even when this extreme opinion was opposite of their original opinion. Students also came up with rationalizations for why members of their reference group had (they were told) strongly endorsed positions such as consideration of sex in job promotions, reinterpreting the question in creative but implausible ways to keep their own attitudes in synch with that of their identity group. Opposite effects were obtained when students were told the attitudes of an out-group they wanted to be disassociated with; given the (fictional) opinion of Ku Klux Klan members, students adjusted their opinions in the opposite direction.

The third, and perhaps most interesting, aspect of social identity and influence is that the salience of particular identity groupings is not equally relevant in every situation and that identities can be "activated," either by circumstance or by design, to achieve different effects. In one typical study, simply reminding Catholic high school students of their religious identity caused them to be more resistant to ideas perceived as anti-Catholic than other students who had not been reminded.[18]

Because most individuals have multiple identity affiliations, and because these identity groups carry different associations, activating different associations can have powerful effects on influence campaigns.

Anti-insurgent messages propagated by American forces in Iraq often stressed that violence was terrorism and was "against Iraq." This simple message attempted to evoke two identity-based associations. First, it created an association between resistance violence and terrorism, an association attractive to few people. More importantly, it evoked "Iraq" as the relevant identity group. This was at a time when much of the violence pitted Sunnis against Shiites. The message evoked a superordinate, unifying identity group, Iraqis. The message could plausibly have argued that violence was against "Islam," also superordinate to Sunni and Shiite. However, "Iraq" was the preferred identity reference because of the United States' goal of creating a unified, modern, secular state. Also, any message about Islam from the United States might have been rejected out of hand.

Psychological operations practitioners are often in the position of trying to influence people who are very aware that their messages are coming from an out-group, if not an outright enemy. An experienced World War II and Cold War propagandist gave this advice:

> I think our propaganda to Russia should be done on the supposition that we are talking to Communists (leadership). If you talk to the Russian peasant as though he is your friend and is opposed to the [Soviet] regime, he may patriotically turn off his wireless because he is a patriot as well as a peasant. . . . There were many Germans who did not like the [Nazi] regime, but if you appealed to them as traitors, they did not take the propaganda. They were much more content to

> overhear you talking to a Nazi. . . . nor is it any good
> pouring out the most moving exhortations to become
> a democrat because the enemy knows that this comes
> from the enemy, and writes it off as propaganda.[19]

Note how certain identity groups are carefully evoked and implied. Associations with the label "traitor" are assiduously avoided. By addressing the country's leadership as an identity group separate from the country's people, the propagandist attempts to drive a wedge between these groups. This propagandist also did not attempt to make an argument based on a superordinate "democrat" identity, which might have been effective in another context but not coming from an enemy in wartime.

DIFFUSION OF INNOVATION

The study of insurgencies as social movements and the applicability of diffusion of innovation dates back at least to the Special Operations Research Office's work in the 1960s.[20] Interpersonal communication, more so than mass media, typically determines whether an innovation is adopted. The process of diffusion can be considered "the acceptance, over time, of some specific item—an idea or practice, by individuals, groups or other adopting units, linked to specific channels of communication, to a social structure, and to a given system of values or culture."[21]

Researchers have also tried to understand how ideas move from individuals and small groups to large-scale social changes. Everett Rogers's 1962 book *Diffusion of Innovation*, referred to at the beginning of this chapter, was a landmark in this area of study and was ahead of its time in understanding the importance of social factors and social network structures.

Rogers proposed the now-famous curve shown in Figure 9-2. The curve divides the population into segments based on when they are likely to adopt a new innovation. The first people to try something new are usually *innovators*, who are likely to be younger, wealthier, and less risk averse than others. Innovators are not, however, the most influential members of society; they are in fact often regarded with bemusement or suspicion. Most people will not adopt a new idea or innovation because of the actions of a few innovators. More change happens if innovators are followed by a somewhat larger group of *early adopters*. This segment is more cautious than innovators and more concerned with losing face through making a foolish choice, but they also see an advantage in being on the forefront of something important. Early adopters also tend to be younger and more affluent than the general population. Diffusion does not really take hold, however, until a group of people known as *opinion leaders* embraces change. Opinion leaders are the most influential members of the society with regard to innovations in their domains of expertise. Unlike innovators, opinion leaders have a reputation for prudence, which they are careful to maintain, and are closely observed by many segments of society. When opinion leaders are observed embracing a new idea, innovations become widely known throughout the opinion leaders' large social networks, and the status of opinion leaders makes the innovations palatable to a much larger percentage of the population. Opinion leaders are not necessarily the highest-status individuals in a society, or the formal leaders, so identifying them often takes some investigation. Some opinion leaders will remain hidden until a social structure is very well understood. In rural regions, a person may be more likely to become an opinion leader by virtue of their character and personal relationships. In more urban, information-dense environments, the crucial characteristic of opinion leaders may be intelligence and more specifically their ability to monitor and process large amounts of information from different sources.[22] In more complex societies where information specialization is necessary, there is also more differentiation as to the type of influence an opinion leader wields; one person may be very influential in arts and culture, for example, but have very little influence on politics.

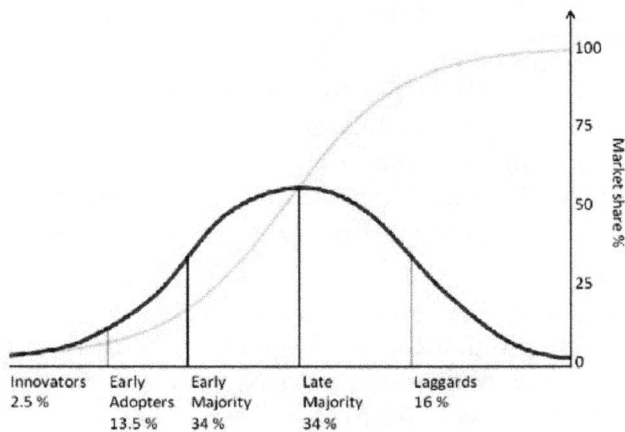

Figure 9-2. Everett Rogers's diffusion of innovation model, role distribution and market saturation. Image from Wikimedia Commons: http://en.wikipedia.org/wiki/File:Diffusionofideas.PNG.

Rogers defines one more important role beyond the opinion leader: the *change agent*. This is a person who is intentionally trying to bring about a large-scale change, possibly as a representative of some larger organization (political, corporate, or ideological). This person is often an outsider, such as the social worker Nelida in the water-boiling anecdote from the beginning of this chapter. They may also be political activists or members of an underground. Successful change agents must be able to identify and work with opinion leaders if their goal is to bring about large-scale change.

INFLUENCE AS AN INDIVIDUAL PROCESS

Most people are exposed to much more information than they remember and many more opinions than they decide to believe in or act upon. Effective influencers must understand how their audience will evaluate and accept or reject the content of their messages. Messages are evaluated on a number of levels: their personal relevance, the nature of the source, the factual content of the message, and the emotional content of the message. The study of rumors is a good introduction to the basic processes of message transmission and acceptance.

Rumors as an Example of Message Transmittal

The study of rumors has provided an interesting method to understand how people receive, reject, and pass on messages. Rumors are unconfirmed information that is widely spread from person to person. Sociologist Tamotsu Shibutani described rumors as "improvised news,"[23] which is especially relied upon when either information is not available through formal channels or formal channels are not trusted and becomes particularly important in times of conflict and war.

Rumors usually have strong emotional components. Knapp's study of wartime rumors in the United States during World War II[24] identified three prevalent types, each with a different emotional theme:

- The Pipe-dream or Wish rumor, which addresses a popular hope: "The Japanese do not have enough oil to last 6 months"

- The Bogie rumor, expressing fear: "The entire Pacific Fleet was destroyed at Pearl Harbor"

- The Wedge-driving or Aggression rumor: "Churchill blackmailed Roosevelt into provoking war with Japan"

Rumors should also have personal relevance; misinformation without personal implications for the listener will probably be forgotten rather than widely distributed. Wartime rumors centered on things such as rations, war bonds, combat stories that contradicted official reports, secret enemy actions, etc. Even seemingly irrelevant rumors, such as urban legends about bizarre household accidents, usually survive only if they hold some moral or practical lesson for the listener.

There are some exceptions to the personal relevance rule, however, such as salacious folklore about celebrities.

Rumors also tend to be distorted over time as they are passed between people. Rumors tend to be shortened and simplified over time; detailed information in original rumors (specific numbers or dates) tends to be forgotten; precise circumstances tend to be lost.[25] At the same time, rumors can pick up new (false) details, often serving the purpose of confirming the listener's preconceived ideas or stereotypes.

More recently Stephanie Kelly of the Naval Postgraduate School conducted a study of rumors prevalent among Iraqi citizens in Baghdad in the years after the United States' invasion in 2003.[26] Rumors were prevalent in this area even before the invasion and were also known as "whispering campaigns" or "souk-telegraph," emphasizing their role as an alternate news source.[27] Rumors about the multinational forces and their intentions were prevalent and were believed to hinder U.S. efforts, so a central database for rumor tracking called the "Baghdad Mosquito" was set up to allow both tracking and rapid response. Kelly found that Knapp's typology of rumors categorized these rumors fairly well; Kelly also added a category of "curiosity" rumors that have less-immediate implications but survive because they involve topics or individuals of interest.

The military value of rumors has long been recognized. As noted by Knapp, "In France, German agents disrupted morale by alternately spreading optimistic and pessimistic rumors in such rapid succession that the French people were almost too confused to act effectively in their own defense. The uses of rumor for propaganda are as varied as ingenuity can devise. Rumors have been used to force the enemy to issue information. They have been used to discredit enemy news sources that take the bait and leave themselves open to 'disclosure.'"

The recipe for creating a successful rumor is much simpler than the recipe for dispelling one. Common advice from experts on responding to rumors includes the following:

- Rapid response to potentially damaging rumors can limit impact and spread. Rumors should be countered with facts from authoritative sources quickly before the rumors change form. These official disclaimers are necessary but not sufficient.

- Combining public statements with visible public actions is more powerful than either words or actions in isolation.

- It pays to spend time developing relationships with opinion leaders and media sources before rumors arise; it may be too late to develop these relationships afterward. Reputation matters a great deal, for both individuals and organizations.

- Rumors are best gathered and countered in person-to-person communication, where feasible.

- An awareness of the length of the local "news cycle," which may be determined by media broadcasts, market schedules, holidays, etc., is useful.

Promoting the Understanding and Retention of Messages

Going beyond rumors, much communications research is directed toward the question of how to get persuasive messages about political issues, innovations, or products to be accepted by an audience.

Personal Relevance

As with rumors, personal relevance is critical to message acceptance. Psychological warfare practitioners have long understood the importance of target audience analysis, as did Maoist Agitprop teams, who interviewed informants and researched local grievances before entering an area.

Under stress, individuals' ability to receive and process messages can be reduced or distorted. People stressed by danger, or simply by high cognitive demands, experience a "narrowing of attention" onto matters most immediately salient or apparent. After a stressful event, the types of messages that can be received roughly follow Maslow's hierarchy of needs.[28] In 1994, a large contingent of Haitian refugees were interdicted at sea while trying to reach the United States and placed in temporary camps in Guantanamo Bay, Cuba. Immediately after the initial trauma of flight, interdiction, and unexpected resettlement, the refugees were preoccupied with basic physical survival needs of food, water, and shelter. Concern with physical safety followed. Later, as they became confident about their immediate survival and how they would be treated, they could pay attention issues related to broader considerations of migrant versus refugee status and political groupings tied to Haitian politics (John Shissler [former J-2A, Joint Task Force 160, Operation Sea Signal], personal communication, January 2012).

Under less stressful conditions, individuals' judgment about what is personally relevant can extend to wider social identity groups. Political scientists have studied what is called "economic voting" and have found that individuals do "vote their pocketbook" but also pay attention to the pocketbooks of their friends, neighbors, and countrymen. When researchers attempt to determine which is more influential, individual circumstances or identity group circumstances, conditions of the larger identity groups tend to be more important in predicting voting behavior. A few other interesting findings from the study of economic voting[29] are as follows:

- Economic change explains approximately one-third of the change in the vote (between elections) as opposed to other factors (policy, personal opinions about candidates, etc.).
- Voters pay most attention to "the big two" economic variables: unemployment/growth and inflation.
- Voters are myopic and so have a short time horizon. Voters react more to past events than to expected future ones.
- Voters may react more to negative changes than to corresponding positive ones.

While most of the research in this area has been done in the industrialized Western countries, there is evidence that economics matter in other settings. One study in Ethiopia showed that economic considerations did affect voting, although the effects were highly variable by district and economic considerations did seem to be of secondary importance behind ethnicity and patronage.[30] Another study in Ghana[31] found similar results: individuals' positive perceptions about the economy in the past five years corresponded with support for the current government, and optimistic beliefs about the next year had an even larger statistical relationship with support for the government. These opinions seemed to go beyond self-centered concerns and reflected an awareness of and interest in the economic situation of the region or nation.

Another study of eighteen Latin American countries[32] focused specifically on whether economic voting was more "egotropic," or concerned only with one's own pocketbook, or "sociotropic," indicating interest and concern with the larger economy. This distinction between self-centered and nation-centered concern is particularly important in countries struggling with patronage and corruption; self-centered voters (or their key leaders) might be more easily "bought," while nation-centered voters will be more likely to demand sound policies and real development. Research indicates that Latin American voters are more sociotropic, i.e., more influenced by the state of the national economy than their own pocketbook. Egotropic considerations are somewhat stronger in the poorest countries in this sample (e.g., Honduras, Nicaragua) but still less influential. Other factors that made a difference were the political fragmentation of the government (in divided governments voters do not know where to give credit or blame) and a leader's length of time in office.

Besides economic conditions, judgments of personal relevance are made on this basis of religious and ideological values, effects on relevant identity groups, and fears related to personal safety. An example of how personal connections to safety concerns affect behavior comes from a study examining rates of smoking cessation among doctors.[33] While almost all doctors were exposed to the increasingly compelling evidence linking smoking to cancer, many doctors continued smoking. One group, however, had a particularly high cessation rate—radiologists, who were frequently exposed to vivid images of cancerous growths in patients' lungs (see Figure 9-3). While all doctors were presumably exposed to the same evidence, and most believed it, the subset personally exposed to vivid visual evidence were more likely to be persuaded.

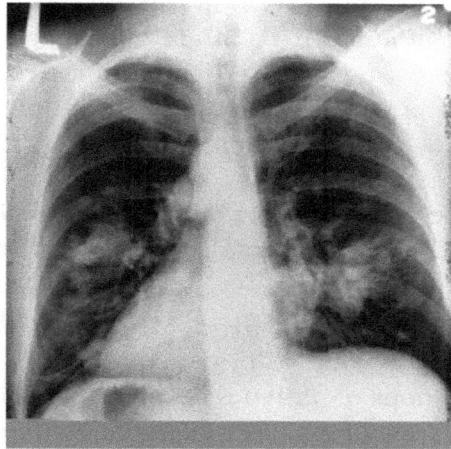

Figure 9-3. X-ray image of a chest showing a growth on the left side of the lung. From Wikimedia Commons: http://www.mediawiki.org/wiki/File:X-ray(Chest)Cancer.jpg.

Evaluating Message Content

Despite a great deal of research on the topic, there are very few universal findings about the content of persuasive messages, and there are many variables that seem to diverge across settings and cultures. Obvious findings are as follows: clear communication is better than unclear communication, authoritative sources increase message acceptance, and support with related evidence improves persuasiveness.

Evaluating evidence is more cognitively demanding than evaluating a source. Because of this, when a message is less relevant to a person, the credentials of the source often become more important. What seems to occur is that people make an implicit decision of how much attention to devote to the matter; when it is less relevant people will use a more "lazy" strategy of evaluating an argument based on the source rather than the evidence.[34]

251

The same seems to be true for "celebrity" endorsements, at least in one study; people were more influenced by the celebrity endorsement when the issues were less personally relevant to them but paid more attention to the arguments themselves when the issues were self-relevant.

Culture can, not surprisingly, have a strong effect on how persuasive a message is. R. S. Zahama[35] describes a set of differences between American and Arab communication styles that often lead to miscommunication. Among other distinctions, Arabs are more indirect especially with criticism, value linear argumentation less, and value the poetic use of language more.

Another study[36] found that culture, particularly the distinction between "collectivist" and "individualist"[a] cultures, predicted different responses to advertising messages. They conducted an experiment with American and Korean students and found that individualistic Americans are more persuaded by ads that promise individual benefits, personal success, and independence, whereas collectivist Koreans are more persuaded by ads that emphasize in-group benefits, harmony, and family integrity.

Epistemology is the study of the nature of knowledge and is often concerned with what "counts" as evidence. Cultures, academic specialties, and identity groups of all types develop their own epistemologies of what is and is not convincing. For example, fundamentalist religious groups usually have well-developed standards for argumentation. Religions that rely on holy writings often have a hierarchy, where there are pre-eminent sources (e.g., the Bible in Christianity, the Koran in Islam) and then other holy writings or commentaries of lesser authority. Arguments based on the Bible or the Koran generally trump arguments based on commentaries or any contemporary writings. A second type of evidence is the ruling of contemporary religious leaders; if they are respected as scholars, their words may be given the same credence as direct evidence from the holy writings because it is presumed that the leader is more learned than the reader. In Islamic scholarship, the academic lineage of the scholar is also very important; clerics inherit credibility from their teachers. Each religious group also develops particular ways of reading and using these writings as evidence. For an outsider, attempting to enter these discussions is a risky proposition because if it

---

[a] The contrast between collectivist and individualist countries has proved to be robust and has a surprisingly broad range of effects. Han and Shavitt[37] explain the distinction this way: "Individualistic cultures are associated with emphasis on independence, achievement, freedom, high levels of competition, and pleasure. Collectivist cultures are associated with emphasis on interdependence, harmony, family security, social hierarchies, cooperation, and low levels of competition."

is done clumsily or with obvious manipulative intent, it may provoke the opposite reaction intended. However, sincere curiosity and concerted attempts to study and understand a religion are often rewarded and can lend an outsider some credibility.

Message Repetition

Repetition of a consistent message is also a key to effective marketing of ideas. An interesting take on repetitive messaging comes from this World War II propagandist:

> We had to order our journalists (it was agony to them!) to repeat the same news day after day in forms in which it did not look the same. . . . 'News creation' does not mean invention . . . assume that you have been instructed to create the impression that typhus is prevalent in the enemy army. If you see that the fourth item in each news bulletin (not too high up, because if you put it as the No. 1 item it is obvious propaganda) is about a case of typhus,.. if you see that over three weeks the bulletin never lacks one true item about typhus, if . . . your leaflet newspaper reports a new case of typhus every day—and if they are all true— then you are doing good psychological warfare.[38]

Being exposed to simple messages repeatedly can sometimes have effects even when the message being heard is not the focus of attention. In one experiment, students were asked to evaluate the sound quality of audio tapes that contained arguments in favor of raising the price of their local newspaper. The more times a student was required to listen to the taped message, the more agreeable they were toward the price increase when asked about it by a different experimenter eight to fourteen days after the experiment.[39] For more complex arguments, however, more deliberate processing is usually required.

Presenting One Versus Both Sides of an Argument: Message Inoculation

There is some debate as to whether it is wise to try to present both sides of an argument or whether it is better to present only your own side. There is some good evidence that "inoculation" strategies, which present your own side along with a synopsis and warning about the other side's argument, can be quite effective under the right circumstances.[40] In one study, students were exposed to an anti-legalization argument about marijuana and then a pro-legalization message. They were less influenced by the second message when the first contained this inoculation statement: "Advocates for the legalization of marijuana claim the substance provides important medical benefits for people, especially for cancer patients who suffer from nausea following chemotherapy treatments. However, many studies show that marijuana provides only short-term antiemetic relief, and only for a minority of users . . ." Inoculation messages have been applied to such varied contexts as preemptive political attack ads, teenager anti-smoking campaigns, and competitive advertising campaigns. These experiments have shown that inoculating people to ideas beforehand is more effective than trying to counter ideas afterward. Inoculation even provides resistance to opinions on the same topic supported by different arguments than the ones "inoculated" against.[41,b]

COMBINING WORDS AND ACTIONS

During periods of uncertainty, people tend to be suspicious of what they hear and rely on their personal experiences, believing only those things which they have seen with their own eyes. When action immediately follows promises, this is called action propaganda. There are two common forms of action propaganda by insurgent groups: one focuses on specific actions that alleviate hunger and suffering among the people and demonstrate the insurgents' ability to accomplish set goals, while the other focuses on military acts, violence, sabotage, and punishment of traitors among the local population. Both show that the insurgents are powerful in spite of being outnumbered. It is a truism of psychological warfare but bears repeating: the combination of words and actions, when applied with planning and consistency, is more powerful than either used in isolation. This tactic, though universally acknowledged, can become harder to achieve as organizations grow bigger, and especially as military and communications functions are divided between specialized units.

Examples of effective coordination of messages and actions include:

- Statements of concern for local grievances coordinated with public works, charity, or new supplies. A directive from the North Korean guerrilla bureau argued that words alone cannot persuade the masses that their lot will improve under Communism; therefore, deeds, however small, are needed to convince the people of the superiority of the Communist system.[42] In Malaya and the Philippines, insurgents were reported to have used a wider variety

---

[b] One U.S. document produced for information operations discusses this concept using different terminology and makes an additional distinction between forestalling (countering possible lines of persuasion prior to the release of propaganda) and conditioning (preemptively shaping target audience vulnerabilities prior to exposure to propaganda).

of topics in their propaganda than did the government forces. Rather than giving ideological reasons for the support they requested, they offered tangible advantages, such as more land, more food, and other material acquisitions, to their supporters.[43]

- Threats preceding military action by insurgent groups. Outnumbered forces can nevertheless make an impression by promising specific actions before then taking action, creating an greater impression of strength than that which would be created if the action were taken in isolation. Many insurgencies have used the practice of posting warnings or "death sentences" against named collaborators before attacking them; the Afghani Taliban's use of "night letters" fits this pattern.

- Defection messages and "free passes" for defectors timed for release at the high water mark of counter-offensives, to coincide with points of maximum demoralization and greatest freedom of movement across contested lines.[44]

Cognitive Dissonance

Some techniques make use of the message recipient's own actions, often to powerful persuasive effect, driven by a phenomenon known as cognitive dissonance. Cognitive dissonance is a process in which people are made uncomfortable by an apparent conflict either between two beliefs that they hold or between their beliefs and their actions. People often resolve this conflict by changing their beliefs, sometimes in surprising ways. A failed experiment in 1953 illustrates the effect. (Hundreds of follow-up variants have confirmed the basic effect.)[45] Experimental subjects were pressured into making speeches supporting positions they did not believe in. Some were paid a small amount of money and some a larger amount of money for the same action. After giving the speech, the subjects were found to have also changed their private opinion somewhat to be in line with the speech they had given. The most unexpected part of the experiment was this: subjects who were paid a small amount of money changed their beliefs to a greater extent than those paid more. The experimenter had anticipated the attitude change but thought that more money would lead to greater change in attitude rather than the opposite. The explanation for this accepted by most psychologists is cognitive dissonance. Subjects found themselves in the uncomfortable position of having made a public speech they did not believe in. To ease their discomfort, they adjusted their beliefs. Those who were paid a larger amount of money experienced less dissonance; they could rationalize their actions more easily because they had been paid well. Those who were paid less experienced more uncomfortable dissonance, because the money seemed paltry in comparison with the act. The low-paid subjects therefore made the largest adjustment to their beliefs to bring them in line with their actions.

Leon Festinger documented what he believed to be a real-world example of cognitive dissonance as part of a study of followers of a UFO cult that believed the world was going to end on December 21, 1954.[46] Festinger followed cult members up to and beyond that date and was surprised to find that, after the expected event failed to occur, the group did not disband. Instead, the opposite seemed to happen: the group rallied around a rationalization of the event (God had spared the planet at the last moment) and became more committed to the cause, not less. The group also became more active in seeking converts and more receptive to press attention. Festiner's explanation was that group members were too committed to the cult's ideas, had committed too many public acts, and were too enmeshed in the social network of the cult to change their beliefs. Instead, group members decided to adjust their beliefs (accepting a suspicious rationalization immediately and without evidence) and renew their commitments to the cause, both socially and ideologically. Recruiting new members to the cause also served the purpose of seeking confirmation; the more people believed the more true their beliefs would seem.

The two examples given here represent two of the standard ways of evoking cognitive dissonance. The speech-writing study is an example of what is called "induced compliance," while the UFO cult's experience is an example of "belief disconfirmation." The third common method is called "hypocrisy paradigm," which involves having individuals recall prior actions and then pointing out where those actions failed to match with stated values. (A fourth method, "free choice," is not discussed here.) Communist self-criticism sessions can be understood as examples of the hypocrisy paradigm. Communists of all rank are supposed to engage in self-criticism sessions, where they scrutinize their behavior, public and private, for instances where their behavior failed to match their ideology. Even when the act of self-scrutiny has been coerced (as in mandatory self-criticism sessions), self-identified inconsistencies can be very unsettling, and the motivation to change quite powerful.

There is reason to believe that cognitive dissonance may take different forms in different cultures. In particular, cultures differ on how important it is for individuals to maintain independent and consistent sets of beliefs (most true in Western individualist cultures) versus making sure ideas are in harmony with others (more true in collectivist cultures) or with authorities (characteristic of "vertical collectivist" cultures).[47]

A coercion experiment done in the United States and Poland illustrates possible differences.[48] The researchers asked students from the United States and from Poland whether they would be willing to participate in a forty-minute marketing survey without pay. For each participant they used one of two different means of indirect persuasion. Half were prompted to think of other times they complied with similar requests; this was an appeal to self-consistency ("if you agreed to this kind of thing before, why wouldn't you do it now?"). The other half were asked to look at information about how many other students had complied with the request. Both techniques increased compliance, but their levels of effectiveness differed. In the individualistic West, people were more affected by appeals to maintain self-consistency; in more collectivist Poland, people were more persuaded by the social evidence.

Persuasion Techniques

Some persuasion techniques, most notably the so-called "foot in the door" technique, make use of peoples' desire for consistency. "Foot in the door" describes a process of asking for small concessions as a precursor to larger ones. This technique can be used for underground recruitment. One source[49] describes how Viet Cong troops would start by soliciting very small amounts of money from rural residents. Asking for donations opened up a relationship, and the fact that the amounts were small and semivoluntary set up a cognitive dissonance dynamic whereby villagers might convince themselves that they were sympathetic to the cause. Later, when requests became larger and more insistent, residents would often continue to comply.

In another approach, a person may also be asked to donate funds to the movement. However, if he seems reluctant, the third member of the team may suggest to him that if he collects money from others, he himself need not donate to the movement. The individual may halfheartedly attempt to collect some money. Even if he collects only a token amount, once he has committed himself to this extent, he can be coerced in to performing other assignments.[50]

An example of a foot-in-the-door technique combined with deception and more overt coercion was reported during the Maoist insurgency in Malaya. A rubber tapper's duties took him daily into the jungle to tend the rubber trees. On one occasion, he was approached by three armed men. They were friendly and talked him into bringing them cigarettes the next day. Soon he began smuggling such items as aspirins and flashlights: getting these things past the guards at the village added a little interest to his life. One day, only two of the three men showed up, explaining that the third had been captured and would almost certainly reveal the rubber tapper's smuggling activities. Faced with the choice between returning to his village and facing arrest and prison or joining the Communists who would protect him, he joined.[51]

In Korea, the Communist Party gave conspicuous assistance to impoverished families when no government aid was forthcoming. Cooperative food stores were started in the Philippines and in Korea; community pools of money and goods were set up in both these countries and in Malaya. Villagers in Korea frequently were abducted, exposed to propaganda, and returned the next day well-fed and unharmed.[52]

Armed Propaganda

The concept of armed propaganda illustrates the integrated nature of military and political operations in an insurgency and stresses the relative importance of psychological and political operations. According to General Vo Nguyen Giap's view, the first of three phases of insurgent conflict is the psychological warfare phase. In the second phase, propaganda and agitation continue but armed struggle comes to the forefront. In the final phase, when victory is near, the emphasis returns to political propaganda. Before, during, and after the armed conflict, military units are charged with carrying out propaganda missions. Likewise, the political arm of the movement has armed enforcing units to support its psychological and military missions.[53]

In Vietnam, Viet Cong agitation was carried out by armed propaganda teams and cadre agitprop agents, supplemented by relatively simple village propaganda subcommittees.[54] A Viet Cong indoctrination booklet succinctly described the nonmilitary responsibilities of its soldiers in a political struggle:

> Because of their prestige, the members of the armed forces have great propaganda potentiality. If the fighter with a rifle in his hand knows how to make propaganda, to praise the political struggle and to educate the masses about their duty of making the political attack, his influence may be very great. But if he simply calls on the population to join him in the armed struggle he will cause great damage. He must say, "Those who do political struggling are as important as we who fight with rifles. If you do not take up the political struggle we will be unable to defeat the enemy with our rifles." This will make our fellow countrymen more enthusiastic and will also help to promote the political struggle.[55]

Armed propaganda companies serving each province, although primarily military units, were also responsible for psychological warfare. These units infiltrated by night those villages that by day were controlled by government troops. They sometimes even entered during the day, assembled the villagers, made their appeals, and left before government troops could respond. The armed propaganda units range in size from a squad to a company. Their duties included agitation, recruitment, and selective terrorism through enforcement of death sentences. They made periodic visits to villages to conduct propaganda sessions. When visiting a village, the armed unit might suggest that all villagers attend the meeting and would usually ostensibly excuse one or two reluctant persons. However, if many of the villagers failed to attend, the team might find it necessary—in the name of patriotism of course—to enforce attendance. The tactics of armed propaganda teams varied, according to the local situation, from a simple display of armed strength to disemboweling an uncooperative village chief. In one operation in Vietnam in January 1962, one hundred youth group leaders were captured by a Viet Cong armed propaganda unit. Most were released several weeks later after a period of intensive political indoctrination. Seven incorrigibles, however, were held back and probably murdered.[56]

The possession of weapons gave such teams an air of prestige before a village audience. As a leading figure in the Philippine Hu movement stated: "The people are always impressed by the arms, not out of fear but out of a feeling of strength. We get up before the people then, backed by our arms, and give them the message of the struggle. It is never difficult after that."[57]

Reciprocity and Obligation

Creating a feeling of obligation by doing favors or making small concessions is a parallel technique. Reciprocating gifts and favors seems to be a fundamental aspect of human relationships, and the motivation to comply is strong and can leave one vulnerable to manipulation.[58] A small gift from a salesperson can make one feel obligated to listen to a sales pitch; a small concession on price may create obligation to make a large purchase that one was not intending. Reciprocity can be a powerful technique in extended interrogations, as very small comforts and freedoms are exchanged for information.

Members of underground groups in Indochina were instructed to recruit new members by creating social indebtedness. They sought out families in difficult life circumstances—the propertyless, the unemployed, and the sick—and offered them aid. This enabled the underground worker to enter the family or the neighborhood and gain the attention and loyalty of a large number of people. Pressure was then applied to have the people repay their indebtedness by assisting the movement and eventually joining it. A favorable word from a mother, father, relative, or close friend was judged a more powerful persuader than any impersonal propaganda message.[59]

COERCION AND PRIVATE BELIEFS

Undergrounds and insurgencies make use of more overt coercive techniques, as do authoritarian governments. Constant pressure to conform can produce cognitive dissonance-like effects. Every tyrant has followers who are true believers; and even longstanding methods of social repression such as India's caste system are often supported and reinforced by the people they victimize. However, systematic coercion can also breed a system in which individuals become used to keeping two sets of opinions, one public and one private. Protests and passive resistance (discussed in greater detail in Chapter 10) also provide strong social coercion to influence the undecided or uncommitted.

Timur Kuran has studied the effects of preference falsification and has a book on the topic called *Public Truths, Private Lies.*[60] Kuran argues, first, that authoritarian regimes are different from other societies only in degree, not kind—every socialized adult routinely has opinions that they keep to themselves and makes public statements that do not fully match private feelings. An economist, Kuran describes the decision whether to express an opinion in terms of economic costs. Individuals often judge the benefit of expressing a true preference to be lower than the potential costs in terms of personal influence, opportunity for advancement, and freedom from persecution, leading to systematic falsification even in the absence of direct coercion. An individual may calculate that it is wisest to verbally support a despised government even in situations where there is little to no chance of being overheard. Widespread preference falsification can keep undesirable social structures in place indefinitely, even though they benefit few and harm many. However, when circumstances change, the cost of expressing contrary opinions even slightly, usually when a critical mass of protestors makes dissent suddenly seem less risky, the result can be very sudden, unplanned revolutionary uprisings such as the French Revolution and the fall of the Iron Curtain in Eastern Europe. The rapidity of these changes often comes as a surprise even to their leaders and participants. Kuran refers to coerced support for Middle Eastern dictators frequently, and his book, written in 1995, seems prescient of the "Arab Spring" uprisings less than a decade later.

ROLE OF NARRATIVES IN JUSTIFYING INSURGENT MOVEMENTS

Insurgent groups often employ narratives as a means of communicating grievances, goals, and justifications for their actions within a story-like framework. A narrative has three distinct elements: (1) actors and the physical, social, and informational environment within which they operate; (2) events along a temporal continuum; and (3) causality, which is the attribution of cause and effect relative to the first two elements.[61] A narrative serves as an organizing framework through which individuals can make sense of their world[62] and provides insight into the beliefs, norms, and values of a group.[63] Narratives facilitate sense-making, the process of interpretation, and production of meaning, whereby individuals and groups decipher and reflect on phenomena.[64] Sense-making enables individuals to conceive and formulate their social environment, creating a shared worldview among members of a particular in-group.

Al Qaeda's strength and appeal did not lay solely in its sophisticated theological discourse, but also in its ability to comprehend, co-opt, and exploit modern grievances. This narrative combination resonated with extremists and moderates alike, regardless of whether the individual approved of the means by which Al Qaeda sought to accomplish its goals.[65] Al Qaeda's narrative, for example, was best exemplified in Osama bin Laden's 1996 fatwa entitled "Declaration of War against the Americans Occupying the Land of the Two Holy Places." The 12,000-word statement appealed to the social identity of Muslims and established a clear out-group by saying "the people of Islam had suffered from aggression, iniquity and injustice imposed on them by the Zionist-Crusaders alliance and their collaborators . . . ."[66]

An effective narrative is a simple, unifying, easily expressed concept that organizes subjective experience and provides a framework through which one can perceive his or her environment.[67] To understand why a narrative resonates with a specific target audience, one must understand how members of that audience view themselves, their in-group, and relations among the collective groups comprising society. At the group to societal level, a well-crafted message that appeals to a variety of vulnerabilities (sociocultural symbols, individual affective vulnerability, etc.) within a selected target audience will help create an environment that incentivizes participation.[c] Bin Laden not only appealed

---

[c] This is increasingly so in relatively closed environments in which it is more difficult to gain access to alternative explanations through additional information resources or in environments in which there is a social or historical predisposition (or bias) toward a particular interpretation.

to Muslims' sense of victimization, but also attempted to embolden and empower them to support his movement. He claimed, "The people of Islam awakened and realized that they are the main target for the aggression of the Zionist-Crusaders alliance" then "from here, today we begin the work, talking and discussing the ways of correcting what had happened to the Islamic world in general, and the Land of the two Holy Places in particular."[68] His criticism of Arab leaders, however, was not simply unsupported rhetoric; he provided specific examples of decisions made by (in particular) the Saudi Arabian government that are contradictory to Quranic values such as the subordination of Shari'a to the rule of law and the permission to billet foreign forces (namely the United States) on the Saudi peninsula starting in 1990.

Framing describes the narratives social movements employ to engender support. Frame alignment is the "linkage of individual and [social movement organization] interpretative orientations, such that some set of individual interests, values, beliefs and [social movement organization] activities, goals, and ideology are congruent and complementary."[69] Frames link grievances to specific goals, actions, and justifications for those actions. The 1996 declaration articulated Al Qaeda's grievances, but the 1998 joint fatwa with bin Laden, Dr. Ayman Al-Zawahiri (Egyptian Islamic Jihad [EIJ]), and Rifi Taha (Egyptian Islamic Group [EIG]) provided the religious justification as well as a challenge to Muslims. Under the title "World Islamic Front Against Jews and Crusaders," the fatwa read:

> The ruling to kill the Americans and their allies—civilians and military—is an individual duty for every Muslim who can do it in any country in which it is possible to do it, in order to liberate the al-Aqsa Mosque[d] and the holy mosque from their grip, and in order for their armies to move out of all the lands of Islam, defeated and unable to threaten any Muslim. This is in accordance with the words of Almighty Allah, "and fight the pagans all together as they fight you all together," and "fight them until there is no more tumult or oppression, and there prevail justice and faith in Allah."[70]

Different types of movements will employ different types of narratives. A recent study[71] done in parallel with this book examined the correspondence of common narrative themes with types of movements. Insurgent movements were classified into five types: Revolution to Modify the Type of Government, Revolution Based on Identity or

---

[d] The Al Aqsa Mosque, considered Islam's third holiest site, is located in East Jerusalem, the proposed capital of an eventual Palestinian state.

Ethnic Issues, Revolution to Drive Out a Foreign Power, Revolution Based on Religious Fundamentalism, and Revolution for Modernization or Reform.

Table 9-1. Assessing Revolutionary and Insurgent Strategies (ARIS) dataset.

Type of Movement	Group
Revolution to Modify the Type of Government	New People's Army (NPA)
	Fuerzas Armadas Revolucionarias de Colombia (Revolutionary Armed Forces of Colombia, or FARC) and the Ejército de Liberación Nacional (ELN)
	Sendero Luminoso
	Iranian Revolution of 1979
	Frente Farabundo Martí para la Liberación Nacional (Farabundo Marti National Liberation Front, or FMLN) of El Salvador
	Karen National Liberation Army (KNLA)
Revolution Based on Identity or Ethnic Issues	Liberation Tigers of Tamil Eelam (LTTE)
	Palestine Liberation Organization (PLO)
	Hutu-Tutsi genocides
	Ushtia Çlirimtare e Kosovës (Kosovo Liberation Army, or KLA)
	The Provisional Irish Republican Army (PIRA)
Revolution to Drive Out a Foreign Power	Afghan mujahidin
	Viet Cong (VC)
	Chechen Revolution
	Hizbollah
	Hizbul mujahidin (HM)
Revolution Based on Religious Fundamentalism	Egyptian Islamic Jihad (EIJ)
	Taliban
	Al Qaeda (AQ)
Revolution for Modernization or Reform	Movement for the Emancipation of the Niger Delta (MEND)
	Orange Revolution of Ukraine
	Revolutionary United Front (RUF)
	Polish Solidarity

Experts knowledgeable in both the history and communications of twenty-three insurgent groups (shown by type in Table 9-1) rated each as to the most common narrative elements used by these movements.

There is some subjectivity and overlap in both the matching of movements to types and the identification of common themes, but the overall pattern is nevertheless interesting.

Table 9-2. Distribution of narrative themes across ARIS categories.

Operational Code Category	Revolution to Modify the Type of Government	Revolution Based on Identity or Ethnic Issues	Revolution to Drive Out a Foreign Power	Revolution Based on Religious Fundamentalism	Revolution for Modernization or Reform
Expel foreign power and reestablish status quo		PLO			
Reform (anti-secularism)				EIJ, AQ	
Moral center of gravity	Sendero Luminoso, Iranian Revolution				
Reform (economic)	FARC/ELN				MEND, RUF, Solidarity
Self-interest (material)	FARC/ELN, FMLN				MEND, RUF
Religious exclusionism				Chechen Revolution, EIJ, Taliban, AQ	
Illegitimate state (political difference)	FMLN		VC, Chechen Revolution	EIJ, AQ	
Reform (institutional)	FMLN, Sendero Luminoso				Orange Revolution, Solidarity

Operational Code Category	Revolution to Modify the Type of Government	Revolution Based on Identity or Ethnic Issues	Revolution to Drive Out a Foreign Power	Revolution Based on Religious Fundamentalism	Revolution for Modernization or Reform
Illegitimate state (ideological difference)	NPA, Sendero Luminoso, Iranian Revolution, FMLN	PLO		EIJ	
Illegitimate state (foreign influence on leader)	NPA, Sendero Luminoso, Iranian Revolution			EIJ, Taliban	MEND
Separatism (separate state)	KNLA	LTTE, PLO, Euskadi Ta Askatasuna (ETA), KLA, PIRA			
Heroic dynamics	KNLA	LTTE, ETA, PIRA	VC		
Illegitimate state (lack of security)		LTTE, PLO, ETA, KLA, PIRA, Hutu-Tutsi		Taliban	
In-group self-determination	KNLA	LTTE, PLO, ETA, KLA, PIRA, Hutu-Tutsi			MEND
Martyrdom	Iranian Revolution	LTTE, PLO, PIRA	Hizbollah	EIJ, AQ	
Illegitimate state (lack of services)	NPA, FMLN	PIRA	Afghan mujahidin, VC	Taliban	MEND, RUF, Solidarity
Oppressive state	Iranian Revolution, FMLN, KNLA	LTTE, PLO, KLA, PIRA		EIJ	

Operational Code Category	Revolution to Modify the Type of Government	Revolution Based on Identity or Ethnic Issues	Revolution to Drive Out a Foreign Power	Revolution Based on Religious Fundamentalism	Revolution for Modernization or Reform
Expel foreign power and reform		PIRA	Afghan mujahidin, VC, Chechen Revolution, Hizbollah, HM	Taliban	MEND
Inherent nobility of cause and/or participants	KNLA	ETA, PIRA	Hizbollah, HM	EIJ, AQ	Solidarity

In Revolutions to Modify the Type of Government, such as the Iranian Revolution of 1979 and the FMLN of El Salvador, class arguments, emphasizing the illegitimacy of the state, is a frequent narrative theme. These groups are highly dependent on a large or powerful segment of the population not having its socioeconomics needs being met by the current governmental system as well as a mature ideological alternative system to propose. Class arguments are quite prevalent.

The next type of group, "Revolution Based on Identity or Ethnic Issues," includes groups such as the Liberation Tigers of Tamil Eelam (LTTE), the Palestine Liberation Organization (PLO), and Euskadi Ta Askatasuna (Basque Homeland and Freedom, or ETA). These types of movements often occur where colonialism or war has artificially combined various ethnic or identity groups into the same government, with one side having preferential treatment, over-representation, or dominant political voice. A contributing factor is also the current governmental system's fragility or constraints for accommodating concessions or evolution to a more egalitarian system. The important factors within this type are the creation/use of a narrative that distinguishes the disaffected group from the ruling group as a unique cultural/social unit. As the data suggest, these groups advocate strongly for self-determination and/or preservation of their heritage. The recurrence of the "In-group self-determination" content code in each of the cases and the "Illegitimate state (lack of security)" code appearing in five of the six help account for the degree of similarity.

"Revolutions to Drive Out a Foreign Power," represented by the Afghan mujahidin, Viet Cong (VC), the Chechen Revolution, Hizbollah, and Hizbul mujahidin, result from a segment of the population trying to oust a foreign occupier, colonial ruler, or a foreign military presence. Common narrative themes are the desire for independence, the need for reform, and the illegitimacy of the occupying power.

"Revolutions Based on Religious Fundamentalism," which include Egyptian Islamic Jihad (EIJ), the Taliban, and Al Qaeda, are heavily dependent on religious scholarship, both in terms of motivation as well as legitimacy, and incorporate appropriate themes and symbology into their narratives. They tend to derive legitimacy from their adherence to a particular interpretation (not necessarily a widely or academically accepted interpretation) of their faith. Common themes are the evils of secularism, represented by the secular government, and the need for religious separation and purification. The intensity with which they delegitimize the state is often greater than with other groups, a trend evident in the psychological warfare tactics of both EIJ and Al Qaeda. The willingness to negotiate (at the strategic level) with the state is considerably less and is often on terms deemed unreasonable by the state.

The final category, "Revolutions for Modernization or Reform," such as the Orange Revolution of Ukraine, the Revolutionary United Front (RUF), and Polish Solidarity, is categorically the least similar, but all movements share aspects relating to weakening states and their disintegration of powers and services. These groups tend to employ narratives that address a particular political issue (e.g., "Illegitimate state [lack of services]") to engender support but not necessarily to incite violence. They rarely wish to acquire the responsibility of governance but rather want a degree of policy reform or political inclusion.

The clustering of narratives around a set of themes identifies referent trends; however, this is done so at a level of abstraction for ease of analysis and not operational relevance. The idiosyncratic application of each of these narratives is where comprehension of the local culture, symbology, and customs differentiates the individual movement from their group. To consider the FARC a Marxist group would be accurate but incomplete. The FARC did not rigidly adhere to Marxist-Leninist ideology; their pragmatism in dealing with farmers and narcotraffickers alike was more capitalistic. Their ideological evolution to Bolivarianism indicates an intellectual and cultural adaptability for which a broader classification cannot account. Furthermore, the PIRA was not a religious fundamentalist organization, but their recognition of Catholic symbolic reverence for those willing to sacrifice themselves for the larger good was clearly indicated during the hunger strikes. Organizations/movements such as Hizbollah and Al Qaeda that have a presence (however small) in many locations throughout the world retain a unifying strategic narrative but employ a far subtler and nuanced approach depending on their geographic location and functional responsibility. A Hizbollah operative recruiting in Lebanon will likely make the imminent threat of Israel more salient than a financier soliciting donations from the Lebanese diaspora in the Tri-Border Region of South America. Likewise, the franchised component of the Al Qaeda social movement that emerged from the Maktab al-Khidamat (Services Bureau) during the Afghan resistance explicitly tailored their recruiting and fundraising themes to the local target audience.

SUMMARY

The study of influence encompasses research from a number of domains and includes both social and individual factors. Understanding person-to-person influence requires understanding of charisma, nonverbal communication, and leadership factors. Individuals are also strongly influenced by their social networks—so much so that facing uniform disagreement often induces conformity. Social identity categorizations are also important to understand how individuals synchronize their opinions with groups that they want to identify with, and avoid those that they do not. On the individual level, people are influenced by messages that have emotional appeal and appear to be relevant to their interests. Stress may narrow the attention of message recipients in traumatic situations to focus on basic needs; successful communicators must assess what level of messaging or persuasion a target audience is ready to receive. Research on rumors, particularly wartime rumors, illustrates how self-interest and emotional appeals attract and hold peoples' attention. For those wishing to craft influential messages, the combination of words and actions is particularly powerful.

An understanding of cognitive dissonance, which is created when an individual's desire to have a consistent and positive self-concept is disrupted, also sheds light on proven influence techniques such as the "foot in the door" process and reciprocity. Influence and associated cognitive effects vary by culture; research has shown some intriguing contrasts, particularly between collectivist and individualistic cultures.

Narratives provide a way to contextualize and justify grievances against the government. Narratives also frame a struggle and tie grievances to specific goals and actions. While there are many common themes between groups, different types of insurgent groups rely on different clusters of narrative types that are congruent with their goals.

ENDNOTES

[1] Joseph Henrich, Steven J. Heine, and Ara Norenzayan, "The Weirdest People in the World," *Behavioral and Brain Sciences* 33, no. 2–3 (2010): 61–83.

[2] Everett Rogers, *Diffusion of Innovations*, 4th ed. (New York: Simon & Schuster, 1995).

[3] Tanya L. Chartrand and Amy N. Dalton, "Mimicry: Its Ubiquity, Importance, and Functionality," in *Oxford Handbook of Human Actions*, ed. Ezequiel Morsella, John A. Bargh, and Peter M. Gollwitzer (Oxford: Oxford University Press, 2009).

[4] Sigal G. Barsade, "The Ripple Effect: Emotional Contagion and Its Influence on Group Behavior," *Administrative Science Quarterly* 47, no. 4 (2002): 644.

[5] Caroline A. Bartel and Richard Saaverda, "The Collective Construction of Work Group Moods," *Administrative Science Quarterly* 45, no. 2 (2000): 197–231.

[6] Arik Cheshin, Anat Rafaeli, and Nathan D. Bos, "Anger and Happiness in Virtual Teams: Emotional Influences of Text and Behavior on Others' Affect in the Absence of Non-Verbal Cues," *Organizational Behavior and Human Decision Processes* 116, no. 1 (2011): 2–16.

[7] Ellen S. Sullins, "Emotional Contagion Revisited: Effects of Social Comparison and Expressive Style on Mood Convergence," *Personality and Social Psychology Bulletin* 17, no. 2 (1991): 166–174.

[8] Joyce Bono and Remus Ilies, "Charisma, Positive Emotions and Mood Contagion," *The Leadership Quarterly* 17, no. 4 (2006): 317–334.

[9] Chris L. Kleinke, "Gaze and Eye Contact: A Research Review," *Psychological Bulletin* 100, no. 1 (1986): 78–100.

[10] Robert B. Cialdini, *Influence: Science and Practice*, 4th ed. (Allyn and Bacon, 2001).

[11] Solomon E. Asch, "Studies of Independence and Conformity: I. A Minority of One Against a Unanimous Majority," *Psychological Monographs: General and Applied* 70, no. 9 (1956): 1–70.

[12] Nicholas A. Christakis and James H. Fowler, "The Collective Dynamics of Smoking in a Large Social Network," *The New England Journal of Medicine* 358, no. 21 (2008): 2249–2258.

[13] Nicholas A. Christakis and James H. Fowler, "The Spread of Obesity in a Large Social Network Over 32 Years," *The New England Journal of Medicine* 357, no. 4 (2007): 370–379.

[14] James H. Fowler and Nicholas A. Christakis, "Dynamic Spread of Happiness in a Large Social Network: Longitudinal Analysis Over 20 years in the Framingham Heart Study," *British Medical Journal* 337 (2008): a2338.

[15] Peter Grose, "Vietcong's 'Shadow Government' in the South," *The New York Times Magazine*, 1965.

[16] Donald L. Horowitz, *Ethnic Groups in Conflict* (Berkley: University of California Press, 2000).

[17] Wendy Wood, Gregory J. Pool, Kira Leck, and Daniel Purvis, "Self-definition, Defensive Processing, and Influence: The Normative Impact of Majority and Minority Groups," *Journal of Personality and Social Psychology* 71, no. 6 (1996): 1181–1193.

[18] Harold H. Kelley, "Salience of Membership and Resistance to Change of Group-Anchored Attitudes," *American Sociological Review* 17, no. 4 (1955): 453–465.

[19] William E. Daugherty, "The Creed of a Modern Propagandist," in *A Psychological Warfare Casebook*, ed. William E. Daugherty and Morris Janowitz (Bethesda, MD: Johns Hopkins Press, Operations Research Office, 1958): 35–46.

[20] Charles Windle and T. R. Vallance, "The Future of Military Psychology: Paramilitary Psychology," *American Psychologist* 19, no. 2 (1964): 119–129.

[21] Ibid.

[22] John R. Zaller, *The Nature and Origins of Mass Opinion* (New York: Cambridge University Press, 1992).

[23] Tamotsu Shibutani, *Improvised News: A Sociological Study of Rumor* (Indianapolis: Bobbs-Merrill Inc., 1966).

[24] Robert H. Knapp, "A Psychology of Rumor," *Public Opinion Quarterly* 8, no. 1 (1944): 22.

[25] Gordon W. Allport and Leo Postman, "An Analysis of Rumor," *Public Opinion Quarterly* 10, no. 4 (1946): 501–517.

[26] Stephanie R. Kelly, "Rumors in Iraq: A Guide to Winning Hearts and Minds" (master's thesis, Naval Postgraduate School, 2004).

[27] Ibid., 20.

[28] Abraham Maslow, "A Theory of Human Motivation," *Psychological Review* 50 (1948): 370–396.

[29] Michael Lewis-Beck, "Economic Voting: An Introduction," *Electoral Studies* 19, no. 2–3 (2000): 113–121.

[30] Leonardo R. Arriola, "Ethnicity, Economic Conditions, and Opposition Support: Evidence from Ethiopia's 2005 Elections," *Northeast African Studies* 10, no. 1 (2008): 115–144.

[31] Jeremy Youde, "Economics and Government Popularity in Ghana," *Electoral Studies* 24, no. 1 (2005): 1–16.

[32] Matthew Singer and Ryan E. Carlin, "Are Latin American Voters Campesinos o Empresarios? Economic Perceptions and Government Support 1995–2008" (paper presented at the annual meeting of the American Political Science Association, 2011).

[33] Richard E. Nisbett and Lee Ross, *Human Inference: Strategies and Shortcomings of Social Judgment* (Englewood Cliffs: Prentice-Hall, 1980).

[34] Richard E. Petty and John T. Cacioppo, "The Elaboration Likelihood Model of Persuasion," *Advances in Experimental Social Psychology* 19, no. 1 (1986): 123–205.

[35] Rhonda S. Zaharna, "Understanding Cultural Preferences of Arab Communication Patterns," *Public Relations Review* 21, no. 3 (1995): 241–255.

[36] Sang-pil Han and Sharon Shavitt, "Persuasion and Culture: Advertising Appeals in Individualistic and Collectivistic Societies," *Journal of Experimental Social Psychology* 30, no. 4 (1994): 326–350.

[37] Ibid.

[38] Daugherty, "Creed of a Modern Propogandist," 39.

[39] Petty and Cacioppo, "The Elaboration Likelihood Model of Persuasion," 123–205.

[40] Michael Pfau, Kyle James Tusing, Waipeng Lee, Linda Goldbold, Ascan Koerner, Linda J. Penaloza, Yah-huei Hong, and Violet Shu-huei Yang, "Nuances in Inoculation: The Role of Inoculation Approach, Ego-Involvement, and Message Processing Disposition in Resistance," *Communication Quarterly* 45, no. 4 (1997): 461–481.

[41] W. J. McGuire and D. Papageorgis, "The Relative Efficacy of Various Types of Prior Belief-Defense in Producing Immunity Against Persuasion," *Journal of Abnormal and Social Psychology* 62 (1961): 327–337.

[42] Fred H. Barton, *Salient Operation Aspects of Paramilitary Warfare in Three Asian Areas, ORO-T-228* (Chevy Chase, MD: Operations Research Office, 1953), 25.

[43] Ibid., 90.

[44] William E. Daugherty and Morris Janowitz, *A Psychological Warfare Casebook, Social Forces* 37 (Bethesda, MD: Johns Hopkins Press, Operations Research Office, 1958).

[45] Eddie Harmon-Jones, "A Cognitive Dissonance Theory Perspective on Persuasion," in *The Persuasion Handbook: Developments in Theory and Practice* (Thousand Oaks: SAGE Publications, 2002), 99–116.

[46] Leon Festinger, Henry W. Riecken, and Stanley Schachter, *When Prophecy Fails* (Minneapolis: University of Minnesota Press, 1956).

[47] Harry C. Triandis and Michele J. Gelfand, "Converging Measurement of Horizontal and Vertical Individualism and Collectivism," *Journal of Personality and Social Psychology* 74, no. 1 (1998): 118–128.

[48] Robert B. Cialdini, Wilhelmina Wosinska, Daniel Barrett, Jonathan Butner, and Malgorzata Gornik-Durose, "Compliance with a Request in Two Cultures: The Differential Influence of Social Proof and Commitment/Consistency on Collectivists and Individualists," *Personality and Social Psychology Bulletin* 25, no. 10 (1999): 1242–1253.

[49] Douglas Pike, *Viet Cong: The Organization and Technique of the National Liberation Front of South Vietnam* (Cambridge: M.I.T. Press, 1966).

[50] Ibid.

[51] Juan DeDios Marin, "Inside a Castro 'Terror School,'" *The Readers Digest*, 1964, 119–120.

[52] Barton, *Salient Operation Aspects of Paramilitary Warfare*, 86.

[53] General Vo Nguyen Giap, *People's War, People's Army* (Washington, DC: Government Printing Office, 1962), 76–81.

[54] Douglas Pike, "The Communication Process of the Communist Apparatus in South Vietnam," (unpublished manuscript, Center for International Studies, Massachusetts Institute of Technology, circa 1964), 39.

[55] Ibid.

[56] Bernard B. Fall, *The Two Vietnams: A Political and Military Analysis* (New York: Praeger, 1963), 137.

[57] William J. Pomeroy, *The Forest: A Personal Record of the Huk Guerrilla Struggle in the Philippines* (New York: International Publishers, 1963), 43.

[58] Cialdini, *Influence: Science and Practice*.

[59] Barton, *Salient Operation Aspects of Paramilitary Warfare*, 43.

[60] Timur Kuran, *Private Truths, Public Lies* (Cambridge: Harvard University Press, 1995).

61 Shaul R. Shenhav, "Political Narratives and Political Reality," *International Political Science Review* 27, no. 3 (2006): 245–262.

62 Andrew D. Brown, Patrick Stacey, and Joe Nandhakumar, "Making Sense of Sensemaking Narratives," *Human Relations* 61, no. 8 (2008): 1035–1064.

63 Montgomery McFate, Britt Damon, and Robert Holiday, "What Do Commanders Really Want to Know?," in *The Oxford Handbook of Military Psychology* (New York: Oxford University Press, 2012), 92–113.

64 Brown, Stacey, and Nandhakumar, "Making Sense of Sensemaking Narratives."

65 Jason Spitaletta and Shana Marshall, "Al Qaeda," in *Casebook on Insurgency and Revolutionary Warfare, Volume II: 1962–2009*, ed. Chuck Crossett (Alexandria, VA: U.S. Army Publications Directorate, in press).

66 "Bin Laden's Fatwa," August 1996, reprinted at *PBS Newshour*, http://www.pbs.org/newshour/terrorism/international/fatwa_1996.html?print, accessed February 24, 2012.

67 David J. Killcullen, *Counterinsurgency* (Oxford: Oxford University Press, 2010).

68 "Bin Laden's Fatwa."

69 David A. Snow, E. Burke Rochford, Jr., Steven K. Worden, and Robert D. Benford, "Frame Alignment Processes, Micromobilization, and Movement Participation," *American Sociological Review* 51, no. 4 (1986): 464–481.

70 As quoted in Spitaletta and Marshall, "Al Qaeda."

71 Chuck Crossett, Summer Newton, and Jason Spitaletta, "The Role of Narrative in Insurgent and Revolutionary Warfare: Examples from 24 Case Studies Spanning 1962–2009," *Journal of Cultural Intelligence* (in press).

CHAPTER 10.

NONVIOLENT RESISTANCE

CHAPTER CONTENTS

Jason Spitaletta and SORO authors

Nonviolent resistance has long been a method and corollary of human conflict, playing a role in many underground and revolutionary activities. Although both the role and, particularly, the process of passive resistance have sometimes been obscured by philosophic and religious considerations, its tactics remain much the same, whether practiced by a figure as prominent as Gandhi from moral and religious convictions or by a pragmatic underground as an expedient. Typically, the underground organizes and directs passive resistance techniques and persuades the ordinary citizen to carry them out. These types of action can be described as activism in opposition to radicalism.

Nonviolent resistance and mass mobilization have contributed to significant revolution in the past thirty years: the Solidarity movement in Poland during the 1980s saw disparate labor strikers organize and ultimately unseat the Communist regime, the 2004–2005 Orange Revolution in the Ukraine saw mass mobilization and broad civil disobedience usher in a democratic electoral process, and the January 2011 Lotus Revolution saw a disparate group of Egyptians occupy Tahrir Square in Cairo until President Hosni Mubarak relinquished his presidency after thirty years.

The twenty-first century has witnessed a synthesis of the global technological networks that link computers on the Internet and the social networks that have connected humans since the dawn of the species.[1] Many innovative forms of protest have emerged in the post-Cold War era. Although few have managed to mobilize a sufficient number to displace a regime, they have provided a forum for a youthful demographic to engage in creative, often social-media-directed alternatives to the picketing and chants their elders employed.[2]

OBJECTIVES

Passive resistance implies a large, unarmed group whose activities capitalize on social norms, customs, and taboos in order to provoke action by security forces that will serve to alienate large segments of public opinion from the government or its agents. If the government does not respond to the passive resisters' actions, the resisters will immobilize the processes of public order and safety and seriously challenge the writ of government. Passive resistance rests on the basic thesis that governments and social organizations, even when they possess instruments of physical force, depend on the voluntary assistance and cooperation of great numbers of individuals.[3] One method, therefore, of opposing an established power structure is to persuade many persons to refuse to cooperate with it.

The principal tactic used to induce noncooperation—tacit withdrawal of the populace support of the government—is frequently described as persuasion through suffering. One of the persistent myths of passive resistance is that persuasion through suffering aims only to persuade the opponent and his supporting populace by forcing him to experience a guilty change of heart and a sense of remorse. "The sight of suffering on the part of multitudes of people will melt the heart of the aggressor and induce him to desist from his course of violence."[4] This conception of the role of suffering makes the fundamental error of presuming that only two actors are involved in the process of passive resistance: the suffering resister and the opponent. Actually, passive resistance operates within a framework involving three actors: the suffering passive resister, the opponent (the government and security forces), and the larger "audience" (the population). Every conflict situation is dramatically affected by the extent to which the audience becomes involved.

This "contagiousness of conflict"[5] inevitably occurs in the nature of the conflict when the audience is included as a third actor. The original participants are apt to lose control or, at the least, the outcome is greatly influenced. It is most important in politics to determine the manner and extent to which the scope of conflict influences the outcome, and how to manipulate that outcome. Passive resistance techniques, particularly the function of suffering, provide one insight into the manipulation of "the contagiousness of conflict." More than anything else, the objective of passive resistance is to create situations that will involve public opinion and direct it against the established power structure. When this happens, the position of the passive resisters is legitimized. While not explicitly linked with the ongoing nonviolent civil rights movement, the conflict in Northern Ireland became increasingly violent on January 30, 1972, in Derry when British paratroopers of the First Battalion of the Parachute Regiment fired on a crowd of anti-internment demonstrators, killing thirteen and wounding another thirteen. All the injured were Catholic, seven of whom were under nineteen years of age. The events of "Bloody Sunday" brought strong about reactions from both the Republic of Ireland and the Catholic community in Northern Ireland as well as the international community. The event was followed by a decision by the British prime minister to suspend the Northern Irish parliament and assume direct governance. The Provisional Irish Republican Army (PIRA) responded with an escalation of violence.[6]

When the passive resister suffers at the hand of the government, it demonstrates his integrity, commitment, and courage, while showing the injustice, cruelty, and tyranny of the government. The essential function of suffering is comparable to the interaction that takes place between a martyr and a crowd. The passive resister's token to power in the face of the security force is his capacity to suffer in the eyes of the onlooker. The courage and dedication of an unresisting martyr can have a tremendous impact on the imagination of a crowd.[7]

If the passive resister provokes a response from the security forces or government that can be made to seem unjust or unfair, his charges of tyranny and persecution are confirmed. Should the government fail to act, it abdicates its control over the population, over the enforcement of law, and over the maintenance of order. Passive resistance techniques suddenly thrust upon a government the initiative, and also the responsibility, for uninvited conflict with unarmed citizens. Jujitsu politics (see *Chapter 6: Group Dynamics and Radicalization*) is when a population is mobilized by conflict between two groups in which one (usually the government) has aggressively/excessively responded to the other, promoting sympathy and outrage.[8] In this context, jujitsu politics refers to the use of a state's (or political adversary's) momentum against itself by a less resourced entity. This mechanism is a generalization of group dynamic theory and states that a group seeks to influence potential sympathizers by reframing the out-group threat. By identifying the out-group as a threat to that in-group reference, and capitalizing on the outrage to the recent over-reaction, those potential sympathizers can be swayed and/or mobilized without much expended effort on the part of the radical group.[9] Carlos Marighella, in his *Minimanual of the Urban Guerilla*, advocated this tactic as a highly effective means of cultivating popular support. Instead of compelling the populace to join an insurgent cause, the insurgents would provoke the state to react violently and indiscriminately, thus driving the populace into the arms of the insurgents.[10] Al Qaeda in Iraq and the Taliban in Afghanistan have employed variations of his tactic against U.S. forces.[11]

Demonstrations and festivals, even if not politically motivated, can be co-opted and exploited by nefarious actors. The Tamil New Tigers (TNT) group was the forerunner and vanguard of the Liberation Tigers of Tamil Eelam (LTTE), a group that at its height was among the most innovative, ruthless, and well-organized insurgent movements in recent history. In January 1974, Sri Lankan police attempted to dismantle and disperse the Fourth International Tamil Conference in Jaffna, resulting in the deaths of eleven Tamils—seven were electrocuted and four were trampled to death—and approximately fifty civilians sustaining severe injuries. The incident took place while Naina Mohamed, a distinguished Tamil scholar from India, was speaking to a crowd of thousands. Although the TNT did not instigate the police or incite the riot, they used the incident to rationalize the assassination of the Jaffna mayor in July 1975.[12] The TNT framed the event as an intentional attack on their cultural heritage and commemorated it annually.

The primary function of passive resistance suffering is to redraw the political battle lines in favor of the resister. There are, of course, a number of variables that affect the effectiveness of suffering. One is the attitude and orientation of the opponent; success seems dependent on whether the opponent really cares how a population views him, whether he is attempting to win favor. However, if the opponent is unconcerned with popular opinion, passive resistance tactics are less likely to be effective. Leaders rarely operate with unlimited agency and are thus somewhat constrained (or certainly influenced by) the political environment. Also, the coercive effect depends on whether the opponents are the passive resisters' own countrymen; if they are, common identity and nationalism tend to induce empathy. In some societies, passive suffering is viewed with contempt and is seen as masochism or "an exploitation of the rulers' good-natured reluctance to allow unnecessary suffering, denying attributes of personal courage or virtue to the sufferer."[13]

In addition to alienating public opinion from a government, underground-sponsored passive resistance has two other equally important objectives. The first is to lower the morale of government officials and security forces. This goal is most relevant to occupying forces. Imposing casualties on a military force is but one way to lower morale; forcing soldiers to question not only the legitimacy of their cause but also the viability of success can erode capability from the inside out. Confronting unarmed and nonviolent activists (particularly women and children) often undermines the ethos that serves as a cohesive force among militaries. The resultant stress may lower productivity (and thus increase manpower demands) and eventually cause psychological injuries among the forces. A second objective is to tie down security forces. By organizing and encouraging a citizenry to use techniques of passive resistance, the underground can successfully divert security forces from other tasks.[14] The coordinated operations of a nonviolent wing and guerrillas can be extremely effective in disorienting security forces by preventing them from focusing on any one particular aspect of the insurgency.

TECHNIQUES

Among the most prominent theorists in nonviolent revolution is Dr. Gene Sharp,[a] whose compilation of nonviolent resistance tactics is included in Table 10-1. Operationalizing those who condemned violence from Gandhi to Thoreau, Sharp's key theme is that political power is not derived from the intrinsic qualities of those in positions of authority. Rather, the power of any state is derived from the consent of the governed, and thus they possess the moral and political authority to take it back. Essentially, leaders lack power without the consent of the governed, and because the state may hold the monopoly on the use of force, nonviolent methods are the ideal means for the people to impose their will on the state.[15] The arsenal of the passive resister contains a number of weapons of nonviolence. One reason these weapons may be effective is that the government forces may not know how to cope with nonviolence. Police and soldiers are trained to fight force with force but are usually "neither trained nor psychologically prepared to fight passive resistance."[16]

Actions of passive resistance may range from small isolated challenges to specific laws to complete disregard of governmental authority, but the techniques of nonviolent resistance have be classified into three general types: attention-getting devices, noncooperation, and civil disobedience.

---

[a] As of December 2011, Dr. Sharp is a professor emeritus of political science at the University of Massachusetts, Dartmouth. Sharp has served as a professor of political science at the University of Massachusetts, Dartmouth, since 1972 and has held a research appointment at Harvard University's Center for International Affairs since 1965. He founded the Albert Einstein Institution, a nonprofit organization that studies and promotes the use of nonviolent tactics, in 1983.

Table 10-1. 198 Methods of nonviolent action.[17]

THE METHODS OF NONVIOLENT PROTEST AND PERSUASION

Formal Statements

1. Public speeches
2. Letters of opposition or support
3. Declarations by organizations and institutions
4. Signed public statements
5. Declarations of indictment and intention
6. Group or mass petitions

Communications with a Wider Audience

7. Slogans, caricatures, and symbols
8. Banners, posters, and displayed communications
9. Leaflets, pamphlets, and books
10. Newspapers and journals
11. Records, radio, and television
12. Skywriting and earthwriting

Group Representations

13. Deputations
14. Mock awards
15. Group lobbying
16. Picketing
17. Mock elections

Symbolic Public Acts

18. Displays of flags and symbolic colors
19. Wearing of symbols
20. Prayer and worship
21. Delivering symbolic objects
22. Protest disrobings
23. Destruction of own property
24. Symbolic lights
25. Displays of portraits
26. Paint as protest
27. New signs and names
28. Symbolic sounds
29. Symbolic reclamations
30. Rude gestures

Pressures on Individuals

31. "Haunting" officials
32. Taunting officials
33. Fraternization
34. Vigils

Drama and Music

35. Humorous skits and pranks
36. Performances of plays and music
37. Singing

Processions

38. Marches
39. Parades
40. Religious processions
41. Pilgrimages
42. Motorcades

Honoring the Dead

43. Political mourning
44. Mock funerals
45. Demonstrative funerals
46. Homage at burial places

Public Assemblies

47. Assemblies of protest or support
48. Protest meetings
49. Camouflaged meetings of protest
50. Teach-ins

Withdrawal and Renunciation

51. Walk-outs
52. Silence
53. Renouncing honors
54. Turning one's back

THE METHODS OF SOCIAL NONCOOPERATION

Ostracism of Persons

55. Social boycott
56. Selective social boycott
57. Lysistratic nonaction
58. Excommunication
59. Interdict

Noncooperation with Social Events, Customs, and Institutions

60. Suspension of social and sports activities
61. Boycott of social affairs
62. Student strike
63. Social disobedience
64. Withdrawal from social institutions

Withdrawal from the Social System

65. Stay-at-home
66. Total personal noncooperation
67. "Flight" of workers
68. Sanctuary
69. Collective disappearance
70. Protest emigration (hijrat)

THE METHODS OF ECONOMIC NONCOOPERATION: (1) ECONOMIC BOYCOTTS

Actions by Consumers

71. Consumers' boycott
72. Nonconsumption of boycotted goods
73. Policy of austerity
74. Rent withholding
75. Refusal to rent
76. National consumers' boycott
77. International consumers' boycott

Action by Workers and Producers

78. Workmen's boycott
79. Producers' boycott

Action by Middlemen

80. Suppliers' and handlers' boycott

Action by Owners and Management

81. Traders' boycott
82. Refusal to let or sell property
83. Lockout
84. Refusal of industrial assistance
85. Merchants' "general strike"

Action by Holders of Financial Resources

86. Withdrawal of bank deposits

87. Refusal to pay fees, dues, and assessments
88. Refusal to pay debts or interest
89. Severance of funds and credit
90. Revenue refusal
91. Refusal of a government's money

Action by Governments

92. Domestic embargo
93. Blacklisting of traders
94. International sellers' embargo
95. International buyers' embargo
96. International trade embargo

THE METHODS OF ECONOMIC NONCOOPERATION: (2) THE STRIKE

Symbolic Strikes

97. Protest strike
98. Quickie walkout (lightning strike)

Agricultural Strikes

99. Peasant strike
100. Farm workers' strike

Strikes by Special Groups

101. Refusal of impressed labor
102. Prisoners' strike
103. Craft strike
104. Professional strike

Ordinary Industrial Strikes

105. Establishment strike
106. Industry strike
107. Sympathetic strike

Restricted Strikes

108. Detailed strike
109. Bumper strike
110. Slowdown strike
111. Working-to-rule strike
112. Reporting "sick" (sick-in)
113. Strike by resignation
114. Limited strike
115. Selective strike

Multi-Industry Strikes

116. Generalized strike
117. General strike

Combination of Strikes and Economic Closures

118. Hartal
119. Economic shutdown

THE METHODS OF POLITICAL NONCOOPERATION

Rejection of Authority

120. Withholding or withdrawal of allegiance
121. Refusal of public support
122. Literature and speeches advocating resistance

Citizens' Noncooperation with Government

123. Boycott of legislative bodies
124. Boycott of elections
125. Boycott of government employment and positions
126. Boycott of government depts., agencies, and other bodies
127. Withdrawal from government educational institutions
128. Boycott of government-supported organizations
129. Refusal of assistance to enforcement agents
130. Removal of own signs and placemarks
131. Refusal to accept appointed officials
132. Refusal to dissolve existing institutions

Citizens' Alternatives to Obedience

133. Reluctant and slow compliance
134. Nonobedience in absence of direct supervision
135. Popular nonobedience
136. Disguised disobedience
137. Refusal of an assemblage or meeting to disperse
138. Sitdown
139. Noncooperation with conscription and deportation
140. Hiding, escape, and false identities
141. Civil disobedience of "illegitimate" laws

Action by Government Personnel

142. Selective refusal of assistance by government aides
143. Blocking of lines of command and information
144. Stalling and obstruction
145. General administrative noncooperation
146. Judicial noncooperation
147. Deliberate inefficiency and selective noncooperation by enforcement agents
148. Mutiny

Domestic Governmental Action

149. Quasi-legal evasions and delays
150. Noncooperation by constituent governmental units

International Governmental Action

151. Changes in diplomatic and other representations
152. Delay and cancellation of diplomatic events
153. Withholding of diplomatic recognition
154. Severance of diplomatic relations
155. Withdrawal from international organizations
156. Refusal of membership in international bodies
157. Expulsion from international organizations

THE METHODS OF NONVIOLENT INTERVENTION

Psychological Intervention

158. Self-exposure to the elements
159. The fast
    a) Fast of moral pressure
    b) Hunger strike
    c) Satyagrahic fast
160. Reverse trial
161. Nonviolent harassment

Physical Intervention

162. Sit-in
163. Stand-in
164. Ride-in
165. Wade-in
166. Mill-in
167. Pray-in
168. Nonviolent raids
169. Nonviolent air raids
170. Nonviolent invasion
171. Nonviolent interjection
172. Nonviolent obstruction
173. Nonviolent occupation

Social Intervention

174. Establishing new social patterns
175. Overloading of facilities
176. Stall-in
177. Speak-in
178. Guerrilla theater
179. Alternative social institutions
180. Alternative communication system

Economic Intervention

181. Reverse strike
182. Stay-in strike
183. Nonviolent land seizure
184. Defiance of blockades
185. Politically motivated counterfeiting
186. Preclusive purchasing
187. Seizure of assets
188. Dumping
189. Selective patronage
190. Alternative markets
191. Alternative transportation systems
192. Alternative economic institutions

Political Intervention

193. Overloading of administrative systems
194. Disclosing identities of secret agents
195. Seeking imprisonment
196. Civil disobedience of "neutral" laws
197. Work-on without collaboration
198. Dual sovereignty and parallel government

Attention-Getting Devices

Passive resistance in the early stages usually takes the form of actions calculated to gain attention, provide propaganda for the cause, or be a nuisance to government forces. These dilemma actions force the authorities to choose between allowing such activities to continue and taking the risk that they will build into something significant and imposing harsh punishment on people who are engaged in a seemingly benign activity.[18] Attention-getting devices include demonstrations, mass meetings, picketing, and the creation of symbols: Demonstrations and picketing help advertise the resistance campaign and educate the larger public about the issues at stake. Such activities provide propaganda and agitation for both internal and external consumption. An example of this was the 1963 Buddhist protest in South Vietnam. The self-immolation of a Buddhist monk was strategically timed to ensure that newsmen and photographers—particularly U.S. reporters—would be present to record the event. The leader of the Buddhists, Thich Tri Quang, wanted publicity in the U.S. press and took pains to make U.S. reporters welcome.[19] The upheaval in Tunisia that began

December 17, 2010, had similar, though less intentional, origins. In the central Tunisian town of Sidi Bouzid, a 26-year-old fruit vendor named Mohammed Bouazizi performed an act of self-immolation to protest the humiliating tactics of local officials. The act jolted Tunisians, and dozens (later hundreds then thousands) began protesting in the streets. The protestors ultimately clashed with Tunisian security forces, resulting in the deaths of approximately one hundred individuals throughout the country. The protest resulted in the renunciation of power and exodus of President Ben Ali (who had been in power since November 1987) on January 14, 2011.[20]

Symbolic interactionism[b] holds that individuals act toward things based on the meanings ascribed to those things[21] and thus the creation of symbols is also a common passive resistance device. Gandhi's exhortation for Indians to use the rudimentary spinning wheel to spin their own cloth and not depend on British factories was so resonant a symbol of Indian resistance to the Raj that today it is on the Indian national flag. Besides gaining attention, the Buddhist monks who immolated themselves—particularly the first, whose heart was preserved and displayed in Saigon's Xa Loi Pagoda—also became symbols for the resistance campaign. An example of another kind of symbol is seen in the Danish resistance movement against the Nazis. King Christian became a symbol embodying the spirit of the passive resistance struggle. The King's traditional morning ride through Copenhagen on his statuesque horse, unaccompanied by police or aides-de-camp, even months after the Nazi occupation began, gained national attention. As the Danish poet Kaj Munk wrote then: "It does us good, as if it says to us, Denmark is still in the saddle."[22] The King also kept his royal standard flying both day and night over his palace, indicating he was always ready—either to negotiate with the Germans or lead his people. Protest puppetry (the use of large puppets to represent an issue to bystanders, the media, and others) has been used in demonstrations in Seattle, Washington; New York City; and Windsor, Ontario.[23] Protest (or radical) puppetry is a confluence of performance art and social messaging that seeks to represent social concerns through readily identifiable caricatures[24] Activists seeking to bring attention to environmental issues, nuclear disarmament, racism, substance abuse, peer pressure, sexism, and homelessness have employed this device.[25]

---

[b] There are three underlying premises of the theory: individuals act toward things based on the ascribed meaning of said things, the meaning of such things is derived from the social interaction an individual has with others and the society, and these meanings are handled in, and modified through, an interpretative process used by the individual in dealing with the things he or she encounters.

Few symbols are more resonant than the martyr, particularly in groups and/or cultures where there is a heightened sense of heroism associated with fallen members and where the community supports and rallies around families of the fallen or incarcerated. Families of fallen or incarcerated members are given enhanced social status (as well as financial and/or material support). Voluntary participation—be it religious or politically motivated—in the cause is considered a victory. As with other phenomena, as individuals adopt this view of success, their own self-image becomes more intimately intertwined with the success of the organization.[26] In January 1978, Iranian internal security forces (SAVAK) responded brutally to protests by Ruhollah Khomeini's students in Qom. At least seventy were killed over two days; the incident was tipping point for the movement, which shifted from an anti-Shah movement dominated by the secular opposition to one led by religious leaders (or *ulema*), particularly Khomeini. The cycle of forty-day commemorations consistent with the tenets of Shia Islam[c] started that February to honor those killed in Qom, and thereafter, every time a demonstrator was killed, individuals knew exactly when and where the next event would occur. Riots and protests soon spread to other cities around Iran, and each cycle of the forty-day commemorations saw an increase in participation and potency, with new martyrs being generated at each commemoration.[27]

Humor can facilitate outreach and mobilization, a culture of resistance, and an inversion of oppression.[28] Humor attracts members though its engendered energy, creativity, and enjoyment. This in turn increases in-group cohesion by boosting morale by comically exacerbating in-group/out-group differences. Finally, humor enables the change agent to seize the sociopolitical momentum by provoking, mocking, and/or ridiculing; reduce the fear associated with confronting a regime; and constrain the regime's response options.[29] Humor can also be used to provoke an enemy and demonstrate contempt. This technique was used by most passive resistance movements in Nazi-occupied Europe. Sometimes it took the form of shouting anti-Nazi jokes in a cinema hall showing German films, or of little jokes made in public about Nazi repression policies, like the Danish streetcar conductor calling out, "All saboteurs change here."[30] The Serbian youth movement

---

[c] Arbaeen, which means "forty" in Arabic, is a Shia religious observation that occurs forty days after the Day of Ashura, the commemoration of the martyrdom and beheading of Imam Husayn Ibn Ali (who along with seventy-two supporters died in the Battle of Karbala in 680 CE), the grandson of the Prophet Muhammad. Forty days is the typical mourning period in many Islamic cultures.

Otpor![d] used public theater and satire through various forms of media in concert with more "traditional" approaches to nonviolent resistance such as demonstrations, concerts, electoral politics, general strikes, and even the occupation of government buildings and disruption of traffic. Otpor! created publicity by spreading handbills, posters, and graffiti showing their symbol (a clenched fist) throughout the country and by having political cartoonists incorporate incongruity and absurdity into their products.[31]

Ostracism campaigns, accusations, whispering campaigns, and refusal to speak or be friendly are also frequent techniques. In the anti-Nazi resistance, these occasionally developed spontaneously; later, they were often organized by the undergrounds. In Denmark, these techniques were labeled *Den kolde Skulder* (the cold shoulder) policy, and many people wore buttons initialed DKS or SDU (*Smid dem ud*, or throw them out). Open contempt was displayed: "If a German military band gave a concert in a public place, they did so without a single listener. If Germans entered a cafe, at a given signal all Danes then rose and left."[32] In Belgium, similar activities were organized. One illustration of the type of witticism that helped the Belgian morale and enraged the Germans centered on the proposed German invasion of England: "An attractive housewife entered a store just before a German officer, who of course told the storekeeper to help the lady first. The lady declined and stated she did not wish to delay the officer who was probably in a hurry to catch his ship for England."[33] Also, anti-German inscriptions began to appear on the sides of buildings, on sidewalks, and on streets. In fact, one ingenious Belgian reportedly cut letters spelling "Down with the Boche" into the rims of his automobile tires and then filled the letters with paint, so that the slogan was painted continuously down the middle of the street.[34] Such programs served dual purposes: they lowered the morale of the Germans while at the same time raising the morale of the populace, creating a feeling of defiance and unity that could be later channeled into more significant resistance activity.

---

[d] Otpor! ("resistance" in Serbian) was an influential youth movement in Serbia from 1998 to 2003 that engaged in a two-year-long, successful nonviolent struggle against Slobodon Milosevic. Otpor! was formed in Belgrade in response to repressive university and media laws introduced earlier that year. The group primarily consisted of members of the Democratic Party Serbia (DS) youth wing, members of various nongovernmental organizations that operated in Serbia, and university students (many of whom were veterans of anti-Milosevic demonstrations). The organization quickly gained prominence as anti-regime media outlets started featuring the clenched fist symbol in open defiance of Serbia's information law. In the aftermath of the 1999 NATO bombing, Otpor! demonstrations resulted in nationwide police repression, resulting the arrests of over 2,000 activists, some of whom claimed to have been beaten in custody. After Milosevic's 2000 resignation, the organization became an international resistance cause célèbre and eventually (2003) transformed into a political party. It eventually merged back into the DS in 2004.

Nuisance activities vary greatly. They may be offensive personal acts against the opponent, such as the Algerian children publicly spitting on French soldiers. If a soldier struck a child, public opinion against the French would solidify all the more. The soldier felt humiliated and was clearly shown how the populace felt about his presence. Similar accounts exist with pro-PIRA children on the streets of Belfast. Another nuisance device is to overload the government security system with reports of suspicious incidents and persons. By following government instructions, large numbers of people can turn in false alarms or make unfounded denunciations of people who are suspected of aiding the enemy and in this way so overload governmental authority that valid reports cannot be handled. This technique has frequently been used against the block-warden surveillance system of countering underground activities.

The confluence of performance art, political theater, and social activism in the early twenty-first century has seen many of these tactics employed in pursuit of rather specific objectives. FEMEN is a Ukrainian activist group that seeks to publicize and eradicate the sex tourism trade among other social issues. Their provocative (and public) use of eroticism has brought them vindication and condemnation.[35] Their tactics typically employ disruption using topless activists (engaging in what Sharp refers to as "protest disrobing") who engage in a form of public theater documented by their internal media team and reporters, photographers, and videographers. The themes of these events have ranged from the corrupt Ukrainian political system, to sex tourism, to students' rights, to a variety of governmental policies/actions deemed unfair to the masses. While FEMEN's approach has been criticized as more of a novelty sideshow, their efforts have helped to catalyze a nascent feminist movement that emerged from the post-Soviet era society.[36] While their efforts have brought some attention to their issues, the group has indicated their intention to evolve into a political party (much like Otpor!) and formerly participate in the political process.

Noncooperation

Techniques of noncooperation call for a passive resister to perform normal activities in a slightly contrived way, but not so that police or government can accuse him of breaking ordinary laws. Activities such as "slowdowns," boycotts of all kinds, and various forms of disassociation from government are all examples of noncooperation. There are numerous examples of noncooperation in the anti-Nazi resistance movements, including falsification of blueprints and deliberate errors in adjustment of machine tools and precision instruments.[37] Workers in shipping departments of Nazi factories addressed shipments to the wrong address or conveniently forgot to include items in the shipments. Feigned sickness was widespread.[38]

These acts of noncooperation impeded the war effort while appearing simply to be honest mistakes. In Yugoslavia railroad workers used a particularly effective noncooperation technique: during an Allied air raid, they deserted their jobs and, after the raid, they stayed away for twenty-four hours or more because of "feigned fear." This seriously delayed railway traffic.[39]

Noncooperation is a principal tool of passive resistance and has been shown to be most effective in disrupting the normal processes of society and severely hampering and challenging the writ of a government—all in a way that is difficult for the government and its security forces to challenge. Many individuals altering their normal behavior only slightly can add up to a society behaving most abnormally.

Civil Disobedience

Mass participation in deliberately unlawful acts, though generally misdemeanors, constitutes civil disobedience. This is perhaps the most extreme weapon of passive resistance; the boundary between misdemeanors and serious crimes can be considered the dividing line between nonviolent and violent resistance. Forms of civil disobedience include the breaking of specific laws, such as tax laws (nonpayment of taxes), traffic laws (disrupting traffic), and laws prohibiting meetings, publications, free speech, and so on. Civil disobedience can also take the form of certain kinds of strikes and walkouts, resignations en masse, and minor destruction of public or private property.

In Palestine, after the Haganah raided the British and hid in a nearby village, passive resistance by the Jewish population was effective in preventing their capture. When the police began a search, people vigorously refused them entrance to their homes, stopping only short of using arms; hand-to-hand fighting with bricks and stones often broke out, and first-aid stations were set up to treat the injured. At the first sign of a British cordon, a gong or siren would sound, and at this signal villagers from nearby settlements would rush into the area, flooding it with "outsiders" and effectively preventing the British from recognizing which "outsiders" had taken refuge in the village following the raid and which had come simply to create confusion.[40]

Civil disobedience is a powerful technique, but to be effective, it must be exercised in large numbers. There is a calculated risk involved: the breach of law automatically justifies and involves punishment by the government and security forces. However, the more massive the scale on which civil disobedience is organized, the less profitable it is for the government to carry out sanctions. For example, during a Huk-led strike in the Philippines, as police were attempting to arrest the leaders, Luis Taruc used the tactic of demanding and forcing the government to arrest everyone participating in the strike. "We must crowd the prison with our numbers," he said. "If there is no room for us [in the police vans], we will walk to jail." It quickly became unfeasible for the security forces to use the threat of jail.[41]

During the Indian independence movement, Gandhi effectively used the same tactic. He led so many millions in the breach of law that it proved impractical, if not impossible, for the British to jail all offenders. As British officials saw, such widespread disrespect for a law makes its enforcement ridiculous and counterproductive.[42] Yet, if a government cannot enforce its writ, it must abdicate authority. As the jails became impossibly full, Gandhi's position in pressing his demands on a government searching for ways to pacify the population was increasingly enhanced.[43]

Organizers of passive resistance are selective about the laws that are to be broken. The laws should be related in some manner to the issues being protested or the demands being made. Examples are Gandhi's selection of the salt tax in India, which was considered a hardship tax on the peasants and representative of unjust British rule; the civil rights sit-ins in the United States, which were directly related to discrimination in public places; and the Norwegian teachers' strike against the Nazi puppet-government's demands that teachers join a Nazi association and that Nazi socialism be taught in the schools. If the laws that are broken have little or nothing to do with the issues involved, it is difficult to persuade a citizenry to risk government sanctions by taking provocative actions.

Cyber (or virtual) activism refers to normal, nondisruptive use of the Internet in support of an agenda or cause. Also called online organizing, electronic advocacy, e-campaigning, and e-activism, operations in this area include web-based research, website design and publication, transmitting electronic publications and other materials through e-mail, and using the web to discuss issues, form communities of interest, and plan and coordinate activities.[44] Hacktivism, the exploitation of computer systems (hacking) for a political purpose, brings methods of civil disobedience to cyberspace.[45] Hacktivist tactics include a litany of constantly evolving techniques often at the leading edge of information security. Included among them are virtual sit-ins, automated e-mail bombs, web hacks and computer break-ins, and computer viruses and worms. A virtual sit-in is the cyberspace equivalent of a blockade where the objective is to disrupt normal operations, thus calling attention to the perpetrator. In 1998, the Electronic Disturbance Theater (EDT) organized a series of web sit-ins against a series of Mexican and U.S. government websites as well as the Frankfurt Stock Exchange to demonstrate solidarity with the Mexican Zapatistas.[46] This variation of a denial-of-service (DOS) attack has supporters visit the specified sites in order to overwhelm the servers and limit accessibility. An e-mail bomb is another form of virtual blockade in which a particular e-mail address (or group of addresses) is inundated with messages, preventing the effective use of a particular account or server. In 1998, a Tamil group sympathetic to the LTTE swamped Sri Lankan embassies with thousands of e-mail messages from the Internet Black Tigers.[47] Website defacement is a form of hacking that does not necessarily seek to exfiltrate information or corrupt a network but rather seeks to replace existing public content with a political message. Also in 1998, a group of Portuguese hackers modified the sites of forty Indonesian servers to add a "Free East Timor" banner.[48] Hacktivists have used computer viruses, worms, and other malicious code to disseminate propaganda and damage target computer systems. On October 16, 1989, computer systems at NASA's Goddard Space Flight Center in Greenbelt, Maryland, were infected by the WANK (Worms Against Nuclear Killers) worm. The attack was executed by a loosely affiliated group of anti-nuclear weapons activists.[49] The WANK worm and other Hactivist tactics are migrations (with some mutation) of techniques used in the physical domain to the infosphere.

In summary, the underlying consideration in most passive resistance techniques is whether they serve to legitimize the position of the passive resister while alienating or challenging the government.

ORGANIZATION

The success of passive resistance rests largely on its ability to secure widespread compliance within the society. A government cannot be robbed of the popular support upon which it depends if only a few individuals act. A boycott, for example, requires participation by great numbers. The relative distribution of the components within a social movement differs on the basis of operational requirements. Experienced and highly effective organizers during the Orange Revolution in the Ukraine were able to mobilize hundreds of thousands of people (often despite frigid temperatures) against the sitting government while simultaneously avoiding the use or provocation of violence. On November 22, 2004, the day after a fraudulent vote, approximately 500,000 people (many dressed in orange) gathered in Independence Square in Kiev and marched to the headquarters of the Ukrainian parliament while carrying orange. This scene was broadcast globally, sending an unambiguous message to the members of parliament who would vote a few days later to void the election results.[50] Figure 10-1, showing a time series component model of the Orange Revolution, depicts how the Orange Revolution represents a departure from both the traditional and contemporary insurgent models in that the armed component was subsumed by the auxiliary and served essentially as a personal security element to protect prominent members of the political component from government attacks. The Yushchenko camp anticipated and prepared for violence. Viktor Yushchenko's personal security detail was led by Yevhen Chervonenko and included fifty-five former military special operations and Interior Ministry security experts. Chervonenko claims that they had an "elaborate system of reconnaissance, intelligence, and physical protection."[51] Chervonenko also built up a team of hundreds of "battle-ready" individuals, including many athletes. "All were armed. Many of them legally held various weapons, including hunting weapons."[52] These paramilitaries were backed by 4,000 volunteers who, according to Orange Revolution organizer Taras Stetskiv, were "ready for everything and only waited for a signal"[53] to storm the presidential administration. While the underground was initially prominent, the relatively rapid expansion of the Orange Revolution was almost entirely in the public component.[54]

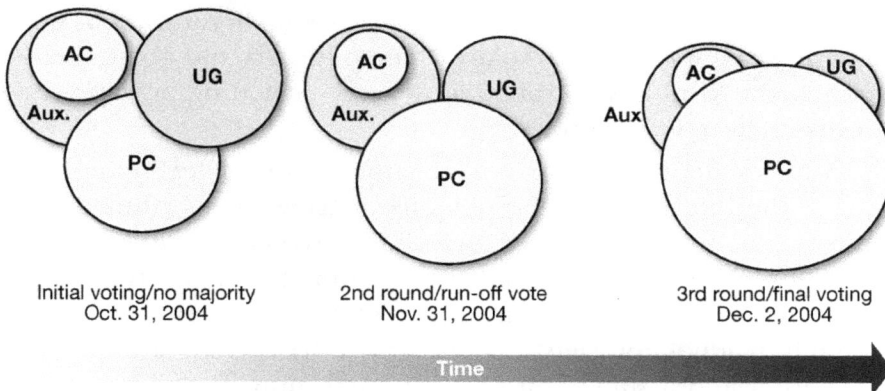

| Initial voting/no majority | 2nd round/run-off vote | 3rd round/final voting |
| Oct. 31, 2004 | Nov. 31, 2004 | Dec. 2, 2004 |

Time

Figure 10-1. Time series component model for the Orange Revolution. AC, armed component; Aux., auxiliary; UG, underground; PC, public component. The Orange Revolution represents a departure from both the traditional and contemporary insurgent models in that the armed component was subsumed by the auxiliary and served essentially as a personal security element to protect prominent members of the political component from government attacks. While the underground was initially prominent, the relatively rapid expansion of the Orange Revolution was almost entirely in the public component.

The mobilization of civic youth organizations under the well-trained Otpor! youth movement leaders from Serbia and Georgia provided a level of maturity and nonviolence that was critical for the effective presentation of a united front against the sitting administration.[55] Solidarity's loose national coordination enabled a decentralized control mechanism, allowing local organizers to gain recognition among their immediate constituency. The movement relied heavily on the organizational resources of seasoned activists and at one point initiated a strike with a single phone call. Once the movement was forced underground during the period of martial law, it had to re-establish its structures and leadership in accordance with the new conditions. Eighty percent of the Solidarity leadership was incarcerated during the initial sweeps, yet those in prison still heavily influenced the underground, with the leaders who avoided capture acting as intermediaries. The underground movement became even more nonhierarchical, local, and autonomous. Strategy and communications were the most centralized functions for this stage, with the agreement that a passive resistance campaign should be followed, and many publishing outlets were established to spread messages across the nation. A Provisional Coordinating Committee was established at the national level to develop and promulgate their strategy as well as issue demands to the government. Regional structures emerged in places where the crackdown had not completely decimated the organizational structure or caused the leadership to flee. Some of these regional structures also set up organizations within the largest factories and workplaces. The underground resisted efforts mostly at the local level through loosely connected organizations and networks that re-formed often.[56]

Advances in information and communications technology (particularly the Internet) and the ubiquity of access to social media have rendered the requirement for physical infrastructure and face-to-face interaction moot. Unlike traditional organizational constructs, these networks do not require physical instruction, geographic collocation, or even individual notoriety.[57] Political pressure no longer requires the aggregation and assimilation of committed individuals into an organizational structure. Activist groups can advance their concerns far more rapidly through the spontaneous formation of distributed networks of concerned individuals. These ad hoc, location-independent, and medium-agency networks self-organize around functional, not geographic, concerns. While the rapidity with which these groups can coalesce and act far surpasses that of an underground movement, an organizer sacrifices control and message discipline for speed. With virtual communities, there is not always a vetting and/or acculturation process and thus there is not the same discipline or unity of effort seen in small, clandestine organizations.[58]

Politically divisive issues often catalyze the formation of virtual networks who organize to discuss, plan, and sometimes take action. An example is the Tibet Autonomous Region, which presents challenges with official relationships with the People's Republic of China (PRC). Because of the geographic isolation of Tibet and the relative inability of Tibetan activists to voice dissent from within the PRC, an online "Free Tibet" network has emerged to support the independence movement. The conglomeration of websites in English, Chinese, and Tibetan are hosted on servers outside China (often in Europe and the United States), are linked to one another, and have similar pro-independence sentiments; however, they do not have a common style or format or common themes, messages, or strategies.[59] While there is conceptual commonality among these activities, there is no explicit command and control apparatus. This affords freedom of action; however, it comes at a cost because loosely affiliated networks can be more easily co-opted, distracted, and/or delegitimized.

Organization is of critical importance to passive resistance. Although a few individuals can launch a passive resistance movement, in order to succeed, thousands—whose participation must be strategically channeled—must join them. How is widespread social compliance secured? What forces and factors induce people to practice passive resistance?

LEGITIMACY

Obtaining legitimacy is integral to the success of any revolutionary movement, regardless of the means by which the organization presents itself to the general population, their opposition, and external actors; a movement must be taken seriously as a legitimate actor within the political realm.[60] The following sections detail the factors that nonviolent movements employ to not only assert but also advertise their legitimacy in the eyes of the government, their constituency, and the international community. Those methods include conditions of normative and mystical factors, and consensual validation.

Normative Factors

One method by which leaders of passive resistance movements secure widespread compliance is by cloaking their movement and techniques in the beliefs, values, and norms of society—those things people accept without question. For example, the clergy led the earliest stages of the Norwegian resistance against the Nazi occupation and Quisling puppet government. From the pulpit and parsonage, the religious leadership of Norway coalesced public opinion against the Nazis by invoking the voice of the church. "When the Nazis established a new ecclesiastical leadership, the bulk of the old established Church ignored the orders of the new hierarchy. Through nonviolent action it preserved its integrity," simply refusing to cooperate in religious affairs with the Nazi occupation.[61]

Because the institutions of religion were held in high esteem, and because the clergy appealed within the framework of religious values, the Nazis never were able, even in the later stages of the occupation, to break the church's resistance. In India, women were used on the front lines of demonstrations, making it awkward for British forces to break up the crowd without inflicting injuries on the women and further stirring up public opinion.[62]

Poland was the only Eastern Bloc state in which the Church remained an independent actor and continued to exert a tremendous influence over the people. The symbols and rituals of the Roman Catholic Church pervaded the scene of strikes during the Solidarity movement. Primarily because of the astute leadership of Cardinal Wyszynski (later Pope John Paul II) and Catholic intellectuals, the Church in Poland began to champion basic human rights and political and civil liberties, providing the non-Catholic intelligentsia and the workers in Poland with a common vocabulary with which to oppose the state. The presence of priests saying Mass and giving Communion during demonstrations further validated the righteousness of the cause among Catholic laborers.[63]

The events in Egypt during January 2011 are an example of long-standing resentment manifested over generations reacting to a catalyst (Tunisia) and mobilizing (with the intent of removing President Mubarak). The turmoil in Egypt is an ideological extension of what transpired in Tunisia and is reminiscent of the 1979 Islamic Revolution in Iran; however, the unique geopolitical circumstances within Egypt differentiate it from both.[c] This distinction is evident in the

---

[c] The dissatisfaction with Mubarak can be traced back to the Free Officer's Movement in 1952 that deposed King Farouk. Nasser, followed by Sadat (assassinated in 1981), and Mubarak represent a series of technocratic military dictators whose relationships with the West alienated the religious elite and the bulk of their populace. Seminal events that contributed to this dissatisfaction were the surprisingly swift defeat by Israel in the 1967 Six-Day War (although a semblance of pride was recovered by Egypt's performance in the 1973 Yom Kippur War with Israel), the 1979 Camp David Accords (which has been arguably the most geopolitically significant policy change in the Middle East since the dissolution of the Ottoman Empire), and the 1981 Sadat assassination, which brought Mubarak to power. This series of leaders increased the prominence (and governmental condemnation) of the Muslim Brotherhood (an Islamic sociopolitical organization established in 1928 that has served as a counterweight to the colonial then the secular government). The continual friction between the various regimes and the Brotherhood gave rise to more militant groups, including the Egyptian Islamic Group (EIG) and Egyptian Islamic Jihad (EIJ). The former emerged as an umbrella organization for the proliferation of Islamic

juxtaposition of typical protest violence and the erudite articulation by the blocked elite who accompanied the younger university students in Cairo's Tahrir Square.

Mystical Factors

Rare or extraordinary factors such as charisma play an important part in mobilizing public opinion in a passive resistance movement. Gandhi's leadership of India's independence struggle verged on the mystical. Thousands of villagers from rural India, who perhaps could not be touched or aroused by any modern means of communication or organized population pressure, were mobilized into action by Gandhi's fasts and his religious mystique. Hundreds of thousands of peasants gathered to meet Gandhi, although they often did not understand his language and could barely see him. It is difficult to estimate the role of Gandhi's mystique in coalescing public opinion against the British, but it is clear that resistance to colonial rule had never appeared on such a large scale before Gandhi. Jean-Bertrand Aristide, the former Catholic priest and Haiti's first democratically elected president, was another charismatic leader who exerted near-mystical influence over the Haitian people. In the wake of the 1991 coup d'état that left Aristide in exile, many Haitians fled the country by sea. In the summer of 1994, the U.S. Coast Guard and Navy retrieved over 20,000 migrants in a three-week period. Debriefings of migrants by U.S. intelligence personnel confirmed that the opposition underground had disseminated rumors that "*Ti dede*" wanted his people to go out to the sea as migrants to influence the Americans. The United States made a deal with Aristide to restore him to power if he would tell the migrants to cease, and within forty-eight hours of Aristide making a broadcast to Haiti (transmitted over via traditional media and U.S. psychological operations platforms), the mass exodus halted.[64]

Although not a nonviolent movement, the PIRA relied heavily on support from Catholics in Northern Ireland who disapproved of the violent tactics (particularly terrorism) but supported the Republican cause. Counterintuitively, Catholic support for the PIRA would wane when the Provos were relatively strong militarily yet would increase

student groups that emerged in the 1970s, while the latter was a more secretive and elitist organization that considered itself the ideological vanguard of the global Salafist jihad movement of the late twentieth and early twenty-first centuries (ultimately merging with Al Qaeda in 2001). The increasingly close political–military relationship between Egypt and the United States, the heavy-handed reactions of the government to perceived internal threats from the militant Islamists against those advocating freedom of speech, and the epidemic underemployment of the educated class are but some of the grievances that have contributed to the widespread dissatisfaction with President Mubarak.

when oppression and the concomitant suffering was at its most fierce. Perhaps nowhere was the confluence of cultural, religious, and political factors more manifest than the engagement in self-immolative behavior as a means of addressing grievances in the 1917, 1920, and 1981 hunger strikes.[65] By March 1981, the conditions at Long Kesh prison had reached another standoff, and another hunger strike was initiated that ultimately led to the death of ten volunteers. Bobby Sands, the Officer-Commanding within Long Kesh, led the effort. The strike generated very high levels of public attention because four days into the protest, Sands's name was nominated to be a member of Parliament from Northern Ireland, and on April 9, he was elected by a narrow margin. The election brought heightened attention to the struggle as the international community debated whether the British government would allow a member of Parliament to succumb to hunger in defiance of their internment policy. On May 5, 1981, his sixty-sixth day on hunger strike, Sands died. The ensuing international publicity generated a substantial increase in support. Sands's death revitalized the PIRA's motivation and prompted the more overtly political role of Sinn Féin.[66] Furthermore, his leadership of the Provisionals during his internment and steadfast adherence to the Republican cause continue to resonate among Republican and revolutionary circles. Although the PIRA was suspicious of politics and equated political participation with compromise, some of the leadership began to see its value.

Consensual Validation

The technique of "consensual validation"—in which the simultaneous occurrence of events creates a sense of their validity—is often used to coalesce public opinion. For example, if demonstrations take place at the same time in diverse parts of a country, the cause which those protests uphold appears to be valid simply because a variety of persons are involved. A minority group can organize a multitude of front organizations so that seemingly widely separated and diverse organizations simultaneously advocate the same themes and give the impression that a large body of opinion is represented. Passive resistance organizers effectively use the psychology of "consensual validation" to rally public opinion.

The same techniques used by passive resisters against the government can be used to ensure widespread social compliance within the resistance movement. Ostracism is frequently used to apply pressure on individuals not participating in the passive resistance campaign. Instances of organized ostracism of collaborators can be found in all the underground resistance movements in Nazi-occupied countries during

World War II. In Denmark, the underground published blacklists that were "feared by all those who acted in the interest of the enemy."[67] Informal, everyday pressures of conformity also help secure widespread compliance. The fact that passive resistance generally appears during times of crisis or of popular unrest indicates that at such times there is often a greater sense of nationalism, of a particular "we" arraigned against "they." There are then strong pressures demanding conformity.

Like the previous Polish Solidarity strikes in 1970 and 1976, the strikes of 1980–81 were precipitated by an economic crisis but were different in that they were more political, and larger. The 1980–81 strikes evidenced a more notable concern for political rather than economic well-being. The scope of these strikes dwarfed those in the past, with nearly 10 million members at the onset of martial law in 1981. An estimated 8,000,000–9,000,000 workers participated in the strikes, effectively halting economic and political life in Poland.[68] The enormity of these efforts not only drew domestic and international attention but also lent legitimacy to the movement.

COMMUNICATION AND PROPAGANDA

As noted earlier in this chapter, the first phase of passive resistance is characterized by a period of attention-getting propaganda activities: parades, demonstrations, posters, newspapers, and other forms of communication, either clandestine or open. Once the resistance movement is launched, there must be continuing means of "spreading the word." No movement can operate without some form of communication between the leaders and the led. The underground press in Denmark played an important role; the first illegal newssheet appeared on the very first day of the German occupation. Within a year, nearly 300 illegal newspapers were being published once or twice a month. They had an estimated circulation of 70,000, and each copy was read and passed on by large numbers of people. Ironically, "as the German domination became sharper, so did the free press become more and more powerful."[69] Similar examples can be found in Norway, Poland, France, and The Netherlands during the Nazi resistance: Besides providing "objective" information on the course of the war, using Allied news-broadcast reports, the clandestine papers instructed the populace in passive resistance techniques and procedures.[70]

Unlike most revolutionary organizations, the Solidarity movement in Poland initially conducted its meetings openly and disseminated its messages through overt networks; it was then forced underground as martial law was imposed. Within Poland, during the last half of 1982, the underground circulated more than 1500 regular publications that bypassed the government censor. Underground publishers also produced vast reprints such as George Orwell's *1984* and similar works. Radio Free Europe was also crucially important to the spread of the opposition's version of events and messages to the Polish population.[71] Informal communication networks were invaluable in countering previously effective strikebreaking tactics. Workers throughout Poland were kept abreast of strike activity and government concessions to the strikes, and as news spread of government concessions, the strikes escalated rather than diminished.[72]

The use of media, particularly nontraditional forms, has been effective in disseminating the organizer's message and in many cases inadvertently serving as a source for mainstream media. Instead of more using elaborate attention-getting tactics to attract mainstream media, some activist groups have created and organize media capability. This "coverage" is often disseminated through sympathetic physical or virtual networks and at times picked up by mainstream media. In one study, activist-based Internet sources reported a higher proportion of protest events at the local, national, and international levels.[73] The use of media and communication technology is essential for the collaborative planning and simultaneous exaction of distributed transnational protest events. A distinct advantage of the Internet over traditional communications is the ability to circumvent government attempts at censorship. Increasingly technologically sophisticated activist groups have employed organic capabilities and contracted information-technology support from third parties to penetrate denied areas and expose sequestered populations to their message.[74] FEMEN rarely conducts an operation without numerous media outlets accompanying their internal photographers and videographers to document the event. The images and/or videos and accompanying commentary are posted to their organizational website and blog within hours. Followers post comments, include links to other media coverage, and share their experiences with members of the network.

Communications technologies have been used to facilitate clandestine organization and planning; however, the government capabilities (although not necessarily their agility) may exceed those of the organizers, and thus redundant systems and plans are necessary. The Tunisian Internet Agency (ATI) allegedly hacked passwords of Facebook sites advocating the removal of Ben Ali and ultimately (under orders of the government) restricted access to the Internet from all IP addresses outside governmental facilities.[75] Tunisia's National Computer Security Agency was engaged in a limited cyber conflict with some activists—many of whom were members of sympathetic groups outside Tunisia—who attempted to overload government websites with distributed denial-of-service (DNS) attacks.[76] The Egyptian government restricted access to the Internet and mobile networks[f] at the height of the tensions (January 28, 2011). The protests attracted worldwide attention thanks to the increasing integration of social media that has enabled activists and spectators to communicate, coordinate, and document the events as they occur. As the level of publicity increased, the Egyptian government accelerated efforts to limit Internet access, especially access to social media, securing essentially all Internet access and mobile phone service on the eve of major planned protests on Friday, January 28. Limited mobile service resumed on January 30.[77]

Clandestine methods of communication resembling those used by espionage organizations were also developed. One of the earliest and most successful acts of resistance in Norway was the dramatic resignation in 1941–42 of 90 percent of Norway's teachers in response to Nazi pressure to teach National Socialism and join a Nazi teachers' association. This dramatic rebuke to the Nazi regime was accomplished by communicating the plan to all of Norway's teachers by disseminating the message through boxes of matches that contained a statement the teachers were required to make to the Nazis. The communication was obeyed, even though the leaders were never known, because each communicator was trusted and people were assured others were going to take similar action.

TRAINING

Once organizational steps are taken to secure widespread social compliance, an effort must be made to instruct and train passive resisters. The idea is to erect a mental barbed-wire fence between resisters and authority. This instruction often takes the form of codes of "do's and don'ts." Many undergrounds have found that it is easier to tell people what not to do than what to do.[78]

Training is particularly critical when positive, not just negative, actions are desired. Noncooperation and civil disobedience are positive acts that necessarily involve training, organization, and solidarity on the part of the resisters, whether they operate in the open or clandestinely.

---

[f] U.K.-headquartered Vodafone Group PLC, which is contracted by the Egyptian government to provide mobile services to the country, released a statement on January 28 that claimed, "All mobile operators in Egypt have been instructed to suspend services in selected areas. Under Egyptian legislation the authorities have the right to issue such an order and we are obliged to comply with it. The Egyptian authorities will be clarifying the situation in due course."

Gandhi placed great emphasis on nonviolent training, not only because he looked upon nonviolence as a moral creed, but also because he understood that it was essential for effective passive resistance. He required his followers to swear an oath, and he developed a code for volunteers. When individual suffering is involved, and when individuals must invite suffering through civil disobedience, considerable discipline is required. The American civil rights movement followed the Gandhi example and applied it in planning the 1958 "sit-in" movement. Special schools were established to train young people to withstand physical violence and tolerate torment without responding with violence that might negate the entire stratagem. Any violent response on the part of the passive resisters might provide a justification for the use of violent procedures in quelling demonstrations.

Organizations like Otpor! have compiled literature and lessons learned from their protest experience and packaged them in print, video, and electronic media. These free materials are readily available on the Internet and often come with points of contacts for additional information, thus building the virtual network of nonviolent activists. Institutions like the Albert Einstein Institution (AEI)[g] and the U.S. Institute for Peace (USIP)[h] have not only served as a repository for this information but also funded research (as well as the translation and publication of nonviolent resistance materials) and hosted events promoting nonviolent resolution to political problems. The works of Gene Sharp (mentioned earlier) have been translated into over thirty languages and employed from the Ukraine to Egypt.[79] There are even organizations such as the International Center on Nonviolent Conflict[i] that have developed video games to serve as interactive training tools.

RIOTS AND DEMONSTRATIONS

The manipulation of crowds and civil disturbances is just one of the means used to accomplish the objective of seizing power. This section introduces some of the strategies and tactics employed in riots and demonstrations and includes two brief case studies.

The internal security forces, which bear a major share of the burden of maintaining order, should understand that the control of subversively manipulated crowds requires special considerations. The security forces of a nation may be composed of paramilitary and military units and civil police. The function of maintaining internal security

---

[g] Online at http://www.aeinstein.org/.

[h] Online at http://www.usip.org/.

[i] Online at http://www.nonviolent-conflict.org/.

may be performed by one or any combination of these forces. Standard priorities of force may be adequate for dispersal of ordinary civil disturbances, but in dealing with a subversively controlled riot, internal security forces must be alert to situations or acts that compel them to respond in ways that the subversives can politically exploit. The underground use of subversively manipulated crowds and civil disturbances adds a new dimension to the problem of maintaining internal security. The difference between civil disturbances that are subversively manipulated and those that are not can be expressed in terms of objectives. Strikes, riots, and demonstrations usually have limited goals, such as better working conditions or social changes.

The subversive manipulation of crowds and civil disturbances involves a relatively small number of underground members who try to guide and direct "legitimate" protests. They attempt to direct the crowd toward emotional issues and arouse them against authority. The U.S. Central Intelligence Agency (CIA) has employed subversive manipulation of crowds to facilitate political objectives. Shortly after President Dwight D. Eisenhower was inaugurated in 1953, CIA officers in Iran involved in anti-Soviet operations shifted their focus to undermining Iranian President Mohammed Mossadeq's rule. Agents orchestrated a pro-Tudeh Party demonstration to push the military and civilian populations toward General Fazlollah Zahedi. The following day, agents again helped incite an anti-Mossadeq crowd that was later joined by police and army units who, after wreaking havoc on government buildings and pro-Mossadeq institutions, marched on toward the president's house, where he surrendered to General Zahedi.[80]

There are several theories for explaining crowd psychology and the ways in which the behavior of the crowd differs significantly from the psychological dispositions of those individuals within it. Freud's crowd-behavior theory, dated but nonetheless valid, holds that people in a crowd tend to act as though their psyches have merged into a single entity whose collective thought process is dramatically different from those comprising the crowd. As the collective enthusiasm increases, individuals tend to lose their inhibition and become less risk averse.[81] This behavior theory, derived from the behavior of animals in herds, flocks, and schools, has been applied to the study of human conduct during activities such as stock market bubbles and crashes, street demonstrations, sporting events, religious gatherings, and episodes of mob violence, as well as decision making, judgment, and opinion forming.[82] A riot is typically the result of an unorganized violent mob that is neither controlled by a leader nor organized into units or another hierarchical structure. Riots have, however, been intentionally incited and/or directed by organized political activists and have been at least partly directed. Although major insurrections have been initiated by riots, such as the French Revolution of 1789 and the Russian Revolution of February 1917, the spontaneous street violence was not part of an orchestrated plan to overthrow the respective governments.[83]

The emotional perceptions and beliefs of the crowds that participate in civil disturbances often do not coincide with objective reality, and the individuals involved do not realize that their grievances are being manipulated in politically subversive ways. The following case studies exemplify the use of protests (both violent and nonviolent) to achieve a political objective. The first is an example of a demonstration subversively manipulated by Communist organizers in Colombia; the El Bogotazo incident had a profound effect on Fidel Castro, who is among the most prominent revolutionaries in the latter twentieth century. The second is a more contemporary example of networked nonviolent warfare with ideologically diverse groups uniting against a common adversary. The 1999 Battle in Seattle is arguably one of the most influential single demonstrations in recent history.

Colombia (1948)

El Bogotazo (Blow at Bogota) demonstrated the effectiveness of the riot technique. It is particularly significant in that the techniques for provoking street crowds into rioting learned in the demonstration are used as training material for revolutionaries. A Venezuelan Communist defector from Castro's training school—the Tarara Training Center in Cuba—reported that the classroom materials included diagrams of the tactics and descriptions of how the crowds were manipulated in the three-day riot that wrecked the Ninth Inter-American Conference and left the Colombian capital in ruins.

The Ninth Inter-American Conference had been called for March 30, 1948, to discuss the pledge among member nations to the mutual defense and resistance to the threat of international Communism. According to testimony before the Judiciary Committee of the 86th U.S. Congress, intercepted Communist communiqués reveal that the party immediately went into action to disrupt the conference and the combined forces of the Latin American Communist Party apparatus set about making plans. A high-ranking official of the Colombian Communist Party (which claimed at that time a membership of 10,000 out of a total population of 11 million) said that the Inter-American Conference must be blocked but that this action was not to be known as a Communist activity; he admonished the party to refrain from open activity so as not to jeopardize or curtail party functions.

The Communists devoted considerable effort to preparing for any proposed riot; they seldom relied exclusively on spontaneity or accidental occurrences, even though they attempted to capitalize on such events. By January 29, 1948, arms and explosives had already been stored in seventeen houses. A Communist dispatch dated February 2 included the information that plans called for organization of mass public meetings, organization of sixteen meetings of cells in outlying districts, recruitment of new members to the party, organization of fifteen syndicates and unions, further organization of cells within the syndicates, and distribution during the conference of 50,000 handbills and 3,000 posters. A committee of the Communist Party was assigned to supervise these arrangements. A dispatch dated March 30 laid out the program of agitation and attacks upon the U.S., Chilean, Brazilian, and Argentine delegations, all of which were especially anti-Communist. During the middle of the first week in April, the Communist-controlled Latin American Conference (CTAL) adopted resolutions in Mexico City condemning the conference.

On April 9, a well-known figure, Dr. Jorge Gaitan, was killed by four bullets from a revolver fired by an unidentified person. Dr. Gaitan, a 47-year-old lawyer, was the leader and former presidential candidate of Colombia's liberal movement. Although it was reported that the Communist Party had supplied money through an intermediary for the support of Dr. Gaitan and his movement, he had maintained an independent attitude toward the Communists. Rumors of Communist plans to disrupt the conference caused Dr. Gaitan to publicly repudiate all acts against the conference, saying that these were acts against democracy and the unity of the Americas. Although his assassin was never identified, both the personality of Dr. Gaitan and the circumstances surrounding his death inspired at least one observer to say that the Communists, needing an appropriate victim whose death could prevent the holding of the conference, targeted a prominent person. Colombian President Ospina Perez also suggested that the man who killed Dr. Gaitan apparently had Communist affiliations and that the entire affair was a Communist maneuver.

A wave of mass violence was triggered by the assassination, which took place within the sight of thousands. The Communists channeled the high emotions into anti-U.S. feelings and acts of violence against U.S. property and individuals. Within fifteen minutes of the attack on Dr. Gaitan, radio-broadcasting stations in Bogota had been taken over and the Communists were issuing instructions and inciting the people to revolt against the government, the conference, and Yankee imperialism. Orders were given to plunder arms depots, hardware stores, gunsmith shops, department stores, government buildings, police precincts, and army barracks, and to organize a "popular militia." The radio also transmitted orders to specific individuals to assault specific places and gave locations where additional weapons could be obtained. Instructions were given on how to manufacture Molotov cocktails. During the broadcasts, fighting—reportedly between the Communists and a group of students—for control of the radio facilities could be heard in the background; the armed Communists forced the students out of the station. By controlling communications, the Communists could incite attacks against the symbols and instruments of power within the government. Within each group of demonstrators in the crowds were organized agitators chanting similar slogans. Prompted by the Communist agitators, a crowd entered the parliament building where the Inter-American Conference was being held and destroyed most of the interior. The rioters concentrated on destroying offices of the Chilean and U.S. delegations. Mob action almost completely suspended transportation, created obstacles for the police, and made the crowd that much more difficult to control.

Led by the Communists, less than 5 percent of the population carried on the riot for three days. Shops, churches, public utilities, and institutions of public service were attacked. Red flags were evident throughout the crowd, and in every group orders could be heard directing the mobs. The word "*abajo*" (down with) was heard frequently. There was heavy sniper fire.

The first burnings of buildings may have been simple, random gestures of protest, but a consideration of which buildings were burned showed a subversive pattern. On the afternoon of April 9, the Ministry of National Education, the Ministry of Justice, the Ministry of Foreign Affairs, the Palace of Justice, the Ministry of Government, the Episcopal Palace, the detective headquarters, and the Identification Section for natives and foreigners were all attacked and burned. The Confederation of Colombian Workers, which represented 109,000 organized workers, called for a general strike throughout Colombia.

After declaring a state of siege and imposing martial law, President Perez eventually restored order. But the Communists had achieved their tactical objective of disrupting the conference and, in the process, had effectively demonstrated the practicability of their methods.

Seattle, Washington (1999)[j]

The incident colloquially known as the "Battle in Seattle" is an example of the effective use of nonviolent resistance techniques to disrupt a high-profile political event. On November 30, 1999, approximately 50,000 activists representing more than 200 organizations[84] from across the globe collaborated in Seattle to demonstrate during the World Trade Organization (WTO) Ministerial Conference held at the Washington State Convention and Trade Center in Seattle, Washington. The purpose of the conference was to initiate "The Millennium Round" of trade negotiations. However, delegates failed to agree on the agenda amidst the disagreements between industrialized and developing nations.[85]

People for Fair Trade/Network Opposed to WTO (PFT), Direct Action Network (DAN), and the American Federation of Labor and Congress of Industrial Organizations (AFL-CIO) were among the coalitions that helped set the conditions for the coming events. All three established bases in Seattle in the months leading up to the WTO Ministerial Conference, and each attempted to build a coalition from diverse organizational members. DAN represents an emerging species of political organization—one based on loosely affiliated networks rather than institutions. The primary networked organizations in DAN were a coalition of such groups as the Rainforest Action Network, Art & Revolution, and the Ruckus Society. Through DAN, these groups coordinated nonviolent protest training, communications, and collective strategy and tactics through a decentralized process of consultation/consensus decision making. DAN organized training sessions for approximately 250 people that were conducted by the Ruckus Society in the weeks prior to the protests.[86] These groups were fairly heterogeneous, with representation across ethic and socioeconomic lines. Many were active in (or supportive of) the environmental movements gaining increasing attention at the time, while others came to be involved through human bridges. Organizers and participants alike derived a

---

[j] Methodological note: The majority of the information presented in this section was derived from sources in the scholarly literature. That information was augmented by the subjective experience of an individual who was on the periphery of planning and coordinating events leading up to the demonstration and who witnessed (but did not participate) in the events. The interviewee (heretofore referred to as "HC") was a recent college graduate at the time and was employed and resided in downtown Seattle. HC was (and is) politically informed and involved in environmental activism but has no criminal record or a history of involvement with violent extremist groups. Although not formally affiliated with any particular group, HC's social network included a number of individuals who were active members of environmentalist and/or anti-capitalist movements (all of which were nonviolent). The interview took place over the phone on December 13, 2011, and lasted for approximately ninety minutes. HC was not compensated for the time.

great deal of pride and satisfaction from the group's diversity of membership and interest yet singularity of focus.[87]

The second principal WTO opponent was the AFL-CIO, a hierarchical institution that employed the traditional labor organization model of unitary top-down command with relatively little decision-making input from rank and file members. The AFL-CIO's policy goals were directed more at American politics and less at international issues. Their target was supporting and legitimizing President Bill Clinton's actions at the conference through symbolic displays. The AFL-CIO helped attract thousands of people to Seattle with the intent of holding a rally at the Seattle Center and then marching downtown to the Paramount Theater. Its main adherents had little interest in joining with DAN's, but during the second and third days of the protests, a spillover from the AFL-CIO crowds into DAN's street actions added to a third wave of protest that ultimately overwhelmed the police.[88]

Although neither numerous nor operationally integral to the WTO protests, a third group (anarchists mostly from Eugene, Oregon)[89] received a disproportionate amount of media attention.[90] The groups of anarchists, or Black Blocs, varied in size from a few dozen to a few thousand and typically appeared on the occasion of a rally and dissolved at its conclusion.[91] A Black Bloc consisted of protesters who wore black, carried anarchist flags and banners, and took a more confrontational approach to protest.[92] The Black Bloc tactic was disseminated mainly via the ultra-left counterculture network through periodicals, touring music groups, and social networks of travelling activists. The first observed Black Bloc tactic in North America was a January 1991 anti-war rally in Washington, DC.[93] The total number of Black Bloc participants in Seattle was estimated between one and two hundred people; however, their tactics continue to be disputed. The vandalism and anarchist graffiti targeting the corporate symbols of Nike, Starbucks, and the like were Black Bloc hallmarks; however, their message was confounded by the presence of nonaffiliated groups looting and vandalizing the shops. While police shadowed Black Bloc operations, some DAN members attempted to stop the vandalism and/or incitement.[94] This direct intervention suggested the degree to which DAN was attempting to preserve the integrity of their protest.

DAN's goal was to halt the WTO meeting by preventing the attendees from gaining access to the venue. The main instrument of doing so was the infiltration of a few dozen affinity groups (small, self-organizing units). DAN's first wave of 200–300 protestors comprised supporters who sought nonviolent civil disobedience and were willing to be arrested. Their job was to penetrate the area immediately surrounding the conference site, seize and hold the dozen strategic intersections that controlled access to the site, and wait for the second wave. They began staging by 5:00 a.m. local time, before the police arrived to establish the initial security cordon. The second wave of several thousand included other affinity groups and supporters who also opted for nonviolent civil disobedience but were unwilling to risk arrest. Their task was to establish street blockades in the vicinity of the WTO conference to inhibit the movement of responding police forces.[95] The first and second waves were organized around a dozen simultaneously converging affinity groups, overwhelming the protest target from multiple directions. Each affinity group blockaded a specific intersection. The first and second waves had control of the decisive ground around the Convention Center by 8:00 a.m. DAN anticipated that the blockade would last only until police had arrested sufficient demonstrators to regain control of the streets. By 9:00 a.m., the police began reporting difficulty with their crowd control efforts, resulting in an alert to potential law enforcement entities in western Washington. Shortly after 10:00 a.m., the first round of tear gas was deployed against DAN in an attempt to expand the police's security perimeter around the Olympic hotel.[96] DAN utilized communications channels—from cellular phones, to portable computers with an Internet connection, to pagers, police scanners, and two-way radios—to command and control certain nodes and maintain a degree of tactical cohesion. In addition to the organizers' all-points network, protest communications were leveraged with individual protesters using cell phones, direct transmissions from roving independent media feeding directly onto the Internet, personal computers with wireless modems broadcasting live video, and a variety of other networked communications.[97]

Simultaneously, another crowd (which would reach approximately 20,000 people by 11:00 a.m.) began staging for the AFL-CIO rally at the Memorial Stadium at the Seattle Center. The assistant police chief was still resisting calls from tactical units to declare a state of emergency; the decision was not made until 3:24 p.m. This third wave arrived in such high numbers that the mass arrests did not start occurring until later Tuesday evening. The third wave was a march by several thousand additional people composed mostly of environmental and human rights groups who elected to participate in the street protests instead of the labor parade. This group entered downtown from the south at about 1:00 p.m. local time and marched to the Paramount Theatre inside the protest zone. By 5:30 p.m., the police reinforcements began arriving, and their tactics became more aggressive with the deployment of mounted police, dismounted police in full riot gear, and the use of tear gas, stun grenades, and nonlethal projectiles. Many protestors began to withdraw from the Convention Center after the establishment of a 7:00 p.m. curfew.[98] By this time, residents were frustrated by their inability to travel effectively amidst the large crowd and emerging police barricades. There were violent incidents and looting during this period. Some commercial establishments had been ransacked, and there were burning cars, tires, and garbage on certain streets. However, the disturbances and confrontations with police were not explicitly linked to DAN or their affiliates but attributed to unaffiliated individuals heretofore not involved in the demonstrations. Some racial tension erupted in some predominantly African-American sections of Capitol Hill, with much of the anger directed toward the police. The aftermath of the day's events provided an odd juxtaposition for many activists: satisfaction with a well-executed preventative action yet disappointment (even victimization and regret) with the destruction and violence. While many of the local environmental activists disagreed with the WTO and its policies, they possessed civic pride in and identified closely with the Seattle community and intended to disrupt the event, not destroy the city.[99]

President Clinton was scheduled to address the WTO Ministerial on December 1 and, by then, order had largely been restored. As the presidential motorcade departed, the police again deployed nonlethal force. They planned a 4:00 p.m. offensive to clear the area (an order about which the Washington governor had not been informed) and began a demonstration of force by sending patrol cars on high-speed patrols through cleared areas. At approximately 7:00 p.m., several hundred demonstrators began withdrawing from downtown area to Capitol Hill to continue their nonviolent protest. At approximately 9:00 p.m., police and National Guard elements arrived and quickly came under attack by members of the crowd. Most accounts indicate that the agitators were Capitol Hill residents and not DAN members. The crowd finally dispersed around 2:00 a.m. on December 2. By this point, DAN's success in disrupting the WTO Ministerial was well acknowledged. Their tactical actions, which limited delegate attendance, coupled with President Clinton's announcement of U.S. policy initiatives essentially rendered consensus impossible. DAN's focus shifted to providing support for detained protestors and it commenced a two-day vigil outside the Public Safety Building.[100]

The anti-WTO demonstration was the largest left-wing protest action in the United States since the Gulf War. Despite causing an estimated $3 million in damage, DAN achieved its objectives through a disciplined adherence to nonviolent principles and a tactical flexibility that readily accepted diverse tactics. Their success was also unintentionally facilitated by the large AFL-CIO presence as well as intelligence failures and tactical mistakes on the part of the Seattle police.

Nevertheless, the 1999 WTO demonstration served as an exemplar of a networked approach to swarm tactics using nonviolent means.

Strategically, the "Battle in Seattle" is viewed a tactical success in the anti-globalization movement[k] because it validated the capability to disrupt powerful economic and governmental process through nonviolent means. The event emboldened nonviolent activists, encouraged many others to mobilize, and infused the aforementioned cottage industry of protest training. Tactically, the 1999 WTO protestors' incorporation of the information domain into their strategy —from tactical command and control networks to real-time global messaging using cellular and mobile computing technology—has been emulated with varying degrees of success in numerous demonstrations throughout the world. What might future actions hold? The logical extrapolation of tactical communication networks will include the use of all conceivable means of technological communication variable. This will likely include sophisticated military-grade hardware, with demonstrators employing unmanned aerial vehicles (UAVs) to provide indications and warnings of police activity, to serve as a command and control node, and/or to serve as a communications relay to disseminate images and accounts of ongoing operations.[102]

SHADOW GOVERNMENT STRUCTURE[l]

One method that is frequently used to both undermine public confidence in a government and secure population compliance with passive resistance is the establishment of parallel structures of government. If a population must depend on an underground-sponsored "government," it will be forced to comply with the underground's passive resistance program and therefore withdraw its support from the regular government.

In India, for example, Gandhi felt that the highest form of passive resistance would be the establishment of parallel institutions of government—not only because they would be a potential weapon against the British colonial government, but also because they would provide

---

[k] The paradox of globalization (the economically driven phenomenon that increases financial interdependency and obscures traditional national boundaries) is that while the underlying theoretical logic is intends to secure peace and prosperity for the world, the result is often political and social instability resulting from the shift in patterns of political authority from the centralized, hierarchical state-based model to more decentralized conglomerates. The result is the proliferation of stretches of territory indefinitely existing in a sort of limbo between sovereignty and subordination to existing juridical states.[101]

[l] For a more comprehensive discussion of insurgent use of shadow governments, refer to the appropriate chapter in the companion to this work, the second edition of *Undergrounds in Insurgent, Revolutionary, and Resistance Warfare.*

positive value in creating a sense of unity and community within the diverse Indian population. Indeed, many observers assert that the passive resistance movement in India was important not so much because of what it did against the English but because of what it did for the Indians: It shaped a new Indian nationalism and provided an opportunity for Indians to repair the wounds to their national self-esteem that had been inflicted during more than one hundred years of outside rule.[103, 104]

Similarly, in Poland, the underground passive resistance movement helped the Polish people maintain a sense of national identity and unity in the face of Nazi harshness. The establishment of a "secret state"—of underground courts, schools, and civil government—maintained a continuity and identification with nationalism, thus denying loyalty to the occupier. During the Algerian War for Independence in the 1950s, passive resistance served to solidify the Arab community. Although unfavorable environmental, social, and political conditions had existed for generations, there was no insurgency until Arab grievances crystallized into national consensus. In mid-1940s Palestine, the Zionist insurgent-led passive resistance campaigns did as much to develop a strong feeling of unity and nationalism as they did to oppose the British forces. The main effect was to lead people away from any form of support for the official government. Thus, a consensual validation of the values and objectives of the insurgents was provided.

The techniques and societal values capitalized upon to undermine popular support of the government also serve the positive function of solidifying public opinion around a larger sense of community and national identification.

SUMMARY

Nonviolent resistance continues to play a prominent role in many underground and revolutionary activities. The twenty-first century has witnessed a synthesis of the global technological networks that link computers on the Internet and social networks to result in innovative forms of protest. Though few have managed to mobilize a sufficient number to displace a regime, they have provided a forum for a youthful demographic to engage in creative, often social-media-directed alternatives to the picketing and chants their elders employed.[105] This trend is likely to continue, as the opportunities for the voiceless to find their voice are numerous and growing as is discontent with the political status quo and the concomitant passivity.

ENDNOTES

[1] Jon Kleinberg, "The Convergence of Social and Technological Networks," *Communications of the ACM* 53, no. 11 (2008): 66–72.

[2] Ellen Barry, "Sound of Post-Soviet Protest: Claps and Beeps," *New York Times*, July 14, 2011, http://www.nytimes.com/2011/07/15/world/europe/15belarus.html?_r=1&pagewanted=all.

[3] Mulford Q. Sibley, ed., *The Quiet Battle* (Garden City, NY: Anchor Books, 1963), 9.

[4] Leo Kuper, *Passive Resistance in South Africa* (New Haven, CT: Yale University Press, 1957), 84.

[5] E. E. Schattschneider, *The Semi-Sovereign People* (New York: Holt, Rinehart and Winston, 1960), chapter I.

[6] Chuck Crossett and Summer Newton, "The Provisional Irish Republican Army: 1969–2001," in *Casebook on Insurgency and Revolutionary Warfare, Volume II: 1962–2009*, ed. Chuck Crossett (Alexandria, VA: U.S. Army Publications Directorate, in press).

[7] C. M. Case, "The Social Significance of Non-violent Conduct," *The Quiet Battle*, ed. Mulford Q. Sibley (Garden City, NY: Anchor Books, 1963), 57.

[8] Clark McCauley and Sophia Moskalenko, "Mechanisms of Political Radicalization: Pathways Toward Terrorism," *Terrorism and Political Violence* 20, no. 3 (2008): 415–433.

[9] Chuck Crossett and Jason Spitaletta, *Radicalization: Relevant Psychological and Sociological Concepts* (Ft. Meade, MD: Asymmetric Warfare Group, 2010).

[10] Ariel Merari, "Terrorism as a Strategy of Insurgency," *Terrorism and Political Violence* 5, no. 3 (1993): 213–251.

[11] Crossett and Spitaletta, *Radicalization: Relevant Psychological and Sociological Concepts.*

[12] Ibid.

[13] Kuper, *Passive Resistance*, 85.

[14] Tadeusz Bor-Komorowski, *The Secret Army* (London: Victor Gollanez, 1950), 79.

[15] Gene Sharp, *Sharp's Dictionary of Power and Struggle: Language of Civil Resistance in Conflicts* (New York: Oxford University Press, 2011).

[16] Feliks Gross, *The Seizure of Political Power in a Century of Revolutions* (New York: Philosophical Library, 1958), 51.

[17] Gene Sharp, *The Politics of Nonviolent Action* (Boston: Portor Sargent, 1973), reproduced by The Albert Einstein Institution, accessed December 11, 2011, http://www.aeinstein.org/organizations103a.html.

[18] Barry, "Sound of Post-Soviet Protest: Claps and Beeps."

[19] Denis Warner, "Vietnam's Militant Buddhists," *The Reporter*, December 3, 1964, 29.

[20] Mike Elkin, "Tunisia Internet Chief Gives Inside Look at Cyber Uprising," *Wired*, January 28, 2011, http://www.wired.com/dangerroom/2011/01/as-egypt-tightens-its-internet-grip-tunisia-seeks-to-open-up/.

[21] Michael P. Arena and Bruce A. Arrigo, *The Terrorist Identity: Explaining the Terrorist Threat* (New York: New York University Press, 1996).

[22] Lt. Col. Th. Thaulow, "King Christian X," *Denmark During the German Occupation*, ed. Borge Outze (Copenhagen: The Scandinavian Publishing Co., 1946), 135–136.

[23] Leslie J. Wood, "Breaking the Wave: Repression, Identity, and Seattle Tactics,"*Mobilization: The Quarterly* 12, no. 4 (2007): 377–388.

[24] Rikki Asher, "Radical Puppets and the Language of Art," *Art Education* 62, no. 3 (2009): 6–12.

[25] Ibid.

[26] Crossett and Spitaletta, *Radicalization: Relevant Psychological and Sociological Concepts.*

27 Chuck Crossett and Summer Newton, "1979 Iranian Revolution," in *Casebook on Insurgency and Revolutionary Warfare, Volume II: 1962–2009*, edited by Chuck Crossett (Alexandria, VA: U.S. Army Publications Directorate, in press).

28 Majken Jul Sorenson, "Humor as a Serious Strategy of Nonviolent Resistance to Oppression," *Peace & Change* 33, no. 2 (2008): 167–189.

29 Ibid.

30 Ronald Seth, *The Undaunted: The Story of the Resistance in Western Europe* (London: Frederick Muller, Ltd., 1956), 99–100.

31 Sorenson, "Humor as a Serious Strategy," 167–189.

32 Seth, *The Undaunted*.

33 George K. Tanham, "The Belgian Underground Movement 1940–1944" (unpublished Ph.D. diss., Stanford University, 1951), 64.

34 Ibid.

35 Marian J. Rubchak, ed., *Mapping Difference: The Many Faces of the New Ukraine* (New York: Berghan Books, 2011).

36 Sarah D. Phillips, *Women's Social Activism in the New Ukraine* (Bloomington, IN: Indiana University Press, 2008).

37 Guillain de Benouville, *The Unknown Warriors* (New York: Simon and Schuster, 1949), 197.

38 John G. Williams, "Underground Military Organization and Warfare" (unpublished master's thesis [thesis 452], Georgetown University, Washington, DC, February 1950), 89–90.

39 David Martin, *Ally Betrayed* (New York: Prentice-Hall, 1946), 177–179.

40 R. D. Wilson, *Cordon and Search* (Aldershot, England: Gale and Polden, Ltd., 1949), 33.

41 Luis Taruc, *Born of the People* (New York: International Publishers, 1953), 41–42.

42 *India in 1930–31: A Statement Prepared for Presentation to Parliament* (Calcutta: Government of India Central Publications Branch, 1932), 73.

43 See Joan V. Bondurant, *Conquest of Violence: The Gandhian Philosophy of Conflict* (Princeton, NJ: Princeton University Press, 1958), 91ff.

44 Martha A. McCaughery and Michael D. Ayers, *Cyberactivism: Online Activism in Theory and Practice* (Bristol, PA: Taylor & Francis, Inc, 2003).

45 Dorothy E. Denning, "Activism, Hactivism, and Cyberterrorism: The Internet as a Tool for Influencing Foreign Policy," in *Networks and Netwars: The Future of Terror, Crime, and Militancy*, ed. John Arquilla and David Rondfelt (Santa Monica, CA: RAND Corporation, 2001).

46 Ibid.

47 Ibid.

48 Ibid.

49 Ibid.

50 Jerome M. Conley, "Orange Revolution (Ukraine): 2004–2005," in *Casebook on Insurgency and Revolutionary Warfare, Volume II: 1962–2009*, ed. Chuck Crossett (Alexandria, VA: U.S. Army Publications Directorate, in press).

51 Ibid.

52 Ibid.

53 Ibid.

54 Ibid.

55 Ibid.

56 Chuck Crossett and Summer Newton, "Solidarity," in *Casebook on Insurgency and Revolutionary Warfare, Volume II: 1962–2009*, ed. Chuck Crossett (Alexandria, VA: U.S. Army Publications Directorate, in press).

[57] McCaughery and Ayers, *Cyberactivism*.

[58] Laura Illia, "Passage to Cyberactivism: How Dynamics of Activism Change," *Journal of Public Affairs* 3, no. 4 (2003): 326–337.

[59] Tianjun Fu and Hsinchun Chen, "Analysis of Cyberactivism: A Case Study of Online Free Tibet Activities," *Intelligence and Security Informatics* (2008): 1–6.

[60] Crossett and Newton, "The Provisional Irish Republican Army: 1969–2001."

[61] Sibley, *The Quiet Battle*, 156–157.

[62] *India in 1930–31*, 660.

[63] Crossett and Newton, "Solidarity."

[64] John Shissler (former J-2A, Joint Task Force 160, Operation Sea Signal), personal communication, January 2012.

[65] James Dingley and Marcello Mollica, "The Human Body as a Terrorist Weapon: Hunger Strikes and Suicide Bombers," *Studies in Conflict and Terrorism* 30, no. 6 (2007): 459–492.

[66] Crossett and Newton, "Solidarity."

[67] Halfdan Lefevre, "The Illegal Press," *Denmark During the German Occupation*. ed. Borge Outze (Copenhagen: The Scandinavian Publishing Co., 1946), 64.

[68] Crossett and Newton, "Solidarity."

[69] Lefevre, "The Illegal Press," 63–64.

[70] Ernest K. Bramstedt, *Dictatorship and Political Police: The Technique of Control by Fear* (New York: Oxford University Press, 1945), 210.

[71] Crossett and Newton, "Solidarity."

[72] Ibid.

[73] Paul D. Almeida and Mark Irving Lichbach, "To the Internet, From the Internet: Comparative Media Coverage of Transnational Protests," *Mobilization: An International Journal* 8, no. 3 (2003): 249–272.

[74] Denning, "Activism, Hactivism, and Cyberterrorism."

[75] Elkin, "Tunisia Internet Chief."

[76] Ibid.

[77] Charles Levinson and Matt Bradley, "Egypt's Regime on the Brink," *Wall Street Journal Online*, January 29, 2011, http://online.wsj.com/article/SB10001424052748703956604576109323492986438.html?mod=WSJ_World_LeadStory.

[78] Bramstedt, *Dictatorship and Political Police*, 210.

[79] Ruaridh Arrow, "Gene Sharp: Author of the Nonviolent Revolution Rulebook," *BBC News*, February 21, 2011, http://www.bbc.co.uk/news/world-middle-east-12522848.

[80] Crossett and Newton, "1979 Iranian Revolution."

[81] Sigmund Freud, *Group Psychology and the Analysis of the Ego* (New York: Liveright Publishing, 1922).

[82] James Surowiecki, *The Wisdom of Crowds: Why the Many Are Smarter Than the Few and How Collective Wisdom Shapes Business, Economies, Societies and Nations* (New York: Doubleday, 2004).

[83] Merari, "Terrorism as a Strategy of Insurgency."

[84] Paul de Armond, "Netwar in the Emerald City: WTO Protest Strategy and Tactics," in *Networks and Netwars: The Future of Terror, Crime, and Militancy*, ed. John Arquilla and David Rondfelt (Santa Monica, CA: RAND Corporation, 2001).

[85] Larry Elliot and John Vidal, "Week of Division On and Off Streets," *The Guardian*, December 3, 1999, http://www.guardian.co.uk/world/1999/dec/04/wto.johnvidal1.

[86] de Armond, "Netwar in the Emerald City."

[87] Interview with HC.

[88] de Armond, "Netwar in the Emerald City."

89 Ibid.

90 Ibid.

91 Francis Dupuis-Deri, "The Black Blocs Ten Years after Seattle," *Journal for the Study of Radicalism* 4, no. 2 (2010): 45–82.

92 de Armond, "Netwar in the Emerald City."

93 Margaret Levi and Gillian H. Murphy, "Coalitions of Contention: The Case of the WTO Protests in Seattle," *Political Studies* 54 (2006): 651–670.

94 de Armond, "Netwar in the Emerald City."

95 Ibid.

96 Ibid.

97 Ibid.

98 Ibid.

99 Interview with HC.

100 de Armond, "Netwar in the Emerald City."

101 Mark Duffield, "Globalization and War Economies: Promoting Order or the Return of History?" *Fletcher Forum of World Affairs* 23, no. 2 (1999): 21–38.

102 Stuart Ackerman, "Occupy the Skies! Protesters Could Use Spy Drones," *Wired*, November 17, 2011, http://www.wired.com/dangerroom/2011/11/ows-drones/#more-63792.

103 Bor-Komorowski, *The Secret Army*.

104 Jan Karski, *The Story of a Secret State* (New York: Simon Publications, 2001).

105 Barry, "Sound of Post-Soviet Protest: Claps and Beeps."

CHAPTER 11.

TERRORISM

CHAPTER CONTENTS

Jason Spitaletta and SORO authors

Terrorism has proven to be one of the more effective forms of psychological warfare. Terrorism is a tactic by which a group seeks to impose its will on a selected target audience.[1] Most liberation or underground movements have at least dabbled in "the unlawful use of violence or threat of violence to instill fear and coerce governments or societies. Terrorism is often motivated by religious, political, or other ideological beliefs and committed in the pursuit of goals that are usually political."[2] Despite often relying on illegal acts and/or criminal networks, terrorism is not traditionally used for financial gain but rather for psychological gain.[3] Insurgents have seldom relied solely on the attractiveness of their appeals or on the persuasiveness of their goals to secure popular support; they have generally assumed that people never entirely pursue idealistic goals or do what logic might tell them is most beneficial. Coercive means are therefore used to focus public attention on the goals and issues identified as important by the insurgents. The Chinese saying "Kill one, frighten ten thousand" summarizes the objective of terrorism: maximize the psychological effect of politically motivated violence.[4] Creating uncertain conditions is often an operational objective of insurgent groups because it supports their narrative that the existing government is powerless and/or lacks legitimacy.[5]

Terror is also used to support other insurgent techniques and operations, such as propaganda and agitation. Because it has limited resources, an underground needs to consider economy of force and thus terrorism can be seen as a prudent course of action and not an irrational or ill-considered act of madmen. Terrorism is the confluence of violence and propaganda, where the former seeks to modify behavior through coercion and the latter through persuasion.[6] This chapter is about the use of terrorism by insurgent and revolutionary movements. The focus will be the psychological rationale of the decision to use terror as well as the effect of the use of terror on the state, other groups, and the civilian populace writ large. The chapter begins with a discussion of the considerations of employing terrorism as a strategy and concludes with a discussion of the psychological effects of terrorism including individual and group responses.

OBJECTIVES OF TERRORISM

The modern use terrorism as a political tool can be considered in a set of four forty-year periods dating back to the late nineteenth century. The first period started coincidently with the discovery of dynamite in the 1880s. Anarchists who believed traditional forms of intellectual rebellion were insufficient to change the political status quo employed terrorism in this wave. The 1920s saw the start of an anticolonial wave of terror that was used in the wars of national liberation. These conflicts saw the transition to a wave of New Left[a] terrorism from which emerged the concept of fourth generation warfare.[b] The New Left wave of terrorism emerged in both Europe and the United States as a rebellion against the political and social intuitions of their parents' generation. Groups such as the Brigate Rossi (Red Brigade) in Italy, the Red Army Faction in Germany, and the Weather Underground in the United States were indicative of these politically informed but discontented groups. The current wave of religious extremism began in 1979, before the wave of New Left terrorism came to its conclusion. Nevertheless, the advent of, particularly Islamist, extremist groups has come to dominate the political landscape.[8] Egyptian Islamic Jihad (EIJ) emerged from the larger Islamic movement in Egypt in the latter half of the twentieth century and was considered the first element of the modern global Salafist jihad.[9] The prominence of this movement grew with the 1979 Islamic Revolution in Iran and its export to the Lebanese resistance, giving birth to Hizbollah. The second element was the mobilization of many across the Muslim world to help the Afghan mujahidin expel Soviet forces from Afghanistan during a decade-long war. From that emerged the third element, Al Qaeda, a group of committed veterans of the Afghan conflict who sought to exploit the global Islamic mobilization to reestablish the caliphate that was dissolved in 1922 with the fall of the Ottoman Empire.[10]

---

[a] "New Left" was a term popularized in the United Kingdom and United States in reference to activists in the 1960s and 1970s who sought to implement a broad range of political and social reforms in contrast to doctrinaire Marxist movements, which focused principally on labor unionization and questions of social class.

[b] The phrase "fourth generation warfare" was coined in a 1989 *Marine Corps Gazette* article[7] that proposed a generational model of modern warfare the coincided with advances in technology dating back to the 1648 Peace of Westphalia. First generation warfare (1GW) was characterized by tactics of line and column that developed in the age of the smoothbore musket. 1GW consisted of tightly ordered soldiers with top-down discipline and rigid, centralized command and control. The advent of the machine gun and rifled barrel saw the transition into second generation warfare (2GW) and its tactics of linear fire and movement, with reliance on indirect fire. 2GW saw the emphasis shift from the offense to the defense and from direct to indirect weapons systems. The carnage of World War I saw a rethinking in the latter stages of that conflict and the beginnings of third generation warfare (3GW). Also called maneuver warfare, 3GW (aided by advances in aviation, armor, and tactical communications as well as a more decentralized command and control philosophy) employs defense in depth and offensive tactics of infiltration to bypass enemy surfaces and exploit gaps. The goal of 3GW is to shatter an adversary's will to fight instead of attritting the forces through firepower alone. Fourth generation warfare (4GW), or asymmetric war, is form of conflict that blurs the distinction between war and politics as well as between combatant and noncombatant and is characterized by terroristic tactics.

Disruption of Socioeconomic and Political Status Quo

A common aim of terrorism is to destroy existing organization of governments, social structures, and/or the socioeconomic status quo. Terrorist acts are often directed toward governmental officials and key supporters, making it unsafe to be a government official and, through systematic assassination, crippling the actual functioning of government. The utility of terror for a subversive movement is multifarious: it disrupts government control of the population, demonstrates insurgent strength, attracts popular support, suppresses cooperation with the government by "collaborators" and "traitors," protects the security of the clandestine organization, and, finally, provokes counteraction by government forces.

The application of this theory can be seen in South Vietnam, where the Viet Cong weakened government control in certain regions by killing and kidnapping province chiefs, police officials, village guards, and landlords. It has been estimated that in the five-year period between 1959 and 1964, the Viet Cong murdered more than 6,000 minor Vietnamese officials. Schoolteachers, social workers, and medical personnel were also favorite targets. The allegiance of the people is the chief prize in an insurgency, and because schoolteachers, as one observer noted, "form young minds and educate them to love their country and its system of government, to close such schools or to cow the teachers into spreading antigovernment propaganda can be a more important victory than to defeat an army division."[11] Almost 80,000 South Vietnamese schoolchildren had been deprived of schools by 1960 because of terrorist action: 636 schools were closed, approximately 250 teachers were kidnapped, and another 30 were killed. The Viet Cong also disrupted the South Vietnamese social welfare and medical program; a highly successful malaria-eradication program, for example, was stopped in 1961 because of high casualties among its personnel caused by terrorists.

To underscore the unprofitability of being, or becoming, a government servant, the Viet Cong carried out assassinations by unusual, brutal, or mysterious methods. Vietnamese village headmen suspected by the Viet Cong "of cooperating with the government or guilty of 'crimes against the people' were disemboweled and decapitated, and their families with them."[12] The attention value of such acts in the press and through word of mouth was great and the implied threat to others obvious. When, through the simple process of attrition, the machinery of government in one area came to a virtual standstill, the Viet Cong reestablished social order by setting up a shadow government with its own official apparatus to collect taxes, operate schools, and implement population control measures.

319

Demonstration of Strength

In a terrorist campaign, the individual citizen lives under the continual threat of physical harm. If government police are unable to curb the terrorists' threats, the citizen tends to lose confidence in the state whose inherent mission it is to guarantee his safety.[13,14] The use of terror, when effective, convinces the people of the movement's strength. Captured Viet Cong documents indicate that this was one of its primary objectives in South Vietnam. Through a demonstration of strength by effective assassination of government and village leaders, it attempted to convince the rural population that the regime in Saigon could not protect them.[15]

In order to publicize the movement's strength, some terrorist activities are conducted publicly. The National Liberation Front (Frente de Liberación Nacional, or FLN) in Algeria used this tactic. Witnesses of terrorist acts were not eliminated but were spared in order to confirm the FLN success. Muslims who supported the French were warned by letters bearing the FLN crest to desist from cooperating with the French; if a Muslim refused, the FLN execution order was attached to the victim's dead body. The FLN murdered approximately 16,000 Muslims and kidnapped 50,000 others[16] and thus silenced "its opposition, weakened the . . . French by depriving them of the support of . . . Muslim leaders, and at the same time, assassination enhanced the prestige of the FLN . . . by affording tangible proof of the organization's effectiveness and intrepidity."[17]

However, in pursuing a similar tactic, the Communist insurgency in Malaya got itself into a difficult position. Unable to carry out and win a guerrilla war, the Communists attempted by mass terror to demonstrate their strength and neutralize the members of the population who were supporting the government. They soon received complaints from their political arm, the Min Yuen, that indiscriminate terror was alienating the voluntary support upon which their long-range success depended.[18]

In response to Republic of Ireland pressure, the Provisional Irish Republican Army (PIRA) considered extending its operations to mainland Britain. In early 1973, the Army Council formally approved action in Britain, to include strikes on economic, military, political, and judicial targets. The operations were designed to capture the attention of the British public, generate a weariness of the problem in Northern Ireland, and increase pressure on their respective Members of Parliament (MP) to withdraw. Attacks within Great Britain would also allow the PIRA to avoid Irish civilian casualties and provide needed relief to Belfast and Derry units. The first mainland operation involved four car bombs in London detonating and injuring 180 people. The next team planted nine bombs in mid-England and a series of small incendiary devices in popular shops around London. Innocent civilians were supposed to be avoided in these mainland-bombing attacks, but the bombings were conducted indiscriminately and without consideration of civilian casualties. Pubs frequented by British soldiers were occasionally targeted; however, the typical targets were those associated with influential members of the British government and business. Residences of the members of parliament, high-end hotels, clubs for the wealthy, and similar locations were favorite targets. The highest-profile bombing was perhaps the attempt on Prime Minister Margaret Thatcher's life in 1984 at the Brighton Grand Hotel. The PIRA planted a large Semtex bomb 24 days before it detonated using a sophisticated long-delay timer. The cabinet and Thatcher were attending a conference at the hotel, and the blast came very close to Thatcher's suite. She was unharmed, though the Trade Secretary was badly hurt and several senior politicians were killed. Five people died in the blast and significant visible damage inside and outside the hotel showed the reach and power of the PIRA.[19]

Strategic Risks

Terror can politically boomerang if the target is unwisely chosen or the assassination unwisely timed. An example of this occurred in the Philippines when a contingent of Huk insurgents ambushed and killed Aurora Quezon, wife of the Philippine president, along with her daughter and other distinguished citizens. Because she was widely known and respected by the people, Mrs. Quezon's death was a serious setback to the Communists. Reacting to the nation's feeling of condemnation, the Huk leaders declared that the terrorists acted without orders.

In Algeria, terrorism in the name of nationalism largely won support from or cowed the Muslim population, but it had an opposite effect on the French community. Although thoroughly unnerved by the FLN terrorist offensive of 1956, the French never reached the point of surrendering. In itself, the dramatic nature of the terrorist challenge ensures a dramatic response to the call for counteraction. The effect of FLN terrorism on French policy was to strengthen the resolve to stamp out the rebellion, indeed to make it politically impossible to follow any other course.[20]

In the early 1990s, the PIRA adopted the proxy bomb, a tactic similar to suicide bombing whereby individuals were forced to drive bombs into military checkpoints or other targets. The individual was typically a suspected enemy collaborator whose family was held hostage until the mission was completed. In 1990, the PIRA Derry Brigade forced Patsy Gillespie, a Catholic cook for the British army, to drive a car bomb into an army checkpoint, killing him and five soldiers. His family in Derry was held hostage as insurance. The proxy bomb experiment was an utter public relations disaster for the PIRA and Sinn Féin. Anger in Derry, a nationalist stronghold, reached unprecedented levels. Disgust and revulsion over the use of human bombs also led to more internal dissension within the Republican movement.[21]

In November 1997, operatives from Egyptian Islamic Group (EIG) and an element of EIJ called Jihad Talaat al-Fath or "Holy War of the Vanguard of the Conquest" killed sixty-three tourists at the Temple of Hatshepsut, Egypt. During the operation, later dubbed the "Luxor Massacre," six assailants dressed in police uniforms systematically shot and stabbed to death fifty-eight foreign and four Egyptian tourists. The day after the attack, despite evidence of the systematic executions, EIG claimed that the attackers intended only to take the tourists hostage. Others denied Islamist involvement completely: Omar Abdel-Rahman (the spiritual leader of EIG) blamed the Israeli Mossad, while Dr. Ayman Al-Zawahiri (then the emir of EIJ) maintained that the Egyptian police were responsible. This operation was to be the last kinetic action attributed to either EIG or its members. The audacity and depravity of the attack stunned Egyptian society, ruined tourism for years, and destroyed much of the remaining popular support for EIG and EIJ in Egypt.[22]

The Weather Underground[c] was an American radical leftist organization that emerged out of the 1960s antiwar group Students for a Democratic Society (SDS). The organization was formed in the wake of the killing of Black Panthers Fred Hampton and Mark Clark in a December 1969 Chicago police/FBI raid. A small cadre of SDS members perceived this event to be a signal that nonviolent resistance to the U.S. government was a futile endeavor and that terrorism was necessary in order to achieve their objectives. The "Days of Rage," which commenced on October 9, 1969, was their first public demonstration; a riot in Chicago was timed to coincide with the trial of the seven individuals who were charged with conspiracy, inciting to riot, and other charges related to protests at the 1968 Democratic National Convention. The Weathermen believed that the American public had become desensitized to the status quo of political activism and that it was only through violence that the group could draw sufficient attention of the

---

[c] The Weather Underground (or Weathermen) described themselves as a revolutionary organization of Communist men and women. Their goal was to create a clandestine revolutionary party for the violent overthrow of the U.S. government and the establishment of a dictatorship of the proletariat. The Weather Underground and the Black Panther Party for Self-Defense were priority intelligence targets of the U.S. Federal Bureau of Investigation (FBI)'s counterintelligence program (COINTELPRO); however, many tactics used under that program were eventually ruled illegal.

U.S. government to the group's political goals. The group's intent was to foment public chaos and mobilize the American public in opposition to the U.S. government's policies toward Southeast Asia. In 1970, the group issued a Declaration of a State of War against the U.S. government and began a campaign of bombings through the mid-1970s. The bombings targeted government buildings along with several banks. Each target was carefully selected for its symbolism and tied to a specific political objective. For the bombing of the United States Capitol on March 1, 1971, they issued a communiqué saying it was in protest of the U.S. invasion of Laos, whereas the bombing of the Pentagon on May 19, 1972, was in retaliation for the U.S. bombing raid in Hanoi. The Weathermen largely disintegrated after the United States withdrew from Vietnam, which also marked the beginning of the general decline of the American New Left.[23]

Punishment and Retaliation

The threat of punishment against collaborators and informers within the general populace is a common feature of insurgent terrorism. In fact, many insurgencies against occupying powers have taken a higher toll on indigenous citizenry than on the occupying forces. If the group adopts an increasingly paranoid defensive posture, including the intimidation, expulsion, even killing, of suspected traitors, there is an increased risk of further violence.[24] For example, the Greek Cypriot terrorist organization, EOKA (Ethniki Organosis Kyprion Agoniston, or National Organization of Cypriot Fighters), killed more Cypriots—as "traitors" or "collaborators"—than it did British security forces or government officials. During the early days of the Cypriot insurgency, the population, easy-going by nature and tradition, seemed not to be taking the rebellion seriously. EOKA immediately began an offensive to punish "collaborators"; the slogan "Death to Traitors" was scrawled on walls to make the threat visible, and selected murders of Greek Cypriots drove the point home. A concrete example of this occurred on October 28, 1955, when Archbishop Makarios publicly called for the resignation of all Greek village headmen. The date was symbolically chosen, for October 28, known as Okhi ("No") Day, is the anniversary of the Greek refusal to submit to the Italian ultimatum in 1940. Only about a fifth of the headmen had responded by the end of the year, however. EOKA then went into action by murdering three headmen and within three weeks, resignations reached 80 percent.

Luis Taruc, the leader of the Huks in the Philippines, bragged that by using old women in the town markets, young boys tending carabaos in the fields, and small merchants traveling between towns, every traitorous act, every puppet crime, every betrayal through collaboration was known to the Huks. After a warning of impending punishment, the Huks would blacklist offenders, and agents were authorized to arrest or liquidate them.[25]

The Polish underground also used techniques of terror propaganda against collaborators. Initially, specific collaborators were morally condemned in the underground press; next, they were blacklisted and death sentences were published. Further, the underground press frequently lumped the crimes of collaborators with the names of Nazi officials, listing the names of German officers who would be brought before war crimes tribunals after the war. In most cases, the underground later executed the condemned individuals.

Terror can be used in retaliation against or as a counterbalance to terror. The Liberation Tigers of Tamil Eelam (LTTE) relied heavily on assassinations of rival Tamil separatists, politicians, police informants, and high-ranking politicians. The first assassination for which the LTTE claimed responsibility was that of the Tamil mayor of Jaffna and SLFP (Sri Lankan Freedom Party) member Alfred Thambirajah Duraiappah, who was killed in 1975. Many assassinations attributed to the LTTE were impossible to verify since its practice was to neither claim nor deny involvement in specific activities, instead pointing to continued paramilitary and police violence against Tamil civilians as contributing to specific assassinations. Some of the suspected LTTE assassinations included the death of TULF (Tamil United Liberation Front) Vice President Neelan Thiruchelvam; the attempt on then-President Kumaratunga in 1999; and the deaths of Rajiv Gandhi, President Ranasinghe Premadasa, Lt. Gen. Sarath Fonseka, and Lakshman Kadirgamar.[26]

Maintenance of Security

To bring about the failure of government countermeasures, the security of the clandestine organization must be maintained. There are two ways in which an underground movement can protect its own security: "It can police the loyalty of its members and take steps to see to it that a complete picture of the movement is held only by a limited few, or it can employ the threat of terror against informers."[27] Terror is used by insurgent organizations against their own membership in order to protect the security of their operations. It is always implicit, and often made explicit, that they who defect or betray the cause will be severely punished. To demonstrate the reality of the threat, undergrounds characteristically have organized terror squads that enforce threats and punish traitors.[28] Most frequently, punishment means death.

Terrorist Oaths

In Kenya in 1952, a Kikuyu tribesman who was being admitted to the select terrorist cadre of the Mau Mau had to swear himself to a brotherhood of murder:

- If I am called upon to do so, with four others, I will kill a European.

- If I am called upon to do so, I will kill a Kikuyu who is against the Mau Mau, even if it be my mother or my father or brother or sister or wife or child.

- If I am called upon to do so, I will help to dispose of the body of the murdered person so that it may not be found.

- I will never disobey the orders of the leaders of this society.[29]

Rituals involving bestial and degrading practices, the object of which was to make the initiates become outcasts who shrank at nothing, often accompanied this murder oath. The acts performed were intended to be so depraved that, by comparison, the mere disemboweling of pregnant women, for instance, would seem mild. Furthermore, this degradation would alienate the initiate from the Kikuyu community, insuring that he could never fully return to normal life and betray Mau Mau secrecy. But where such tribal superstitions proved inadequate, the gap was filled by fear of personal violence or death.[30]

Members of the LTTE's elite Black Tigers, whose acceptance into the unit was ritualized and presided over by Vellupillai Prabhakaran, carried laminated cards with a warning (in English and Sinhala) that, "I am filled with explosives. If my journey is blocked I will explode it. Let me go."[31]

Osama bin Laden, the leader of Al Qaeda from its inception in the late 1980s until his death at the hands of U.S. Special Operations Forces in May 2011, recruited perspective members to swear an oath of allegiance (or bayat) to him before they were admitted.[32]

Enforcing Squads

An example of underground enforcing squads was the "Traitor Elimination Corps," which was one of the first units organized by the Communist underground in Malaya. The very existence of this unit caused recruits to the underground to be quickly impressed with the importance of "discipline" and with "the severity of the Party's means of enforcing discipline."[33] However, one source cites evidence that the

Malayan Communist Party also used lesser threats, hoping to avoid liquidating agents who broke discipline yet always reserving the implied "escalation" of the threat of death.

Interviews with numbers of surrendered Malayan insurgents revealed that although 80 percent said they feared bodily harm, the real reasons for their fear were more subtle threats. Nearly all said they feared most "the Party's practice of disciplining its members by depriving them of their firearms; . . . they claimed this was something to be dreaded because, without their weapons, they would be defenseless."[34] Coupled with social ostracism, physical defenselessness threatened the very basis of the man's sense of personal security.

The Huk underground in the Philippines also used terror to maintain the security of its operations, executing its threats through a special terror squad. [35] Similarly, the FLN in Algeria enforced discipline in its urban underground networks by summarily executing "traitors" when discovered. The security of the Yugoslav Communist underground was maintained by a secret police unit called the Department for the Defense of the People (O.Z.Na.). This group provided intelligence on the behavior of underground comrades and was authorized to liquidate those who were disloyal to the partisans.[36] During the Moroccan independence insurgency, the counterterrorist "Red Hand" guarded its secrecy by eliminating defectors through mysterious accidents.[37]

The PIRA employed counterintelligence units (called "nutting squads") to be responsible for internal security. These units killed at least twenty-four suspected informers. The Provisionals were sensitive to the effect the execution of informers had on their popular support. The bodies were generally found with no shoes, a folded bank note in the pocket, masking tape over the eyes, and a bullet in the back of the head. Several times, the PIRA attempted to demonstrate their capacity for forgiveness, as in 1982 when they offered amnesty to all those who came forward and admitted their crime. Provisional punishment was harsh and swift for informers, but so was the community's castigation directed at known informers and their relatives. Informing, becoming a "tout" as informers are called, was viewed as a severe betrayal in the Republican-sympathetic community.[38]

Provocation

Insurgent movements frequently utilize terrorism to provoke a counteraction that may be strategically useful. The Organisation de l'Armée Secrète (Organization of the Secret Army/Secret Armed Organization, or OAS), the illegal secret French army organization in Algeria, used this tactic to provoke the nationalist FLN into upsetting the cease-fire upon which negotiations between the French government and the FLN were based.[d] By carrying out indiscriminate terrorist attacks on Arab civilians, "the OAS leadership evidently believed it would so exacerbate French–Algerian relations that the Algerians would be provoked into massive countermeasures, that full-scale war would be resumed, and that no settlement would be possible."[39]

Provocation often attempts to exploit the "accidental guerrilla" phenomenon, where violent movements gain popular support by default after a heavy-handed response by the state.[40] The Iraqi resistance after the initial surge of Operation Iraqi Freedom (OIF) in 2003 was particularly fierce in Al Anbar province and was initially the most hostile toward the American presence. The predominantly Sunni population of Al Anbar was historically resistant to outside influence, regardless of ethnicity, nationality, or religion. The Anbar resistance was fueled by the "accidental guerrilla" syndrome, the process[e] by which heretofore-noninvolved individuals determine it is in their best interest to aid an insurgency because it provides the best opportunity for them to achieve their desired goals. The process began with Al Qaeda in Iraq (AQI)[f] infecting the area by establishing a base of operations outside the control of U.S. or Iraqi authorities. From this safe haven, AQI operatives began contaminate the zone by expanding into neighboring areas, often marrying into local tribes. In response to this, the state or occupying authority (in this case the U.S. military) intervened by launching

---

[d] The OAS was a French far-right nationalist organization during the Algerian War (1954–62). The OAS used asymmetric tactics (including terrorism) to prevent Algeria's independence from France. It was officially formed in Madrid in January 1961 by French politicians and military officers in response to the January 8, 1961, referendum on Algerian self-determination organized by General Charles de Gaulle. The OAS was an amalgamation of existing, albeit unofficial, resistance networks to the FLN. The OAS carried out attacks on the FLN and their perceived supporters throughout the war.

[e] Killcullen's model entails four phases: "infection," "contagion," "intervention," and "rejection."

[f] In this case, AQI is used as a composite of multiple foreign-fighter organizations operating in Iraq who were ideologically affiliated with, but not explicitly commanded and controlled by, Al Qaeda. Among those groups was Jama'at al-Tawhid wal-Jihad, the group led by Abu Musab Al-Zarqawi (who became the de facto "face" of AQI). The stated goals of AQI were to force a withdrawal of United States-led forces from Iraq, topple the Iraqi interim government, assassinate collaborators with the occupation, marginalize the Shia population and defeat its militias, and establish an Islamic state. Initially, the various insurgent groups primarily targeted coalition forces (with a minimal amount of targeting directed at Iraqi civilians). AQI developed an expansive militant network, including some of the remnants of Ansar al-Islam (a Kurdish militant group) and a growing number of foreign fighters, to resist the coalition occupation forces and their Iraqi allies. Unlike the former Baath regime elements or the Shia groups, AQI employed asymmetric tactics (suicide bombings, car bombings, and the use of improvised explosive devices [IEDs] as counter-mobility weapons), along with more traditional guerrilla tactics such as ambushes and small-scale raids.

military operations against AQI. The consequence of such action, rejection of the coalition by the Anbari people, served to reinforce AQI's narrative, and the combination of psychological preparation by AQI and the reality of circumstances on the ground strengthened the bond between AQI and the Al Anbar Sunnis, thus growing the capability of the in-group and further perpetuating this cycle.[41] AQI exploited U.S. tactics to mobilize the mass of Iraqis who were opposed to both the U.S. and AQI presence in Al Anbar. AQI successfully used coalition force to turn small tactical defeats into operational successes by exploiting the information domain and framing the coalition as oppressive occupiers and themselves as defenders of the Iraqi people.[42]

Governments that seek the consent of the governed, even partially, must be more cautious about proportionate use of force. One use of terror to create a provocative situation is the deliberate assassination in a riot of innocent bystanders in order to create a martyr and provoke the populace into further actions against the government. Carlos Marighella, in his *Minimanual of the Urban Guerilla*, advocated this tactic as a highly effective means of cultivating popular support. Instead of compelling the populace to join an insurgent cause, the insurgents would provoke the state to react violently and indiscriminately, thus driving the populace into the arms of the insurgents.[43] A similar outcome may result whether the casualty was intentional or not. The "Bloody Sunday" attacks in Northern Ireland—in which British paratroopers killed thirteen unarmed protestors during a largely peaceful Catholic demonstration—catalyzed decades of subsequent conflict. This was an unintentional provocation on the part of the British. In retrospect, military leaders regretted assigning a unit of paratroopers known for aggressiveness to crowd control, a mission for which they were unsuited. Prior to this, Catholic protest groups had welcomed the presence of British troops as protection against sectarian violence. However, this event turned the Irish Catholics against the British soldiers and energized the violent factions within the IRA, accelerating a decline into "The Troubles."[44]

CONSIDERATIONS OF TERRORISM

A key feature of terrorism is detailed preparation. To demonstrate insurgent strength and sustain momentum, early success is essential. Targets are selected so that the terrorists are free to choose the time and place that will best insure the success of the mission.[45] Effective use of terror requires a thorough knowledge of localities, people, customs, and habits; it requires extensive and secret reconnaissance activities. Undergrounds have both a technical and social infrastructure that is actively concealed yet requires financial resources to function. Agents must be compensated and guerrilla forces armed; escape-and-evasion networks require money for extra food, for safe houses, and to provide to escapees; psychological operations require funds for products and activities; and headquarters and administrative sections require office supplies. Al Qaeda is often considered the exemplar of underground networks. Dating back to the mid-1980s, the Maktab Al Khidamat (Services Office) or MAK served as a clearinghouse to recruit, finance, train, and employ foreign fighters from across the globe in support of the Afghan resistance to the Soviet Union. Using methods acquired from the Muslim Brotherhood, both MAK and Al Qaeda used a sophisticated, complex, and resilient money-generating and -transfer network.[46]

Unity of Effort

Operational discipline is a necessity if terrorism is to be an effective component of a psychological warfare campaign. Normally unorganized terrorism involves unplanned acts against unselected targets. Such acts are the incidental result of more generalized attacks. The distinguishing feature of unorganized terrorism is that individuals in large units that do not have terror as their sole function commit it. These acts are often against targets of opportunity and prosecuted by independent cells with either vague mission-type orders or the autonomy to attack at their discretion. Examples may include AQI cells deploying a vehicle-borne improvised explosive device (VBIED) against a U.S. patrol that had fallen into habitual patterns. Instead of risking exposure of the network by requesting permission, a cell leader would exploit the vulnerability in the patrol's tactics and attempt to cause as much damage to the U.S. personnel as possible. The risk with this lack of organizational direction, however, is the perceived indiscriminate nature of the attacks due to civilian casualties.

Undergrounds usually are careful to avoid wanton acts of terrorism against the populace. Such notable experts as China's Mao Tse-Tung and North Vietnam's General Vo Nguyen Giap counseled extreme restraint, advising that great care be taken to avoid bringing undue suffering to the populace and unduly alienating public opinion.[47] As noted earlier, terror squads are frequently used as an enforcing arm for underground political units. For this purpose, terroristic acts are specifically designed to support the underground's political goals and are usually carried out by specially trained and organized squads. An example of this is the organization of the Communist Party underground in Malaya. Just as propaganda and political units were attached to the party's Liberation Army to assist in its "military" activities, terrorist squads were attached to the party's political arm, the Min Yuen, to enforce support of its political activities. Somewhat similar to the "Blood and Steel Corps," these terrorist units consisted mostly of "trusted party thugs who, in addition to perpetrating acts of extortion and intimidation against those designated by the Party, were ordered to strengthen the treasury by engaging in payroll robberies and raids on business establishments."[48]

Another dimension of terroristic activity used in support of insurgent political goals is seen in the plan for urban insurgency adopted in 1961 by the Castro-Communist-supported insurgents (Fuerzas Armadas de Liberación Nacional [Armed Forces of National Liberation, or FALN]) in Venezuela. Here, the insurgents attempted to organize terrorist units or shock brigades to serve as the catalytic agent for urban revolution. The tactic was to induce a state of paralysis and alarm within the urban public through an extended period of urban violence, eventually undermining the support and power of the government and leading to a rapid victory. Terror squads, called Tactical Combat Units (TCUs), were used for robberies, sabotage, arson, murders, and the creation of street violence and riots. The units were usually organized in detachments of about thirty: five to eight men engaged in the bolder terrorist actions while the rest filled lesser supporting roles.[49]

The success of many of the units' terrorist actions was largely attributable to careful advance preparation, including written operations plans. Although much of the street violence appeared to be spontaneous, careful examination reveals a pattern: attacks on Venezuelan-owned properties were usually limited to robbery, whereas those on U.S. properties involved some use of incendiaries or explosives. The terrorist units operated in the mobile hit-and-run style, usually proceeding to and from their targets in stolen automobiles. Although the TCU attempted on numerous occasions in 1963 to induce an atmosphere of mass terrorism in Caracas through stepped-up sniper fire and associated acts of violence, these operations failed. Although they interrupted normal patterns of urban life, they did not succeed in producing mass terrorism of the sort that would immobilize or cripple the functioning of government.[50]

There are times when tactical objectives of subordinate leaders may be inconsistent with the overall organizational strategy. By 2005, the tactics used by Abu Musab Al-Zarqawi (the leader of AQI) to foment secular conflict between Iraq's Shia and Sunni populations drew criticism from Al Qaeda's leadership. Zawahiri advised Zarqawi to eschew internecine conflict in Iraq because he (Zawahiri) believed the American exit was imminent. He also admonished Zarqawi to cease beheading hostages and posting the executions on the Internet because "We are in a battle and more than half of this battle is taking place in the battle-field of the media."[51] Zarqawi did not heed Zawahiri's advice, and as AQI operations devolved into "armed banditry," Al Qaeda leadership shifted their resources to support alternative fronts where the chances of victory seemed more promising.[52]

Franchising Terror

Rifts within insurgent groups are inevitable, particularly when security requirements demand a decentralized network. A recent trend in organizational structure is the extension of the networked systems of cells to ad hoc or post hoc affiliations. Since late 2001, weakened by U.S. operations in Afghanistan, Pakistan, and elsewhere, Al Qaeda has encouraged affiliates to conduct operations without necessarily possessing significant financial or operational links to the original group. It may be this contagion effect that is Al Qaeda's most lasting legacy. Al Qaeda has transformed into a diffuse global network and philosophical movement composed of dispersed nodes having various degrees of independence. Operations include the London subway bombings (2005), the Madrid train bombings (2004), the nightclub bombing in Bali, Indonesia (2002), and the hostage crisis in a Moscow theater (2002). Al Qaeda-affiliated groups are also blamed for attacks in Morocco, Turkey, Tunisia, Pakistan, Kenya, Saudi Arabia, Madrid, London, Indonesia, and Algeria (as well as failed attempts in the United States). This proliferation of loosely affiliated groups and their high-profile attacks have also provided many authoritarian regimes with free hand to repress legitimate and nonviolent dissent, citing dubious linkages between Al Qaeda and domestic opposition forces.[53] Again, the effect of successful attacks may unintentionally result in an innocent populace voluntarily surrendering their civil liberties to a regime that not only simplifies the problem but also articulates a solution perceived as suitably reciprocal.[54] The linkage between mortality salience and individual political preference is by no means deterministic; however, the pervasive threat caused by the omnipresent threat of small-scale attacks has a profound effect on political will (particularly in the immediate wake of a terrorist event).[55]

Suicide Terrorism

Among the more historically, socially, and psychologically compelling tactics used by undergrounds are suicide attacks. From their modern inception by Hizbollah in Lebanon, to their strategic incorporation by the LTTE in Sri Lanka, to their employment as a strategic special operation by EIJ and Al Qaeda, suicide terrorism has been defined in various manners: (1) as a conscious, logical decision,[56] (2) as a tactic of desperation by a numerically and/or technologically inferior force,[57] and (3) as the exploitation of a vulnerable demographic by more mature, albeit inconsiderate, organizers.[58]

Particularly talented LTTE soldiers were selected for membership in the Black Tigers, a special unit tasked with the most difficult missions including suicide bombings and assassinations. Unlike many insurgent groups, the LTTE integrated suicide bombings and assassinations not out of desperation but as an integral aspect of their campaign against the Sri Lankan government. Their first suicide attack was used to breach a Sri Lankan Army camp. The attack was followed by a much larger operation with a truck bomb, allowing LTTE regular forces to storm the camp in order to halt an impending army offensive.[59]

Although Hizbollah's successful exploitation of suicide attacks contributed to the popular association of militant Islam with suicide missions, 81 percent of suicide attacks during the Israeli occupation (1982–86) were carried out by Christians or affiliates of secular or leftist parties, and only twelve of Hizbollah's attacks involved the intentional death of a party operative.[60] Hizbollah inspired the use of suicide tactics in high-profile terrorist operations. Their 1983 attacks precipitated the withdrawal of both the French and American forces after the deaths of 241 U.S. Marines and 58 French paratroopers.[61] This model proved desirable but difficult to replicate.

In August 1993, EIJ operatives attempted to assassinate the Egyptian interior minister, who was leading a crackdown on Islamic militants and their terror campaign. A bomb-laden motorcycle exploded next to the minister's car, killing the bomber and his accomplice but not the minister. The failed attack marked the first time Sunni Islamists had made use of suicide in terrorism. Several months later, in November, EIJ and EIG used a suicide truck bomb to destroy the Egyptian embassy in Islamabad, Pakistan. Two dismounted operatives approached the embassy and executed the security detail with a combination of small-arms fire and hand grenades. Then, a taxi loaded with 250 pounds of explosives breached the gate where the driver detonated the bomb, destroying both the vehicle and the gate. Minutes later, a second vehicle carrying a larger bomb was detonated alongside the Embassy, destroying the building and damaging the adjacent Japanese and Indonesian embassies and a nearby bank. The attack killed sixteen, injured sixty, and served as a prototype for the 1998 Al Qaeda attacks on the two American embassies in Africa.[62] The near-simultaneous explosions at the U.S. embassies in Tanzania and Kenya on August 7, 1998, were carried out using similar tactics and resulted in the deaths of 223 and injuries to more than 4,000.[63]

Suicide terrorism results in both shock to and victimization of the groups that identify with the target (discussed later in this chapter) as well the groups who identify with the attacker. To be effective, the social construction of martyrdom must rule out the possibility of the martyr being afflicted with some kind of psychological disorder because it would cause the populace to see the suicide as being motivated by something other than selflessness.[g] Rather, higher-status martyrs (those among the in-group with higher social, academic, or hierarchal standing) have more credibility with the masses and thus their sacrifice carries more weight. The study of suicide terrorism has been criticized for methodological problems (selection bias and selection effects), use of unverified and often-contradictory anecdotal evidence, inadequate sample size, and poor internal validity; [64] however, its inclusion of analyses of modern underground movements is nonetheless pertinent because there are unique considerations that differentiate this type of operation from "traditional" terrorism.

Terrorism Planning

Many underground movements organize specialized terror units to conduct special operations. Such units are characterized by well-planned operations carried out by a small, highly trained professional elite, usually organized on a cellular basis. The targets are usually selected individuals of special importance or iconic structures/organizations having symbolic importance (e.g., the U.S. embassies in Kenya and Tanzania, the Marine Barracks in Beirut, the USS *Cole*, and the World Trade Center). Historically, the skill necessary to execute such an attack necessitated highly trained personnel and thus operations were tailored to secure the safe escape of the terrorists. However, with the advent of the religious extremist wave of terrorism, suicide attacks (discussed in the previous section of this chapter) have become more prevalent.

---

[g] Durkheim proposed four subtypes of suicide: egoistic (reflecting a sense of social isolation resulting in apathy and depression), altruistic (being overwhelmed by an integrated in-group's objectives), anomic (a moral confusion resulting from socioeconomic turmoil), and fatalistic (an over-controlled hyper-regulation). In Durkheim's typology, consumption with, and subsequently being overwhelmed by, a group's goals and ideology characterize the altruistic suicide. This phenomenon occurs in organizations (and societies) with high integration, where individual needs are subordinated to that of the whole. Many religious cultures, political extremist groups, and military organizations hold those who sacrifice themselves for the common good in great reverence and thus this mechanism can be considered a perverse exploitation of the dead as an unimpeachable source of credibility.

In Cyprus, one of the first steps taken by George Grivas, the leader of the Greek Cypriot insurgents, was to organize a cadre of specialized terrorists. Never numbering more than fifty, this small group "terrorized half a million people."[65] Significantly, Grivas developed his cadre of terrorists only from the very young. His study of Communist tactics used during the Greek insurgency convinced him that only youths in their late teens or early twenties can be molded into assassins "who will kill on order, and without question."[66] Young men combine youthful daring and, after indoctrination, fanatical conviction and can be made to believe they are behaving in a heroic way. Grivas groomed the youths for their role as specialized terrorists through a process of escalating acts of lawlessness: first they smeared slogans on walls, then they advanced to throwing bombs into open windows or bars. Only after an extended period of testing and training were the youths given their first "professional" assignment of killing a selected target.[67]

Terrorism, like many planning processes, can be considered in a cyclic series of phases: (1) target selection; (2) intelligence, surveillance, and reconnaissance; (3) rehearsal; (4) execution; and (5) escape and evasion. Not all organizations employ rigid processes, and most differ in some details; however, successful operations do tend to be the result of rigor more so than providence alone. Some organizations are more hierarchal; the PIRA's executive council issued the approval order for the 1981 hunger strikes as an operation as well as for each individual participant,[68] while Al Qaeda's leadership did not necessarily exert command and control over tactical planning and operations, instead emphasizing local initiative.[69]

Target Selection

The proper selection of targets is important in any terrorist campaign. Potential targets are compared against intent and assessed based on accessibility, symbolic value, critical infrastructure vulnerabilities, expected casualties, and potential media exposure. A decision to proceed requires continued intelligence collection against the chosen target.

Viet Cong terrorists were particularly directed to seek as first targets government authorities who were corrupt or unpopular; after murdering them they could then boast: "We have rid you of an oppressor." A captured copy of the Viet Cong "Military Plan of the Provincial Party Committee of Baria" specifically orders squads in charge of villages to conduct assassination missions with the principal targets being security forces and "civil action district officials, hooligans and thugs." Relieving the community of undesirables through selective assassination, Viet Cong terrorists sought to win popular support while crippling the operation of government. A 1965 Viet Cong directive was explicit in identifying those who should be punished:

> The targets for repression are counterrevolutionary elements who seek to impede the revolution and work actively for the enemy and for the destruction of the revolution. Elements who actively fight against the revolution in reactionary parties such as the Vietnamese Nationalist Party (Quoc Dan Dang), Party for a Greater Viet Nam (Dai-Viet), and Personality and Labor Party (Can-Lao Nhan-Vi), and key reactionaries in organizations and associations founded by the reactionary parties and the US imperialists and the puppet government.[70]

Intelligence, Surveillance, and Reconnaissance

Undergrounds typically operate from a position of numerical and technological inferiority relative to government forces. Offensive operations therefore require accurate intelligence in order to maximize effectiveness and maintain security of often small, vulnerable guerilla forces. The 1956 version of the IRA Green Book considered information gathering a continuous evolution. Undergrounds provide this critically important military intelligence not through normal military methods but rather through their infiltration of society and networks of clandestine operatives. Careful intelligence work is a prerequisite for all terrorism, particularly professional terrorism. The target is special, and therefore the underground must find concrete answers to "who," "where," and "when." Intelligence must both identify the target and document his modus operandi. A local person may identify the target for outside agents who then execute the plan; although a local person can facilitate identification, it is difficult for him to commit the assassination because of local loyalties and the difficulty of escaping. Another approach is to have outside agents undertake both the intelligence work and the execution.

The difficulties of requiring local agents to assassinate local people are illustrated in an experience of the Red Hand counterterrorists in Morocco. In 1954, the underground's entire security was severely jeopardized after it assigned a local terror squad to murder the editor of the local Moroccan nationalist newspaper, *Maroc-Presse*. One of the members of the Red Hand squad was a close friend of the editor and, learning of the plot, he became remorseful and defected. The result was that the *Maroc-Presse*, "the only important French-language paper in Morocco not in sympathy with the counterterrorist Red Hand, became the best informed about their activities."[71] However, the repentant Red Hand agent soon met with sudden accidental death."

The local populace, or popular antennae (as these sources are described in one Viet Minh manual), may collect information on security force disposition and activities. The Viet Minh used children playing near Japanese (and later French) fortifications as a source of information on troop arrivals and departures, the guard system, and other pertinent details—all of which were easily observable by untrained children—that aided the guerrillas in planning attacks.[72]

PIRA insurgents were successful in preventing British security forces from infiltrating base areas and in gaining intelligence by cultivating close relations with their communities and using both coercion and incentives to encourage loyalty.[h] Eventually, government forces intensified efforts to gather intelligence through informers, surveillance, and interrogation.[73] The struggle for control of intelligence continued throughout the 1970s, and the increasing success of British efforts led to the PIRA changing its organization and practice in favor of greater security. They began to organize in small cells rather than in the larger battalions of the earlier years. They also took greater care in instructing recruits to refrain from discussing operations with anyone. For example, their sophisticated response to British intelligence efforts included training members to avoid leaving forensic evidence after an operation and how to resist interrogation after capture.[74] PIRA counterintelligence served two primary purposes. First, adequate counterintelligence allowed the PIRA to engage the British army with a measure of confidence and mitigate operational uncertainty. Second, successful counterintelligence served important symbolic measures, enhancing the PIRA's ability to maintain internal control, increase volunteer morale, and demonstrate to volunteers that the British could be overpowered. While British intelligence efforts in Northern Ireland would eventually reach mammoth levels, their initial efforts in the early 1970s were inadequate at best. Prior to the large civil rights marches and the ensuring riots in 1969, because of poor resources and British indifference, the Royal Ulster Constabulary (RUC) intelligence on the IRA was scanty and outdated because most of the available intelligence related to the Border Campaign. After the break with the Official IRA and an influx of young Republicans, such information was largely unhelpful in thwarting PIRA operations, preventing even the identification of active

---

[h] However, initial British intelligence efforts were also hampered by lack of knowledge regarding the nascent PIRA insurgent movement. Most intelligence they did have of IRA members was outdated, identifying primarily members of the older, Marxist-influenced Official IRA. The 1975–76 cease-fire provided the British an opportunity to accelerate their intelligence efforts.

PIRA members.[75] The implicit and explicit support of the Catholic population ebbed and flowed throughout the PIRA campaign. There was always support for the cause, but the violence against civilians or the periodic hope for a political solution would often erode that support. Catholics regarded the PIRA members as members of their community and had strong familial and community ties to the men that they knew (or suspected) were members. Some Catholics could turn a blind eye or deaf ear when they saw or heard of an operation. Vehicles might be loaned, or intelligence passed on. Catholics in government administrative jobs provided rich information, such as the home addresses of policemen or loyalist paramilitary members, to the Provisionals. Those who provided information about the PIRA to the security forces were intimidated or executed. The killing of civilians, however, often placed the PIRA on the defensive, and marches and organizations were formed by some Catholics to voice their dissatisfaction with the PIRA's policy of violence.[76]

Underground intelligence networks most often extend beyond the borders of the movement's native country. It is not uncommon for undergrounds to have cells distributed throughout the world among populations sympathetic to the cause. The LTTE, for example, maintained more than fifty offices and cells in foreign countries, especially in countries with large numbers of Tamil expatriates, such as England, France, Australia, Canada, and the United States. The LTTE had communication hubs in Singapore and Hong Kong to facilitate its weapons procurement activities; secondary cells in Thailand, Pakistan, and Myanmar; and front companies in Europe and Africa. From these locales, LTTE operatives coordinated purchases and shipments from Asia, Eastern Europe, the Middle East, and Africa.[77]

As a plan matures, a terrorist cell shifts to preattack surveillance and detailed planning. Members of the operational cells may appear in the vicinity of the objective during this phase. Trained intelligence and surveillance personnel or members supportive of the terrorist cell may be organized to gather information on the target's current patterns, usually days to weeks prior to an operation. The attack team confirms information gathered from previous surveillance and reconnaissance activities.

Rehearsals

As with conventional military operations, rehearsals are conducted to improve the odds of success, confirm planning assumptions, and develop contingencies. Terrorists also rehearse to test security reactions to particular attack profiles. Terrorists use both their own operatives and unsuspecting people to test target reactions.

Rehearsals, like many terrorist or guerilla training evolutions, require physical safe havens that afford sufficient privacy and security. Physical safe havens are important in the early, vulnerable periods of an insurgency, when political objectives are still being formulated and a support base is being established. Operationally, the segregated spaces provide a secure location for launching operations and for training and offer a defensible enclave, while making the group more cohesive.[78] It should be noted the aforementioned Maoist view of safe havens is a physical requirement; there is also the psychological safe haven.[79]

Execution

As indicated through this chapter, terrorist attacks come in many different forms for many different purposes. They can range from targeted assassination of political leaders, such as the PIRA's attempt on Margaret Thatcher, to indiscriminate suicide bombings in marketplaces such as those that seemed proliferate in Baghdad, Iraq, during the height of the insurgency. The first attack attributed to Al Qaeda took place in 1992, when bombs were detonated at the Mövenpick and Goldmohur Hotels in Yemen. Although American soldiers transiting Yemen on a United Nations mission to Somalia were the presumed targets, the attacks killed only two civilians. In 1993, a team of Al Qaeda, EIG, and EIJ operatives parked a rental truck loaded with a 1,500-pound explosive in the parking garage of the World Trade Center hoping to damage the first tower sufficiently to send it crashing into the second tower, with a casualty estimate of about 250,000. Although the blast did not destroy either tower, it shook both, killing six and injuring 1,042. In August 1998, the near simultaneous explosions at the U.S. embassies in Tanzania and Kenya killed 223 and injured over 4,000.[80] In 2000, Al Qaeda operatives attacked the destroyer USS *Cole* while it was refueling at a port in Yemen. The attack, a suicide mission using a skiff packed with approximately 1,000 pounds of explosives, killed seventeen U.S. servicemen and damaged the vessel. Then, in 2001, nineteen Al Qaeda operatives simultaneously hijacked four commercial airliners, flying two into the World Trade Center towers in New York City while another crashed into the Department of Defense in Washington, DC (the fourth, targeting the U.S. Capitol building, crashed in Shanksville, Pennsylvania). The attacks killed nearly 3,000 altogether.[81]

Escape and Evasion

Escape plans, as much as attack sequences, require detailed rehearsal and disciplined execution. Because professional terrorists are usually highly trained and skilled in their craft, they are obviously valuable to any underground. The problem of safe escape after executing a mission receives considerable attention. This is an inherent advantage to a suicide attack, as the escape and evasions plans and network are not required.

The Communist-led underground in Greece in 1945 developed cells of three agents whose identities were secret and who were unknown to each other. These units, called synergeia, were formed any time a professional terrorist job was required. When the mission was accomplished, the members of the group dispersed, changed their addresses, habits, and clothes, and concocted alibis.

In Cyprus, the EOKA adopted a number of techniques to secure the escape of its terrorists. One method used was for a terrorist, posing as a journalist-photographer, to be followed down a street by two or three young girls. When he sighted the man he was to kill, he shot him in the back and immediately threw the revolver to one of the girls trailing him, who slipped it into her purse and vanished. The terrorist remained briefly on the scene under the pretext that he was a journalist.[82]

Terrorists may manipulate social conditions and grievances and execute specific criminal acts (that may be inconsistent with their public ideology) to set the appropriate conditions for the establishment of escape and evasion networks.[83] Often Iraqi tribal customs and courtesies enabled insurgents to flee from coalition forces in the wake of an attack. The tribes provided financial, material, and personnel resources to facilitate escape and evasion.[84]

Rationalizing Terrorism

The indiscriminate nature of terror operations requires organizations to justify and/or explain the purpose of such operations. This rationalization is required for purposes of internal cohesion; the individual required to support and execute such attacks needs appropriate validation from their superiors as well as their accepted sources of authority.[85] EIJ and later Al Qaeda drew its legitimacy from a Salafist interpretation of Islam viewed through a Qutbist lens.[i] EIJ saw its organization not only as the vanguard of Qutb's vision for an Islamic revolution but also as an entity with requisite political and religious authority to declare all those who did not meet their requirements for piety *takfir* (or essentially non-Muslim), regardless of what the individuals professed to believe. EIJ also declared that all able-bodied Muslims were obligated to dedicate their lives to jihad. Farraj (EIJ's founder) took this to its logical extension and proposed violent jihad as an obligation of all pious Muslims.[86] Al Qaeda synthesized Qutbism's sophisticated

---

[i] This particular ideology is described in greater detail in Chapter 5.

theological discourse with a nuanced ability to comprehend, co-opt, and exploit modern grievances. This narrative combination resonated with extremists and moderates alike, regardless of whether an individual approved of the means by which Al Qaeda sought to accomplish its goals. The specific messages within the larger narrative rarely focused on citing authoritative texts (beyond selective interpretations of previous theorists reinforced by Quranic quotes without context), but rather relied on the application of general religious or ethical principles to modern political and social problems.[87] Attacks that resulted in the deaths of innocents (even Muslims) were justified on the basis that the act was sanctified as religious obligation. Sympathetic Imams often issued fatwas as religious justification to insulate the operatives from moral culpability.

While the PIRA was not a Catholic fundamentalist group, they did employ some theological rationale for acts of terror. The PIRA's arguments exploited cultural and not necessarily doctrinally adherent or historically accurate perceptions of the sociopolitical issues at hand.[88] Nevertheless, the deeply entrenched social networks within the Catholic community combined with the long-standing narrative of the Republican movement allowed the PIRA to immediately have an implied legitimacy across the entire Irish Catholic population. By invoking the name and mission of the IRA and harkening the struggle for independence, the Provisionals framed their campaign as historically legitimate and morally just.[89] This support, however, waned as civilian casualties in Northern Ireland and Great Britain increased: the Catholic population may not have actively supported the targeting of British soldiers garrisoned in Northern Ireland, but it was largely viewed a necessity to resist an occupying force. The targeted killings of Protestant prison guards were even tolerated because those individuals were portrayed as being complicit in the systematic abuse of detainees and violators of civil rights. The mainland bombing campaign and targeted assassinations, however, were not well received and alienated the PIRA from their support networks.[90]

Terrorist Threats

Two essential principles of phrasing a threat can be postulated on the basis of general patterns of human behavior under stress. First, there are specific threats in which the demands and consequences are communicated so that they cannot be misunderstood. Use of a specific threat rests on the basic assumption that an individual confronted with persuasively stated, clear-cut demands and imminent harmful consequences will take the path of least resistance, which is compliance. Rather than endanger himself, his family, or his property, the individual will accept, albeit reluctantly and as evasively as possible, the threatener's alternative. The effect of a terrorist threat is maximized when the target audience believes the likelihood of future attacks on themselves and their in-group is increasingly likely.[91]

Specific Threats

In specific threats, the threatener seeks to secure compliance without actually being required to execute the threat. Persuasive communication, leaving little room for misunderstandings, is essential for effectiveness. For example, Viet Cong guerrillas attacking a South Vietnamese fortification during the night call over loudspeakers saying: "We only want to kill the Americans. All the rest can go free if they leave their weapons."[92] Surrounded by superior Viet Cong forces and offered a clear alternative to further resistance, South Vietnamese militia have been known to throw down their weapons and leave the Americans to fend for themselves.[93] Occasionally the Viet Cong varied this kind of threat by distributing leaflets saying they will fire on government troops only if they are accompanied by U.S. military advisers. The intensity of delegitimization is the degree to which the radical group challenges the legitimacy of its opponents—the more intense the delegitimization, the greater the risk for radicalization.[94] If the group increasingly believes that change is not possible in existing society and that a radical change is necessary, they are more likely to radicalize. Additional signs of delegitimizing an adversary include the following: the group calling for violent action against enemies, specifying targets with explicit detail or expanding the spread of targets from specific to general,[j] emphasizing the historical sins of a designated group, and/or characterizing group members as righteous and uniquely empowered to rectify the perceived ills of society. A radical can delegitimize the out-group by idealizing the goals and means of the in-group's revolutionary nation or leaders associated with violence, terrorism, or revolution.[95]

Terrorists have often used the tactic of escalating warnings. A mild first warning is followed by more harshly stated threats and then by an imminent show of action. World War II resistance movements in occupied Europe provide numerous illustrations of specific threats issued to collaborators to make them desist from supporting the Nazis. The threat often began with a blacklisting of the individual's name in an underground newspaper, escalated to a warning note delivered directly to him, and ended when the threat was enforced. Meanwhile, the demand behind the threat was specifically stated. As the Fuerzas

---

[j] The perversion or misapplication of the law of social substitutability (which holds that the killing of any member of a segment is considered a group offense and can be avenged by the killing of any member of the offender's segment) greatly expands the threat assessment associated with the particular group.

Armadas Revolucionarias de Colombia (Revolutionary Armed Forces of Colombia, or FARC) expanded the scope of its operation to include narcotics trafficking and kidnapping for ransom during the 1990s, their first diplomatic foray into an area was typically an unambiguous threat against the local political apparatus to ensure necessary freedom of movement.[96]

If a threat is to affect a large number of people, it must be related to a clearly discernible act, so that all learn the lesson demonstrated by the enforcement of the threat. Viet Cong terrorists in Vietnam, after several implicit or explicit warnings to a prominent villager or government official, sent a signed "death sentence" to that individual. When the threats (usually assassinations) were carried out, the tale of Communist omnipotence was spread by the terror-stricken widows and children who still had the written death sentence hanging over them. The bandwagon effect sought in these cases was to convince the inhabitants of the village that they had better obey the Viet Cong or the same fate may be theirs.[97]

Another requirement for the effectiveness of a specific threat is that the threatener must have visible means of administering punishment if he is to be persuasive. The threatener must be able to determine that the threatened person did not in fact comply with the demands before inflicting punishment; otherwise, the environment created by a specific threat changes to one of general terror.

The phrasing and structuring of specific threats vary according to the kind of compliance that is sought. Essentially, there are three kinds of threat demands. First, a threatener may choose to demand actions toward which the populace is already predisposed. This is the easiest kind—the one for which compliance is most easily secured. It may be of considerable advantage for an underground to use this tactic because its easy success gives the threatener a disproportionate amount of credit for power and influence. For example, in Algeria, the OAS, in order to demonstrate its displeasure with the French government, demanded that all Algerians stay off the streets during the evening hours and turn off their lights and threatened to punish anyone found on the streets during the evening hours. These demands on the populace were consonant with what an individual might do on his own during any kind of disorder. Thus compliance was easy. The LTTE's reliance on Tamil expatriates for financial support (a necessity after loss of control of the Jaffna peninsula in the mid-1990s) often relied on coercive acts. While the vast majority of contributions to the Tamil cause were voluntary, competition among rival groups prompted the LTTE to not only systematize their collection process but also implement punishments for noncompliance.[98]

General Threats

A second kind of demand seeks to induce an individual or group to change specific behavior by demanding alternative actions. This is more difficult than the first kind of demand and requires clearly stated alternatives and persuasively stated (and perhaps demonstrated) consequences for refusing. To make these demands effective, that target audience must "personalize" the event through either a social or geographic connection; this decreases the psychological distance from the event and thus makes the threat more imminent.[99] Peru's Sendero Luminoso ("Shining Path") portrayed themselves as a modern-day Peruvian Maoist version of Robin Hood. They secured geographic areas through assassinating local government officials and taking control of their belongings/property and redistributing it to the local villagers. Sendero Luminoso attempted to frame their efforts as a positive force with the charismatic Abimael Guzmán as a benevolent revolutionary; however, they often resorted to threats and coercion. Students and teachers at the venues where Guzmán (a former university professor) spoke attended because many felt obliged. Those that did not attend (whether out of disinterest or a competing obligation) were threatened, at a minimum, and often verbally and physically harassed. Faculty members who did not support Sendero Luminoso were encouraged to resign their posts and leave the region, even under the threat of death if they chose to stay. Throughout this near-cultish movement, fear was used as the great neutralizer of the political opposition.[100]

The third and most difficult kind of demand is one that orders an individual to refrain from a course of action he is already pursuing. The Taliban have successfully used shabnamah (or "night letters") to communicate threats to Afghan villagers. The documents are typically handwritten (but have been printed pamphlets and/or communiqués) and often address specific individuals, circumstances, and events to demonstrate proximity, credibility, and legitimacy.[101] The purpose is generally to dissuade villagers from supporting or even working with coalition and/or Government of the Islamic Republic of Afghanistan (GIRoA) forces. Night letters have been a traditional and common instrument of Afghan religious figures, jihadists, and rebels to encourage rural populations to oppose both state authority and regulations. Night letters often invoke traditional narratives to exert greater social control by reinforcing in-group/out-group distinctions while reminding the target of the credible threat the Taliban presents.[102]

Generalized or uncertain threat is the second pattern of phrasing. The generalized threat does not delineate behavior or specify demands and consequences; these are left to the imagination of the threatened individual. Uncertain threats are used to create terror among the populace, making them vigilant and sensitive to terrorist suggestions. The threatener captures attention at a point when persons under stress are desperately searching to eliminate uncertainty and ambiguity. He may suggest escape routes and alternatives and make compliance demands that are readily accepted in order to eliminate the uncertainty of the threat and reduce terror. Occasionally, terrorists do not even seek compliance with specific demands but rather hope to cause evacuation or psychological breakdown of popular morale. The South Vietnamese army used uncertain threat in a counterterror campaign dubbed "Operation Black Eye" against the Viet Cong. Selected Vietnamese troops were organized into terror squads and assigned the task of working with rural agents in penetrating Viet Cong-held areas. Within a short time, Viet Cong leaders—key members of the clandestine infrastructure—began to die mysteriously and violently in their beds. On each of the bodies was a piece of paper printed with a grotesque human eye. The appearance of "the eye" soon represented a serious threat. The paper eyes, 50,000 copies of which were printed by the U.S. Information Service in Saigon, turned up not only on corpses but also as warnings on the doors of houses suspected of occasionally harboring Viet Cong agents. The eyes came to mean, "big brother is watching you." The mere presence of "the eye" induced members of the Viet Cong to sleep anywhere but in their own beds. It was an eerie, uncertain threat.[103]

Generalized terror seems to have limited effectiveness over a period of time. Uncertain threat reaches a point of diminishing returns when the populace finally either becomes habituated to the stress of ambiguity or focuses hostility on some objective it perceives, correctly or not, as the source of threat.[104] Once such a "hostile belief" develops, the populace's openness to suggestion ends. The essential distinction, then, between specific and uncertain threat is the difference between threat used to secure specifically stated demands (known and planned in advance) and threat designed to debilitate and/or sensitize a populace to later suggestions.

PSYCHOLOGICAL EFFECTS OF TERROR

Unlike the guerrilla, the terrorist does not occupy territory in the physical domain but seeks to dominate the psychological, social, and information domains by manipulating target audiences, sending implied messages about the cause and meaning of terrorist tactics.[105]

There are two principal audiences: the organization's constituency[k] (in-group) and the enemy (out-group). In-group messages stress violent resistance is necessary to accomplish the desired end state, that negation is acquiescent to tyrannical authority, and that the adversary is vulnerable. Out-group messages stress the likelihood of future attacks, the identification of all those affiliated with the government (including civilians) with the enemy, and the lack of government control over—as well as the inability of the government to protect—civilians.[106]

As Lenin indicated, the purpose of terrorism is to cause terror—an unremitting, paralyzing sense of fear that permeates one's psyche.[107] Fear precipitates the acute secretion of catecholamines and glucocorticoids typical of the human stress response. This response is elicited by four situational characteristics: novelty, unpredictability, threat to survival, and perceived lack of control,[108] all of which have been identified in survivors of terrorist events.[109] Emotional responses to manmade disasters are considerably more stressful than responses to naturally occurring phenomena.[110] Stress is not entirely negative, although it can have harmful consequences if one is not appropriately prepared for it. Moderate stress can have positive effects on behavioral and neuro-endocrine measures; however, prolonged exposure to stress without sufficient protective measures or rest can result in cognitive depletion and emotional disturbances.[111] With terrorism, it is not only the effect of previous attacks but also the anticipation of future attacks that can induce this stress.[112]

The psychological manifestation of terror is neither simple nor straightforward because different audiences will respond with various interpretations of terrorist acts. Observers may identify with either the victims or the aggressor based on in-group bias and not necessarily the subjective morality of the act.[113] People who have experienced or witnessed a terrorist attack may go into a state of acute stress (termed acute stress disorder, or ASD). Symptoms of an acute stress response include recurring thoughts of the incident, irrational fears of previously normal activities, significant deviation from one's daily routine, survivor guilt, a pronounced sense of loss, a reluctance to communicate feelings, and a subjective uncertainty or loss of control. If these symptoms persist for more than thirty days after a traumatic event, an individual may have posttraumatic stress disorder (PTSD). Symptoms of PTSD fall into three main categories: intrusive memories, avoidant behaviors, and arousal. Intrusive memories, or episodes of re-experiencing the event that disrupt daily life, include flashbacks, nightmares, and/or uncomfortable or disturbing reactions to those memories.

---

[k] Neutral (or uncommitted groups) can be considered a third audience, however, with the higher-order goal being to assimilate the neutrals into the in-group.

Avoidance includes emotional numbing, feelings of detachment, the inability to recall the traumatic event, a general malaise (particularly toward activities previously considered pleasurable), avoiding people and/or places that are reminiscent of the event, and an uncertainty regarding the future. Arousal symptoms include difficulty concentrating, startling easily (and with a more exaggerated reaction), hypervigilance, irritability, and difficulty sleeping. Diagnostic criteria for PTSD include a history of exposure to a traumatic event meeting two criteria and symptoms from each of three symptom clusters.[114] The psychological consequences of a successful act of terrorism are both acute and chronic and tend to increase with proximity to the event. Populations exposed to the attack show higher rates of PTSD and those suffering losses as a result of the attack show higher rates of depression.[115]

Terror Management Theory

The psychological effects of terrorism are not simply manifested in individual physiology and psychology but also in a social context. Terror Management Theory (TMT) states that human behavior is mostly motivated by the fear of death and if mortality is made salient, individuals will intensify strivings for self-esteem and will respond positively toward people and ideas that support their worldview and respond negatively toward those people and ideas that undermine that worldview. If self-esteem is lowered or the validity of a cultural worldview is damaged, death anxiety will increase, necessitating an active solution. Mortality salience exacerbates group defenses and reinforces in-group biases, thus achieving simultaneous higher-order effects on both the target audiences. Individuals evaluate in-group members positively because similarly minded individuals are assumed to support, and therefore validate, their own cultural worldview. In contrast, individuals evaluate out-group members negatively because alternatively minded individuals (relative to the in-group) are assumed to threaten their worldview.

Under uncertain conditions, individual cognitive processes are biased toward emotionally evocative events, resulting in an increased estimate of a perceived threat and a tendency toward indecision. This results in a desire for predictability and therefore the tendency to gravitate toward charismatic leaders who simplify (or split) complex problems into binary issues. When individuals perceive they lack the necessary information to come to a judgment, they tend toward negatively valanced information; this is particularly so in the aftermath of a terrorist incident. Social amplification (which is also increased under ambiguous and/or uncertain conditions) further magnifies the negative bias and thus social interaction further compounds the terror effect. Although unintentionally, media coverage of acts of terror exacerbates the aforementioned phenomena and thus increases the effect of an attack, therefore incentivizing such tactics. Given the ubiquity and global reach of modern media outlets as well as social media, this trend is unlikely to cease.

Terror is not a static phenomenon: As threatening acts accumulate or escalate, the degree of terror heightens. A stimulus can be anything from an act of social sanction to threats of physical violence or actual physical attack. The corresponding interpretation of these threatening stimuli is a heightening state of psychological distress. The response may vary from coerced compliance to acquiescence, from physical flight to psychological immobilization and breakdown. The effect of terror upon individuals cannot always be determined from an objective description of the terrorist act. That which threatens or terrorizes one individual may not affect another in the same way. Essentially, however, the process of terrorism can be viewed in the following manner: the stimulus is the threatening or terroristic act, and the response is the course of action, or inaction. If the perception of the threats leads to disorganized behavior or the inability to take appropriate action, the individual is said to be in a state of terror.

Individual Responses

The effect of terrorism upon individuals differs widely. Similar threats or actions affect individuals differently. Behavior patterns are affected to a large extent by personality and previously established behavior habits. Because personality variables affect an individual's perception of threat, the vagaries of perception are the keys to understanding human behavior under stress. It is not the objective character of the threat that determines an individual's behavior so much as his subjective evaluation of the situation.

Human response to threat also varies according to the nature of the threatening situation—whether it is specific or uncertain. The terrorist may wish to have the threatened party perform a particular act and may issue a highly specific threat. Where the threat is clearly defined and specifically communicated to an individual, with demands, alternatives, and consequences apparent and persuasively stated, an individual's reaction is probably based on a relatively clear assessment of known variables and he may comply out of fear of having the threat carried out. However, the terrorist may seek to cause disruptive behavior or panic by issuing an uncertain, generalized threat. The very ambiguity of the situation makes rational decision-making and assessment functions break down and leads to hysteria and panic.[116] The more specific the threat, the more fear inducing it is; the more vague the threat, the more anxiety inducing it is, making an individual hypersensitive to ordinarily neutral situations and causing disruptive behavior.

The relative intensity of threat, regardless of whether it is vague or specific, determines whether a person will be able to take effective action.[117] Thus, unlike the previous "specific versus uncertain" threat theory, where individuals respond rationally and positively to a specific threat and rather hysterically to an uncertain threat, this theory suggests that whenever the magnitude of threat is great, it tends to produce an ineffective or irrational response regardless of the vagueness or specificity of content. Others argue that threats or threatening acts need not necessarily grow in magnitude for terror to intensify; the mere continuance of threats over a period of time is sufficient to intensify the reaction.[118]

Different behavior patterns emerge from situations in which there are conflicting threats. The individual usually succumbs to the threat that appears most imminent or is greatest in magnitude. Regardless of whether a threat is specific or uncertain, or different in magnitude and danger, an individual's vigilance response generally evolves in five phases:

- Recognition: The threat, or threatening situation, is perceived by some cue or message.
- Probability: An estimation of the probability of the threatening event occurring is made; the validity of the threat is checked.
- Assessment: The qualitative nature of the threat is assessed (whether physical pain, loss of loved ones, loss of property, etc.). There is some attempt to define the situation: the nature, timing, and magnitude of the threat are assessed and an estimation is made of the means of coping with it, the probability of success, and the cost to the individual.
- Defense: Commitment to an avenue of escape and adaptation; progressive use of a "lattice of defense," in which the failure of one defense leads to another more extensive defense.
- Reassessment: When first-defense routes fail, other avenues are attempted, but under the stressful or threatening situation, the individual tends to become overcompensatory or excessively sensitive or may exhibit other nonadaptive responses; psychological immobilization or breakdown may occur at this stage.

This sequence can be terminated at almost any point, and the phases may telescope so that they are virtually simultaneous.[119] Where threat demands and consequences are apparent, rational assessment is relatively easy. Where they are ambiguous or uncertain, the likelihood of irrationality and hysteria is increased. The most debilitating factors in human response to threat are uncertainty and ambiguity because the individual tries to resolve the uncertainty before he takes action to escape the threat. The more difficult the perceived difficulty of resolving the uncertainty, the more unnerved the individual becomes.

When uncertain threat leads to a state of hysteria, the individual attempts to remove the ambiguity of the threatening situation by identifying some certain source—even if the "source" has little or nothing to do with the real origin of the threat. In certain cases, this can lead to the individual identifying with the aggressor instead of the victims, particularly if there is an uncertain (or divided) group loyalty, for example, between an ethnic group and a state apparatus. This is particularly so when those individuals have expressed the desire for revenge, have a history of disagreements with the authorities, have a history of violent behavior or experience with weapons (including participation in military training, paramilitary, or other violent organizations or campaigns), or possess specialized skills that can contribute to or facilitate violent action.[120] A corollary human response to hysteria is the predilection to suggestion (what Freud called wish-fulfillment beliefs).[121] In trying to identify the source of the threat and redefine the uncertain situation, an individual is more susceptible to rumors and targeted influence operations that exploit these biases.

Individuals narrow or restrict their span of attention under threat. Becoming hypervigilant, they focus their attention on the threat and the threatener, to the virtual exclusion of other stimuli. Thus, hypervigilance leads an individual to concentrate on the demands and suggestions of the underground threatener and reduces his attention to communications from the government or security forces. In a group context, the prolonged isolation or segregation can foster a sense of humiliation or collective loss of self-esteem. If the group experiences a growing sense of stigmatization or isolation directed against the group, individual members, or their constituents, their propensity for a violent response may increase. Organizations can experience a sense of helplessness and rage in response to collective attacks against the group or other actions designed to demonstrate the group's inferiority.[122]

If an individual can perceive no avenue of escape from a threat, he or she develops a sense of helplessness and this sense increases the stress reaction. If the purpose of a threat is to achieve compliance with certain demands, a threat that leaves the individual with no influence over the outcome may backfire. The individual either breaks down and is unable to comply or pursues an opposite, hostile course. For example, the Nazi policy of threatening reprisals in occupied Greece during World War II tended to operate against the German objectives of population control. Indiscriminate reprisals against the Greek popu-lace left the individual citizen helpless to influence the outcome: guer-rilla band activity near a village, over which the villager had no control, brought the threat of death. "The wanton nature of the retaliation—the picking of victims at random—meant that pro-German Greeks or their relatives suffered as much as anti-German Greeks. Under these circumstances there was little advantage in being a collaborator. As the reprisals continued they tended to give credence and prestige to the guerrillas . . . ."[123] Further, indiscriminate "burning of villages left many male inhabitants with little place to turn except to the guerrilla bands."[124]

Group Responses

Uniting against a common enemy is one of the most powerful acti-vators and unifying factors for identity groups, increasing the group's cohesion, decreasing internal dissent, and increasing a sense of uni-fied purpose. External threats are often the catalyst for identity groups to radicalize toward violent action. The external requirements for this mechanism do not require deprivation or an oppressive out-group but merely the perception of an external threat. In small, face-to-face groups, an out-group threat leads to increased group cohesion, increased respect for in-group leaders, increased sanctions for in-group deviates, and idealization of in-group norms.[125] This results in increased cohesion and a solidification of tighter social networks. The presence of a threat reinforces the need for cooperation and agree-ment, exacerbating the "us against them" mentality and the "we're all in this together" motivation to cooperate and deindividuate. The group can also perceive a serious threat to individual members or their lead-ers after physical attacks (including arrests, torture, and assassination) or catastrophe. The combination of isolation and outside threat makes group dynamics more powerful in the underground.[126] The group experiences fear that the regime or other opponent is attempting to destroy the group as a whole.[127] The underground group, isolated from society, develops tighter cohesion in response to shared danger, provid-ing an exaggerated variant of the fight-flight group.[128]

There are a number of indications and warnings of this mecha-nism that have been derived empirically. Among the observables is the active recruitment from a pool of disenfranchised, victimized, radical-ized, or violent individuals. Another observable signifying a movement toward radicalization is a change in recruitment strategy such as adapt-ing methods to attract personnel with skills and motivations necessary for violence, using more elitist entry requirements, and perpetrating media-coverage-generating events such as demonstrations and open confrontations with police to draw recruits. This is illustrated by some experiences of the Philippine army during the Huk insurgency. To counteract Huk terrorism or to dissuade a village from giving strong support to the Huk movement, the Philippine army gathered the villagers, including the mayor and village policemen, in an open area. Approximately 200 yards away, Philippine troops, in full uniform, would line up a number of "captured Huks." Then they ushered out each Huk blindfolded and executed him by bayonet. As one Philippine officer reported, "While we were killing them, some were shouting out the name of the mayor, the names of the policemen, and . . . the names of their principal suppliers. Seeing the Huks killed before their eyes, hearing themselves named as the supporters of those we had just massacred, these civilians naturally expected to be next on the death lists."[129] In reality, the villagers had witnessed a mock execution of regular Philippine troops equipped with chicken blood and stage presence. But the executions had the desired effect of making the government counter-threat apparent. Afterward, officers talked individually with the villagers, explaining that they now knew everything about the village and that those who confessed or cooperated would not be treated like the captured Huks. To protect the villagers from further Huk threats, the officers established several meeting places that evening where individuals could report to receive protection in exchange for information. The threat of the government was thus made more pressing and real than the Huk terror; effective responses were obtained.[130]

Cultural factors are also a significant variable in human behavior under threat. Unique cultural mores and beliefs frequently affect an individual's sense of threat or subjective experience of terror. One needs only to think of the role voodoo terror plays in certain areas, such as Haiti, where agents of Haitian dictator Dr. Francois "Papa Doc" Duvalier reputedly use the threat of the pin-in-the-doll with some effectiveness. In Angola, it is believed that a mutilated body cannot enjoy an afterlife, a fear exploited by the Angolan administration during the 1961 rebellion. While the tribesmen will occasionally charge fearlessly into a barrage of machine-gun fire, reported one writer, they will think twice about attacking anyone armed with a machete.[131]

Segregation and isolation exacerbate a perceived threat, particularly during times of conflict. Palestinian refugee camps were established after the 1948 Arab-Israeli War to accommodate the Palestine refugees who were forced to leave or chose to do so after the creation of Israel. United Nations General Assembly Resolution 194 grants Palestinians the right to return to their homeland, but Israel has refused to allow the vast majority of refugees to return.[132] The collective

351

humiliation at the hands of and subsequent anger toward the Israelis has served as a unifying theme among the Palestinian people. The clear distinction between the in-group and the out-group is reinforced through the physical security measures of the refugee camps and the strictly controlled lifestyle required of the inhabitants. Isolation creates a unique combination of relative deprivation and frustration-aggression called cramping. Groups feel cramped when their desire for security and social needs is unavoidably interfered with. This negative sensation grows increasingly intolerable and results in violence against those who are perceived to interfere with the aforementioned needs. Interviews with both secular and nonsecular Palestinian terrorists have identified the common trend of being under constant threat by Israel. This shared fear and contempt appears to be a defining characteristic of those engaged in political violence against Israelis. In a series of interviews with thirty-five incarcerated Palestinian operatives, most reported their families had good social standing, but their status and experience as refugees was paramount in their development of self-identity. For the secular terrorists, enlistment was a natural step and it led to enhanced social status. Armed attacks are viewed as essential to the operation of the organization. There was no question that these types of attacks were necessary for the success of the cause; the attacks provided a sense of control or power for Palestinians in a society that had stripped them of it. The hatred socialized toward the Israelis was remarkable, especially given that few reported any contact with Israelis. There was a common theme of having been unjustly evicted from their land, of being relegated to refugee status or living in refugee camps in a land that was once considered theirs.[133] They expressed a fatalistic view of the Palestinian/Israeli relationship and a sense of despair or bleakness about the future under Israeli rule. Few of the interviewees were able to identify personal goals that were separate from those of the organization to which they belonged.[134]

SUMMARY

John Boyd extrapolated Clausewitz's Wondrous Trinity (referenced in the Preface) and proposed that uncertainty pervades everything in life, including warfare. In fact, uncertainty is a fundamental and irresolvable characteristic of war, no matter how good our observations, theories, and/or rationalizations are.[135] Terrorism seeks to exploit the endemic uncertainty in the human condition and use the natural result of violence (fear) to exert control over members, reinforce the biases of supporters, and intimidate adversaries. This uncertainty not only creates a receptiveness to insurgent ideology but also exacerbates preexisting prejudices in the wake of traumatic experiences.

Despite often relying on illegal acts and/or criminal networks, terrorism is not traditionally used for financial gain but rather for psychological gain.[136] Insurgents have seldom relied solely on the attractiveness of their appeals or on the persuasiveness of their goals to secure popular support; they have generally assumed that people never entirely pursue idealistic goals or do what logic might tell them is most beneficial. Coercive means have been used to focus public attention on the goals and issues identified as important by the insurgents. Negative sanctions are used to ensure that recalcitrant individuals comply and do act in their own self-interest. Terror has been used to support other insurgent techniques and operations, such as propaganda and agitation. Terrorism, however, has its inherent risks. If brutality is used to maintain internal cohesion, both recruiting and retention become difficult. Furthermore, the wanton murder of innocent civilians can have a negative effect on the organization's ability to raise money and engender popular support. Nevertheless, the psychological benefit of terrorist action often outweighs the cost and thus has been an effective psychological warfare tactic from antiquity to modernity and will continue to be employed.

ENDNOTES

[1] L. Morgan Banks and Larry C. James, "Warfare, Terrorism, and Psychology," ed. Bruce Bongar, Lisa M. Brown, Larry E. Beutler, James N. Breckenridge, and Philip G. Zimbardo, *Psychology of Terrorism* (New York: Oxford University Press, 2007).

[2] Department of Defense Dictionary of Military and Associated Terms, Joint Publication 1-02 (JP1-02) (Department of Defense, 2012), 342.

[3] Boaz Ganor, "Terrorism as a Strategy of Psychological Warfare," *Journal of Aggression, Maltreatment, and Trauma* 9, no. 1/2 (2004): 33–43.

[4] Alex Schmid, "Terrorism as Psychological Warfare," *Democracy and Security* 1 (2005): 137–146.

[5] Gerard Chalian and Arnaud Blin, *The History of Terrorism: From Antiquity to Al Qaeda* (Berkeley: University of California Press, 2007).

[6] Schmid, "Terrorism as Psychological Warfare."

[7] William S. Lind, Keith Nightingale, John F. Schmitt, Joseph W. Sutton, and Gary I. Wilson, "The Changing Face of War: Into the Fourth Generation," *Marine Corps Gazette* 73, no. 10 (1989): 22–26.

[8] David C. Rapoport, "The Four Waves of Rebel Terror and September 11," *Anthopoetics* 8, no. 1 (2002), http://www.anthropoetics.ucla.edu/ap0801/terror.htm.

[9] Jason Spitaletta, "Egyptian Islamic Jihad," in *Casebook on Insurgency and Revolutionary Warfare, Volume II: 1962–2009*, ed. Chuck Crossett (Alexandria, VA: U.S. Army Publications Directorate, in press).

[10] Jason Spitaletta and Shana Marshall, "Al Qaeda," in *Casebook on Insurgency and Revolutionary Warfare, Volume II: 1962–2009*, ed. Chuck Crossett (Alexandria, VA: U.S. Army Publications Directorate, in press).

[11] Bernard Fall, *The Two Vietnams: A Political and Military Analysis* (New York: Praeger, 1963), 360.

[12] Ibid., 360–361.

[13] Robert B. Rigg, "Catalog of Viet Cong Violence," *Military Review* 42, no. 12 (1962): 25.

[14] Roger Trinquier, *Modern Warfare* (New York: Praeger, 1964), 16–177.

[15] Wesley R. Fishel, "Communist Terror in South Vietnam," *The New Leader* III, nos. 27–28 (July 4–11, 1960): 14.

[16] Ariel Merari, "Terrorism as a Strategy of Insurgency," *Terrorism and Political Violence* 5, no. 3 (1993): 213–251.

[17] Paul Jureidini, *Case Studies in Insurgency and Revolutionary Warfare: Algeria 1954–62* (Washington, DC: Special Operations Research Office, 1963), 98–99.

[18] Lucian W. Pye, *Guerrilla Communism in Malaya* (Princeton, NJ: Princeton University Press, 1956), 104.

[19] Chuck Crossett and Summer Newton, "The Provisional Irish Republican Army: 1969–2001," in *Casebook on Insurgency and Revolutionary Warfare, Volume II: 1962–2009*, ed. Chuck Crossett (Alexandria, VA: U.S. Army Publications Directorate, in press).

[20] Brian Crozier, *The Rebels* (London: Chatto and Windus, 1960), 176.

[21] Crossett and Newton, "The Provisional Irish Republican Army: 1969–2001."

[22] Spitaletta, "Egyptian Islamic Jihad."

[23] Chuck Crossett and Jason Spitaletta, *Radicalization: Relevant Psychological and Sociological Concepts* (Ft. Meade, MD: Asymmetric Warfare Group, 2010).

[24] Jerrold M. Post, Keven G. Ruby, and Eric D. Shaw, "The Radical Group in Context 2: Identification of Critical Elements in the Analysis of Risk for Terrorism by Radical Group Type," *Studies in Conflict and Terrorism* 25, no. 2 (2002): 101–126

[25] Luis Taruc, *Born of the People* (New York: International Publishers, 1953), 134, 137.

[26] Maegen Nix and Shana Marshall, "Liberation Tigers of Tamil Eelam (LTTE)," in *Casebook on Insurgency and Revolutionary Warfare, Volume II: 1962–2009*, ed. Chuck Crossett (Alexandria, VA: U.S. Army Publications Directorate, in press).

[27] Interview with former member of the Polish underground, Washington, DC, April 27, 1965.

[28] Joestan Joachim, *The Red Hand* (London: Abelard Schuman, 1962), 47–49.

[29] Crozier, *The Rebels*, 169–170.

[30] Ibid.

[31] Nix and Marshall, "Liberation Tigers of Tamil Eelam (LTTE)."

[32] Spitaletta and Marshall, "Al Qaeda."

[33] Pye, *Guerilla Communism in Malaya*, 104.

[34] Ibid., 252.

[35] Taruc, *Born of the People*, 134.

[36] See U.S. Senate, Committee of the Judiciary, *Yugoslav Communism: A Critical Study* (Washington, DC: Government Printing Office, 1961), 124.

[37] Joachim, *The Red Hand*, 47–49.

[38] Crossett and Newton, "The Provisional Irish Republican Army: 1969–2001."

[39] James Eliot Cross, *Conflict in the Shadows: The Nature and Politics of Guerrilla War* (Garden City, NY: Doubleday and Co., 1963), 54.

[40] David Killcullen, *The Accidental Guerrilla: Fighting Small Wars in the Midst of a Big One* (Oxford: Oxford University Press, 2009).

[41] Ibid.

[42] Linda Napoleoni, *Insurgent Iraq: Al-Zarqawi and Al-Qaeda's New Generation* (Washington, DC: Seven Stories Press, 2005).

[43] Merari, "Terrorism as a Strategy of Insurgency."

[44] Crossett and Newton, "The Provisional Irish Republican Army: 1969–2001."

[45] Col. de Rocquigny, "Urban Terrorism," trans., *Military Review* 38, no. 2 (February 1959): 93–99.

[46] Spitaletta and Marshall, "Al Qaeda."

[47] See Vo Nguyen Giap, *People's War, People's Army* (Washington, DC: Government Printing Office, 1962), 65.

[48] Pye, *Guerrilla Communism in Malaya*, 88.

[49] Spitaletta and Marshall, "Al Qaeda."

[50] Ibid.

[51] Lawrence Wright, "The Master Plan: For the New Theorists of Jihad, Al Qaeda Is Just the Beginning," *The New Yorker*, September 11, 2006, http://www.newyorker.com/archive/2006/09/11/060911fa_fact3#ixzz1hYvNJLzp.

[52] Ibid.

[53] Spitaletta and Marshall, "Al Qaeda."

[54] James N. Breckenridge and Philip G. Zimbardo, "The Strategy of Terrorism and the Psychology of Mass-Mediated Fear," in *Psychology of Terrorism*, ed. Bruce Bongar, Lisa M. Brown, Larry E. Beutler, James N. Breckenridge, and Philip G. Zimbardo (New York: Oxford University Press, 2007).

[55] Tom Pyszczynski, Abdolhossein Abdollahi, Sheldon Solomon, Jeff Greenberg, Florette Cohen, and David Weise, "Mortality Salience, Martyrdom, and Military Might: The Great Satan Versus the Axis of Evil," in *Psychology of Terrorism*, ed. Jeff Victoroff and Ariel W. Kruglanski (New York: Psychology Press, 2009).

[56] Robert Pape, *Dying to Win: The Strategic Logic of Suicide Terrorism* (Chicago: University of Chicago Press, 2005).

[57] Ivan Arreguin-Toft, *How the Weak Win Wars: A Theory of Asymmetric Conflict* (Cambridge: Cambridge University Press, 2006).

[58] Ariel Merari, *Driven to Death: Psychological and Social Aspects of Suicide Terrorism* (Boston: Oxford University Press, 2010).

[59] Nix and Marshall, "Liberation Tigers of Tamil Eelam (LTTE)."

[60] Shana Marshall, "Hizbullah: 1982–2009," in *Casebook on Insurgency and Revolutionary Warfare, Volume II: 1962–2009*, ed. Chuck Crossett (Alexandria, VA: U.S. Army Publications Directorate, in press).

[61] Ibid.

[62] Spitaletta, "Egyptian Islamic Jihad."

[63] Ibid.

[64] Alex Mintz and David Brule, "Methodological Issues in Studying Suicide Terrorism," *Political Psychology* 30, no. 3 (2009): 365–371.

[65] Dudley Barker, *Grivas: Portrait of a Terrorist* (New York: Harcourt, Brace and Co., 1959), 142.

[66] Ibid., 140–141.

[67] Ibid., 142.

[68] Crossett and Newton, "The Provisional Irish Republican Army: 1969–2001."

[69] Spitaletta and Marshall, "Al Qaeda."

[70] Merari, "Terrorism as a Strategy of Insurgency."

[71] Joachim, *The Red Hand*, 47–49.

[72] Bryan Gervais and Jerome M. Conley, "Viet Cong: National Liberation Front for South Vietnam," in *Casebook on Insurgency and Revolutionary Warfare, Volume II: 1962–2009*, ed. Chuck Crossett (Alexandria, VA: U.S. Army Publications Directorate, in press).

[73] Crossett and Newton, "The Provisional Irish Republican Army: 1969–2001."

[74] Ibid.

[75] Ibid.

[76] Ibid.

[77] Nix and Marshall, "Liberation Tigers of Tamil Eelam (LTTE)."

[78] Crossett and Newton, "The Provisional Irish Republican Army: 1969–2001."

[79] Inge Bretherton, "The Origins of Attachment Theory: John Bowlby and Mary Ainsworth," *Developmental Psychology* 28 (1992): 759–775.

[80] Spitaletta and Marshall, "Al Qaeda."

[81] Ibid.

[82] Barker, *Grivas: Portrait of a Terrorist*, 145.

[83] John B. Wolf, "Organization and Management Practices of Urban Terrorist Groups," *Studies in Conflict and Terrorism* 1, no. 2 (1978): 169–186.

[84] Montgomery McFate, "Iraq: The Social Context of IEDs," *Military Review* 85, no. 3 (May/June 2005): 37–40.

[85] David Grossman, *On Killing: The Psychological Cost of Learning to Kill in War and Society* (Boston: Back Bay Books, 2007).

[86] Spitaletta, "Egyptian Islamic Jihad."

[87] Spitaletta and Marshall, "Al Qaeda."

[88] Timothy Shanahan, *The Provisional Irish Republican Army and the Morality of Terrorism* (Edinburgh: Edinburgh University Press, 2009).

[89] Crossett and Newton, "The Provisional Irish Republican Army: 1969–2001."

[90] Shanahan, *The Provisional Irish Republican Army and the Morality of Terrorism*.

[91] Ganor, "Terrorism as a Strategy of Psychological Warfare."

[92] "What's News: South Vietnamese Communists," *The Wall Street Journal*, CLXIV, August 12, 1964, 1.

[93] *The Evening Star* (Washington, DC), February 14, 1965, A-1.

[94] Jerrold M. Post, Keven G. Ruby, and Eric D. Shaw, "The Radical Group in Context 1: An Integrated Framework for the Analysis of Group Risk for Terrorism," *Studies in Conflict and Terrorism* 25, no. 2 (2002): 101–126.

[95] Post, Ruby, and Shaw, "The Radical Group in Context 2."

[96] Ronald J. Buikema and Matt Burger, "Fuerzas Armada Revolucionarias de Columbia (FARC)," in *Casebook on Insurgency and Revolutionary Warfare, Volume II: 1962–2009*, ed. Chuck Crossett (Alexandria, VA: U.S. Army Publications Directorate, in press).

[97] Fishel, "Communist Terror in South Vietnam," 14.

[98] Nix and Marshall, "Liberation Tigers of Tamil Eelam (LTTE)."

[99] Ganor, "Terrorism as a Strategy of Psychological Warfare."

[100] Ronald J. Buikema and Matt Burger, "Sendero Luminoso (Shining Path)," in *Casebook on Insurgency and Revolutionary Warfare, Volume II: 1962–2009*, ed. Chuck Crossett (Alexandria, VA: U.S. Army Publications Directorate, in press).

[101] Sanaz Mirazei, "Taliban 1994–2009," in *Casebook on Insurgency and Revolutionary Warfare, Volume II: 1962–2009*, ed. Chuck Crossett (Alexandria, VA: U.S. Army Publications Directorate, in press).

[102] Thomas H. Johnson, "The Taliban Insurgency and an Analysis of Shabnamah (Night Letters)," *Small Wars and Insurgencies* 18, no. 3 (2007): 317–344.

[103] Malcolm W. Browne, *The New Face of War* (New York: Bobbs-Merrill, 1965), 119–120.

[104] Neil J. Smelser, *Theory of Collective Behavior* (Glencoe, IL: Free Press, 1963), 83, 101.

[105] Schmid, "Terrorism as Psychological Warfare."

[106] Scott Gerwehr and Kirk Hubbard, "What is Terrorism? Key Elements and History," in *Psychology of Terrorism*, ed. Bruce Bongar, Lisa M. Brown, Larry E. Beutler, James N. Breckenridge, and Philip G. Zimbardo (New York: Oxford University Press, 2007).

[107] Breckenridge and Zimbardo, "The Strategy of Terrorism and the Psychology of Mass-Mediated Fear."

[108] Sonia J. Lupien, "Brains Under Stress," *Canadian Journal of Psychiatry* 54, no. 1 (2009): 4–5.

[109] Sandro Galea, Jennifer Ahern, Heidi Resnick, Dean Kilpatrick, Michael Bucuvalas, Joel Gold, and David Vlahov, "Psychological Sequelae of the September 11 Terrorist Attacks in New York City," *The New England Journal of Medicine* 346, no. 13 (2002): 982–987.

[110] Breckenridge and Zimbardo, "The Strategy of Terrorism and the Psychology of Mass-Mediated Fear."

[111] Charles A. Morgan III, Sheila Wang, Steven M. Southwick, Ann Rasmusson, Gary Hazett, Richard L. Hauger, and Dennis S. Charney, "Plasma Neuropeptide-Y Concentrations in Humans Exposed to Military Survival Training," *Biological Psychiatry* 47, no. 10 (2009): 902–909.

[112] Banks and James, "Warfare, Terrorism, and Psychology."

[113] Schmid, "Terrorism as Psychological Warfare."

[114] American Psychiatric Association, *Diagnostic and Statistical Manual of Mental Disorders*, 4th ed. (DSM-IV-TR) (Washington, DC: American Psychiatric Association, 2000).

[115] Galea et al., "Psychological Sequelae of the September 11 Terrorist Attacks in New York City."

[116] Smelser, *Theory of Collective Behavior*, 83ff.

[117] James C. Davies, *Human Nature in Politics* (New York: John Wiley and Sons, 1963), 67–68.

[118] Withey, "Reaction to Uncertain Threat," 121.

[119] Ibid.

[120] Crossett and Spitaletta, *Radicalization: Relevant Psychological and Sociological Concepts.*

[121] Smelser, *Theory of Collective Behavior*, 84.

[122] Post, Ruby, and Shaw, "The Radical Group in Context 2."

[123] Barker, *Grivas: Portrait of a Terrorist*, 145.

[124] Doris M. Condit, *Case Study in Guerrilla War: Greece During World War II* (Washington, DC: Special Operations Research Office, 1961), 268.

[125] Clark McCauley and Sophia Moskalenko, "Mechanisms of Political Radicalization: Pathways Toward Terrorism," *Terrorism and Political Violence* 20, no. 3 (2008): 415–433.

[126] Ibid.

[127] Post, Ruby, and Shaw, "The Radical Group in Context 2."

[128] Jerrold M. Post, "Rewarding Fire with Fire: Effects of Retaliation on Terrorist Group Dynamics," *Studies in Conflict and Terrorism* 10, no. 1 (1987): 23–35.

[129] Medardo T. Justiniano, "Combat Intelligence," *Counter-Guerrilla Operations in the Philippines, 1946–1953* (seminar, Fort Bragg, NC: June 15, 1961), 47–48.

[130] Ibid.

[131] Ronald Waring, *The War in Angola: 1961* (Lisbon: Silvas, 1961), 27.

[132] Crossett and Spitaletta, *Radicalization: Relevant Psychological and Sociological Concepts.*

[133] Jerrol M. Post, "When Hatred Is Bred in the Bone: Psycho-Cultural Foundations of Contemporary Terrorism," *Political Psychology* 26, no. 4 (2005): 615–636.

[134] Ibid.

[135] Franz P. B. Osinga, *Science, Strategy, and War: The Strategic Theory of John Boyd* (New York: Routledge, 2006).

[136] Ganor, "Terrorism as a Strategy of Psychological Warfare."

GLOSSARY

PART I: ABBREVIATIONS AND ACRONYMS

AD	Acción Democrática (English: Democratic Action Party)
AEI	Albert Einstein Institution
AFL-CIO	American Federation of Labor and Congress of Industrial Organizations
AKE	Greek Agrarian Party
APD	Antisocial Personality Disorder
AQAP	Al Qaeda in the Arabian Peninsula
AQI	Al Qaeda in Iraq
AQIM	Al Qaeda in the Islamic Maghreb
ASU	Active Service Unit
ATI	Tunisian Internet Agency
BBC	British Broadcasting Company
BGF	Black Guerrilla Family
CIA	Central Intelligence Agency
CLO	Congress of Labor Organizations
CNN	Cable News Network
CTAL	Latin American Conference
DAN	Direct Action Network
DCID	Director of Central Intelligence Directive
DDR	Disarmament, Demobilization, and Reintegration
DKS	Den Kolde Skulder (English: The Cold Shoulder)
DNS	Denial of Service
DSM	*Diagnostic and Statistical Manual of Mental Disorders*
EDCOR	Economic Development Corps
EIG	Egyptian Islamic Group
EIJ	Egyptian Islamic Jihad
EOKA	Ethniki Organosis Kyprion Agoniston (English: National Organization of Cypriot Fighters)
EPON	United All Greece Youth Organization
ERP	Ejército Revolucionario del Pueblo (English: People's Revolutionary Army)
ETA	Euskadi Ta Askatasuna (English: Basque Homeland and Freedom)
EZLN	Ejército Zapatista de Liberación Nacional (English: Zapatista National Liberation Army)
FALN	Fuerzas Armadas de Liberación Nacional (English: Armed Forces of National Liberation)

FARC	Fuerzas Armadas Revolucionarias de Colombia (English: Revolutionary Armed Forces of Colombia)
FBI	Federal Bureau of Investigation
FLN	Frente de Liberación Nacional (English: National Liberation Front)
FMLN	Frente Farabundo Martí para la Liberación Nacional (English: Farabundo Marti National Liberation Front)
FOSF	Friends of Sinn Féin
GDP	Gross Domestic Product
GFA	Good Friday Agreement
GHQ	General Headquarters
GIRoA	Government of the Islamic Republic of Afghanistan
GLU	General Labor Unions
HAMAS	Ḥarakat al-Muqāwamah al-'Islāmiyyah (English: Islamic Resistance Movement)
ICC	International Criminal Court
ICP	Indochinese Communist Party
IED	Improvised Explosive Device
INLA	Irish National Liberation Army
IP	Internet Protocol
IRA	Irish Republican Army
ISAF	International Security Assistance Force
ISP	Internet Service Provider
KKK	Ku Klux Klan
KLA	Ushtia Çlirimtare e Kosovës (English: Kosovo Liberation Army)
LTTE	Liberation Tigers of Tamil Eelam
MAK	Maktab Al Khidamat (English: Services Office)
MAR	Minorities at Risk
MCMI	Millon Clinical Multiaxial Inventory
MCP	Malayan Communist Party
MEND	Movement for the Emancipation of the Niger Delta
MIR	Movimiento de Izquierda Revolucionario (English: Movement of the Revolutionary Left)
MP	Member of Parliament
NAFTA	North American Free Trade Agreement
NASA	National Aeronautics and Space Administration
NATO	North Atlantic Treaty Organization
NFLSV	National Front for the Liberation of South Vietnam

NGO	Nongovernmental Organization
NIACRO	Northern Ireland Association for the Care and Resettlement of Offenders
NORAID	Northern Aid Committee
OAS	Organisation de l'Armée Secrète (English: Organization of the Secret Army/Secret Armed Organization)
OEF	Operation Enduring Freedom
OIF	Operation Iraqi Freedom
PCL-R	Psychopathy Checklist-Revised
PCV	Partido Comunista Venezolano (English: Venezuelan Communist Party)
PFT	People for Fair Trade
PIRA	Provisional Irish Republican Army
PLO	Palestinian Liberation Organization
PRC	People's Republic of China
PRP	People's Revolutionary Party
PTSD	Posttraumatic Stress Disorder
RAND	Research And Development Corporation
RIRA	Real IRA (Irish Republican Army)
RPG	Rocket-Propelled Grenade
RUC	Royal Ulster Constabulary
RVN	Republic of Vietnam
SAVAK	Sāzemān-e Ettelā'āt va Amniyat-e Keshvar (English: National Intelligence and Security Organization)
SDLP	Social Democratic and Labour Party
SDS	Students for a Democratic Society
SDU	Smid Dem Ud (English: Throw Them Out)
SLFP	Sri Lanka Freedom Party
SMS	Short Message Service
SORO	Special Operations Research Office
SUV	Sport Utility Vehicle
TCU	Tactical Combat Unit
TKB	Terrorism Knowledge Base
TMT	Terror Management Theory
TNT	Tamil New Tigers
TRC	Truth and Reconciliation Commission
TWA	Trans-World Airlines
U.K.	United Kingdom

UN	United Nations
URD	Unión República Democrática (English: Republican and Democratic Union)
U.S.	United States
USIP	United States Institute for Peace
VC	Viet Cong
WANK	Worms Against Nuclear Killers
WTO	World Trade Organization
WWII	World War II

PART II: TERMS AND DEFINITIONS

Agitation: Immediate, observable action that follows propaganda promises: one form is specific action to alleviate hunger and suffering, thereby demonstrating insurgents' ability to accomplish set goals, and another form focuses on retaliatory acts of violence, sabotage, and punishment of so-called traitors among the local population.

Armed component: The visible element of a revolutionary movement organized to perform overt armed military and paramilitary operations using guerrilla, asymmetric, or conventional tactics.

Auxiliary: The support element of the irregular organization whose organization and operations are clandestine in nature and whose members do not openly indicate their sympathy or involvement with the irregular movement. Members of the auxiliary are more likely to be occasional participants of the insurgency with other full-time occupations.

Beliefs, values, and norms: Beliefs are ideas, knowledge, lore, superstition, myths, and legends shared by members of a society. Associated with each cultural belief are values—the "right" or "wrong" judgments that guide individual actions. Norms are acceptable patterns of behavior that are reinforced through a system of rewards and punishments dispensed within the group.

Cell: The smallest organizational element of an underground formed around a specific process, capability, or activity. Cells are kept small for secrecy, and communication between cells is often limited to limit damage if any one cell is compromised.

Clandestine operations: Activities to accomplish intelligence, counterintelligence, and other similar activities in such a way as to ensure secrecy or concealment.

Coercion: Physical or psychological pressures exerted with the intent to ensure that an agent or group will respond as directed. Coercion is often contrasted with voluntary persuasion, although the two are also often used in combination.

Cognition: The mental processes of attention, memory, learning, language comprehension, problem solving, and decision making.

Command and control: The exercise of authority and direction by a properly designated commander over assigned and attached forces in the accomplishment of the mission. Command and control functions are performed through an arrangement of personnel, equipment, communications, facilities, and procedures employed by a commander in planning, directing, coordinating, and controlling forces and operations in the accomplishment of the mission.

Compartmentalization: Establishment and management of an organization so that information about the personnel, internal organization, or activities of one component is made available to any other component only to the extent required for the performance of assigned duties.

Covert operation: Operations planned and executed so as to conceal the identity of those involved in, or permit plausible denial of, subversive operations. Covert operations differ from clandestine operations in that emphasis in clandestine operations is placed on concealment of the operation rather than the concealment of personal identity.

Deprivation: A state of lacking in psychological, economic, political, or social resources. Relative deprivation theory states that the subjective sense of being deprived of certain needs or freedoms by a domestic or international governing body can result in feelings of frustration, and when individuals can no longer bear this misery or indignity, a rebellion ensues. When these feelings of frustration go unresolved through productive or legal means and are left to fester, they can manifest in acts of violence motivated by, but not always directed toward, the governing body.

DSM-IV and Axis I and II Disorders: Currently in its fourth edition, the *Diagnostic and Statistical Manual of Mental Disorders* (DSM) is published by the American Psychiatric Association and provides a common language and standard criteria for the classification of mental disorders. It is used in the United States and to varying degrees around the world by clinicians, researchers, psychiatric drug regulation agencies, health insurance companies, pharmaceutical companies, and policy makers. The DSM-IV-TR uses a multiaxial or multidimensional approach to diagnoses because rarely do other factors in a person's life not affect their mental health. Axis I (Clinical Syndromes) disorders are those psychological disorders that are the focus of a diagnosis. Axis I disorders are divided identified into fourteen categories, including Anxiety Disorders, Childhood Disorders, Cognitive Disorders, Dissociative Disorders, Eating Disorders, Factitious Disorders, Impulse-Control Disorders, Mood Disorders, Psychotic Disorders, Sexual and Gender-Identity Disorders, Sleep Disorders, Somatoform Disorders, and Substance-Related Disorders. Axis II (Developmental Disorders and Personality Disorders) are long-standing chronic conditions that may affect the clinical syndromes listed in Axis I. Developmental disorders include autism and mental retardation, disorders that are typically first evident in childhood. Personality disorders are clinical syndromes that have enduring symptoms and encompass the individual's way of interacting with the world. They are divided into three clusters: Cluster A (odd or eccentric) includes Paranoid, Schizoid, and, Schizotypal. Cluster B (overly emotional, unstable, or self-dramatizing) includes Antisocial, Borderline, Histrionic, and Narcissistic. Cluster C (tense and anxiety ridden) includes Avoidant, Dependent, and Obsessive-Compulsive. Axis III disorders are medical/physical conditions that play a role in the development, continuance, or exacerbation of Axis I and II disorders or other physical conditions such as brain injury or HIV/AIDS that can result in symptoms of mental illness. Axis IV includes social and environmental stressors that may affect the clinical syndromes listed in Axis I. Events in a person's life, such as death of a loved one, starting a new job, college, unemployment, and even marriage can affect the disorders listed in Axis I and II. These events are both listed and rated for this axis. Axis V represents the highest level of functioning where a clinician rates the individual's level of functioning at both the present time and the highest level within the previous year. This helps the clinician understand how the above four axes are affecting the person and what type of changes could be expected.

Group dynamics: The study of two or more individuals connected by social relationships and how they interact and influence each other. Groups, relevant to the fields of psychology, sociology, and communication studies, comprise two or more individuals who are connected to each other by social relationships. Because they interact and influence each other, groups develop a number of dynamic processes that separate them from a random collection of individuals. These processes include norms, roles, relations, development, need to belong, social influence, and effects on behavior.

Human factors analysis: The psychological, cultural, behavioral, and other human attributes that influence decision making, the flow of information, and the interpretation of information by individuals and groups at any level in any state or organization.

Identity: Identity theory addresses the individual's sense of self and/or their place in the world. The theory posits a distinction among the psychological sense of continuity from the self (ego-identity) to one's distinguishing idiosyncrasies (as the personal identity) to the set of social roles an individual may fulfill (social identity). Identity is a broad term used throughout the social sciences to describe a person's self-concept and expression of their individual and group affiliations.

Ideology: A set of beliefs that constitute one's goals, expectations, values, and actions and form a comprehensive worldview. In insurgencies, a well-developed ideology serves the purpose of *unifying* disparate members of the movement, *organizing* actions around goals and shared values, and *justifying* actions that may include violence against countrymen.

In-group: A social group toward which an individual feels loyalty and respect, usually because of membership in the group based on social or familial ties. Commonly, in-groups include one's family, team, professional organization, and those of the same race, culture, gender, or religion. This affinity often manifests itself as an in-group bias, whereby individuals tend to define their group over against a reference group and look more favorably upon their in-group than members of an out-group.

Insurgency: An organized movement aimed at the overthrow of a constituted government through use of subversion and armed conflict.

Jihad: An Islamic term translated as a noun meaning "struggle" and a religious duty of Muslims. An individual engaged in jihad is called a *mujahid* (plural: *mujahidin*). Jihad is often dichotomized into greater and lesser jihad, although Islamic scholars differ in their belief as to which is indeed is more important. The greater jihad is the striving each Muslim experiences to live a life of piety amid temptation, while the lesser jihad is the armed struggle in defense of the faith. Sunni scholars refer to this duty as the sixth pillar of Islam, although it occupies no such official status. In Shia Islam, however, Jihad is one of the 10 Practices of the Religion.

Mortality salience: Mortality salience is an increased awareness of and fixation upon one's death. It has the potential to cause worldview defense, a psychological mechanism that strengthens people's connection with their in-group as a defense mechanism.

Nonviolent resistance: Methods employed by resistance movements that capitalize upon social norms, customs, and taboos in order to provoke action by security forces that will serve to alienate large segments of public opinion from the government or its agents.

Out-group: A social group toward which an individual feels contempt, opposition, or a desire to compete.

Propaganda: Any form of communication, especially of a biased or misleading nature, designed to influence the opinions, emotions, attitudes, or behavior of any group in order to benefit the sponsor, either directly or indirectly.

Psychodynamics: The theory and systematic study of the psychological forces that underlie human behavior, especially the dynamic relations between conscious motivation and unconscious motivation.

Psychological operations (PSYOP): A set of techniques used by the underground aimed to influence a target audience's value systems, belief systems, emotions, motives, reasoning, or behavior.

Public component: The overt political component of an insurgent or revolutionary movement. Some insurgencies pursue military and political strategies. At the termination of conflict, or occasionally during the conflict, the movement can transition to the sole legitimate government or forms part of an existing government. Thus, the four spheres— armed component, underground, auxiliary, and public component —form a dynamic and evolving relationship changing in response to internal and external drivers. The public component's overt position distinguishes it from the clandestine underground. However, it frequently overlaps with the underground in that the latter's functionality includes the management of propaganda and communications in general.

Radicalization: The process by which an individual, group, or mass of people undergo a transformation from participating in the political process via legal means to the use or support of violence for political purposes (radicalism).

Risk factors: A set of attributes (traits) or observable behaviors that may predispose an individual to or increase the likelihood that they adopt a set of beliefs or engage in politically motivated violence.

Resistance movement: An organized effort by some portion of the civil population of a country to resist the legally established government or an occupying power and to disrupt civil order and stability.

Sabotage: Actions to withhold resources from the government's counterinsurgency effort by acts of destruction. An act or acts with intent to injure, interfere with, or obstruct the national defense of a country by willfully injuring or destroying, or attempting to injure or destroy, any national defense or war materiel, premises, or utilities, to include human and natural resources.

Safe haven: Any space, whether physical, legal, financial, or virtual (e.g., cyber), that enables insurgent organizations to plan, organize, train, conduct operations, or rest with limited interference from enemy or counterinsurgent forces.

Salafism: An Islamic school of thought employed by Sunni theologians since at least the fifth Muslim generation to differentiate the creed of the first three generations (the *Sahabah,* or Companions of the Islamic prophet Muhammad and the two succeeding generations, the *Tabi'un* and the *Tabi' al-Tabi'in*) from subsequent variations in belief system and method. Salafists view those generations as an eternal model for all succeeding Muslim generations, especially in their beliefs and methodology of understanding the texts, but also in their method of worship, mannerisms, morality, piety, and conduct.

Shadow government: A parallel governance structure established by an insurgent group that mimics the functions and attributes of the nation-state. Its functions include one or more of the following: extension of force, provision of social services, national identity and legitimacy, and revenue generation.

Social identity: Social identity is membership in a group that helps to define a person's self-concept and provide self-esteem. An individual has multiple social identities including those of his or her family, sports team, ethnic group, military unit, etc., all of which help define who he or she is relative to the society and provide a particular sense of self-worth through identification with said group.

Social media: A set of Internet-based applications resulting from technological advances in communications technology that allows the creation and exchange of user-generated content. Social media supports informal, usually text-based communication in one-to-one, one-to-many, and many-to-many formats. Examples of three popular services that are considered social media are Facebook, Twitter, and YouTube.

Social network: A social network is a structure composed of individuals or organizations that are connected by one or more specific types of interdependency. Those interdependencies may be friendship; kinship; common interest; financial exchange; group affiliation; dislike; social relationships; or relationships of beliefs, knowledge, or prestige. Understanding these connections can give insight into a group's patterns of influence and decision making; they can also be used to understand organizations' strengths and weaknesses. Social network analysis is an active area of academic research.

Subversion: Actions designed to undermine the military, economic, psychological, or political strength or morale of a governing authority.

Terrorism: Coercive acts of violence utilized by a subversive movement and usually directed toward disrupting government control over the citizenry and creating a state of mind that makes the citizenry acquiesce to subversive demands.

Terror management theory: Terror management theory states that existential anxiety (or the fear of death) is assuaged by adopting a worldview that makes death comprehensible and manageable. Terror management theory focuses on the implicit emotional reactions when individuals are confronted with their imminent mortality. It attempts to provide a rationale for the motivational catalysts of human behavior when life is threatened and advances the idea that a shared worldview provides an individual with a defense against inevitable existential anxiety (the fear of death). This cultural worldview minimizes death anxiety, providing an understanding of the universe that has order, meaning, and standards of acceptable behavior, resulting in increased self-esteem.

Uncertainty reduction theory: An application of communications research, uncertainty reduction theory puts forth the idea that group affiliation is motivated by the desire to alleviate uncertainty. An underlying assumption is most people do not tend to adopt a specific worldview and/or affiliate with a specific identity group unless there is a motive to alleviate uncertainty.

Underground: A clandestine organization established to operate in areas denied to the armed or public components or conduct operations not suitable for the armed or public components.

Unity of command: The operation of all forces under a single responsible commander who has the requisite authority to direct and employ those forces in pursuit of a common purpose.

Unity of effort: Coordination and cooperation toward common objectives, even if the participants are not necessarily part of the same command or organization.

Vanguard: Organizational theory common to Communist doctrines that calls for establishing a "front" or "vanguard" to infiltrate existing liberation or independence movements and orchestrate the overthrow of the incumbent authority. The vanguard is theoretically the most ideologically advanced sector of society not prone to the "false consciousness" infecting mass society. The organizational theory was most clearly articulated by Lenin and adopted by the Bolshevik Party in Russia and has since been used by various theorists, including non-Communists, such as Sayyid Qutb in his influential book, *Milestones*.

www.ingramcontent.com/pod-product-compliance
Lightning Source LLC
Chambersburg PA
CBHW052108020426
42335CB00021B/2677